Daoism and Ecology

Publications of the Center for the Study of World Religions
Harvard Divinity School

General Editor: Lawrence E. Sullivan
Senior Editor: Kathryn Dodgson

Religions of the World and Ecology

Series Editors: Mary Evelyn Tucker and John Grim

Cambridge, Massachusetts

Daoism and Ecology

Ways within a Cosmic Landscape

edited by

N. J. GIRARDOT

JAMES MILLER

and

LIU XIAOGAN

distributed by
Harvard University Press
for the
Center for the Study of World Religions
Harvard Divinity School

Grateful acknowledgment is made for permission to reprint the following:

Tao Song. Copyright © 1975 Ursula K. Le Guin.

Cover art: Tomioka Tessai (1837–1924), "A Pleasant Life in a Gourd," 1923. Light color on paper. 132.7 cm x 32 cm. Collection of the Kiyoshikojin Seichoji Temple.

Cover design: Patrick Santana

Library of Congress Cataloging-in-Publication Data

Daoism and ecology : ways within a cosmic landscape / edited by N. J. Girardot, James Miller and Xiaogan Liu.
       p. cm. — (Religions of the world and ecology)
    ISBN 0-945454-29-5 (alk. paper)
    ISBN 0-945454-30-9 (pbk. : alk. paper)
       1. Taoism. 2. Ecology. I. Girardot, N. J. II. Miller, James, date. III. Liu, Xiaogan, date. IV. Series.

    BL1923 .D36 2001
    299'.514178362—dc21

                                                                    2001028611

Contents

Preface

LAWRENCE E. SULLIVAN

Religion distinguishes the human species from all others, just as human presence on earth distinguishes the ecology of our planet from other places in the known universe. Religious life and the earth's ecology are inextricably linked, organically related.

Human belief and practice mark the earth. One can hardly think of a natural system that has not been considerably altered, for better or worse, by human culture. "Nor is this the work of the industrial centuries," observes Simon Schama. "It is coeval with the entirety of our social existence. And it is this irreversibly modified world, from the polar caps to the equatorial forests, that is all the nature we have" (*Landscape and Memory* [New York: Vintage Books, 1996], 7). In Schama's examination even landscapes that appear to be most free of human culture turn out, on closer inspection, to be its product.

Human beliefs about the nature of ecology are the distinctive contribution of our species to the ecology itself. Religious beliefs— especially those concerning the nature of powers that create and animate—become an effective part of ecological systems. They attract the power of will and channel the forces of labor toward purposive transformations. Religious rituals model relations with material life and transmit habits of practice and attitudes of mind to succeeding generations.

This is not simply to say that religious thoughts occasionally touch the world and leave traces that accumulate over time. The matter is the other way around. From the point of view of environmental studies, religious worldviews propel communities into the world with fundamental predispositions toward it because such religious world-views are primordial, all-encompassing, and unique. They are *primordial* because they probe behind secondary appearances and stray thoughts to rivet human attention on realities of the first order: life at its source, creativity in its fullest manifestation, death and destruction at their origin, renewal and salvation in their germ. The revelation of first things is compelling and moves communities to take creative action. Primordial ideas are prime movers.

Religious worldviews are *all-encompassing* because they fully absorb the natural world within them. They provide human beings both a view of the whole and at the same time a penetrating image of their own ironic position as the beings in the cosmos who possess the capacity for symbolic thought: the part that contains the whole—or at least a picture of the whole—within itself. As all-encompassing, therefore, religious ideas do not just contend with other ideas as equals; they frame the mind-set within which all sorts of ideas commingle in a cosmology. For this reason, their role in ecology must be better understood.

Religious worldviews are *unique* because they draw the world of nature into a wholly other kind of universe, one that appears only in the religious imagination. From the point of view of environmental studies, the risk of such religious views, on the one hand, is of disinterest in or disregard for the natural world. On the other hand, only in the religious world can nature be compared and contrasted to other kinds of being—the supernatural world or forms of power not always fully manifest in nature. Only then can nature be revealed as distinctive, set in a new light startlingly different from its own. That is to say, only religious perspectives enable human beings to evaluate the world of nature in terms distinct from all else. In this same step toward intelligibility, the natural world is evaluated in terms consonant with human beings' own distinctive (religious and imaginative) nature in the world, thus grounding a self-conscious relationship and a role with limits and responsibilities.

In the struggle to sustain the earth's environment as viable for future generations, environmental studies has thus far left the role of religion unprobed. This contrasts starkly with the emphasis given, for example, the role of science and technology in threatening or sustaining the ecology. Ignorance of religion prevents environmental studies from achieving its goals, however, for though science and technology share many important features of human culture with religion, they leave unexplored essential wellsprings of human motivation and concern that shape the world as we know it. No understanding of the environment is adequate without a grasp of the religious life that constitutes the human societies which saturate the natural environment.

A great deal of what we know about the religions of the world is new knowledge. As is the case for geology and astronomy, so too for religious studies: many new discoveries about the nature and function of religion are, in fact, clearer understandings of events and processes that began to unfold long ago. Much of what we are learning now about the religions of the world was previously not known outside of a circle of adepts. From the ancient history of traditions and from the ongoing creativity of the world's contemporary religions we are opening a treasury of motives, disciplines, and awarenesses.

A geology of the religious spirit of humankind can well serve our need to relate fruitfully to the earth and its myriad life-forms. Changing our habits of consumption and patterns of distribution, reevaluating modes of production, and reestablishing a strong sense of solidarity with the matrix of material life—these achievements will arrive along with spiritual modulations that unveil attractive new images of well-being and prosperity, respecting the limits of life in a sustainable world while revering life at its sources. Remarkable religious views are presented in this series—from the nature mysticism of Bashō in Japan or Saint Francis in Italy to the ecstatic physiologies and embryologies of shamanic healers, Taoist meditators, and Vedic practitioners; from indigenous people's ritual responses to projects funded by the World Bank, to religiously grounded criticisms of hazardous waste sites, deforestation, and environmental racism.

The power to modify the world is both frightening and fascinating and has been subjected to reflection, particularly religious reflection, from time immemorial to the present day. We will understand ecology better when we understand the religions that form the rich soil of memory and practice, belief and relationships where life on earth is rooted. Knowledge of these views will help us reappraise our ways and reorient ourselves toward the sources and resources of life.

This volume is one in a series that addresses the critical gap in our contemporary understanding of religion and ecology. The series results from research conducted at the Harvard University Center for the Study of World Religions over a three-year period. I wish especially to acknowledge President Neil L. Rudenstine of Harvard University for his leadership in instituting the environmental initiative at Harvard and thank him for his warm encouragement and characteristic support of our program. Mary Evelyn Tucker and John Grim of Bucknell University coordinated the research, involving the direct participation of some six hundred scholars, religious leaders, and environmental specialists brought to Harvard from around the world during the period of research and inquiry. Professors Tucker and Grim have brought great vision and energy to this enormous project, as has their team of conference convenors. The commitment and advice of Martin S. Kaplan of Hale and Dorr have been of great value. Our goals have been achieved for this research and publication program because of the extraordinary dedication and talents of Center for the Study of World Religions staff members Don Kunkel, Malgorzata Radziszewska-Hedderick, Kathryn Dodgson, Janey Bosch, Naomi Wilshire, Lilli Leggio, and Eric Edstam and with the unstinting help of Stephanie Snyder of Bucknell. To these individuals, and to all the sponsors and participants whose efforts made this series possible, go deepest thanks and appreciation.

Series Foreword

MARY EVELYN TUCKER and JOHN GRIM

The Nature of the Environmental Crisis

Ours is a period when the human community is in search of new and sustaining relationships to the earth amidst an environmental crisis that threatens the very existence of all life-forms on the planet. While the particular causes and solutions of this crisis are being debated by scientists, economists, and policymakers, the facts of widespread destruction are causing alarm in many quarters. Indeed, from some perspectives the future of human life itself appears threatened. As Daniel Maguire has succinctly observed, "If current trends continue, we will not."[1] Thomas Berry, the former director of the Riverdale Center for Religious Research, has also raised the stark question, "Is the human a viable species on an endangered planet?"

From resource depletion and species extinction to pollution overload and toxic surplus, the planet is struggling against unprecedented assaults. This is aggravated by population explosion, industrial growth, technological manipulation, and military proliferation heretofore unknown by the human community. From many accounts the basic elements which sustain life—sufficient water, clean air, and arable land—are at risk. The challenges are formidable and well documented. The solutions, however, are more elusive and complex. Clearly, this crisis has economic, political, and social dimensions which require more detailed analysis than we can provide here. Suffice it to say, however, as did the *Global 2000 Report*: ". . .once such global environmental problems are in motion they are difficult to reverse. In fact few if any of the problems addressed in the *Global 2000*

Report are amenable to quick technological or policy fixes; rather, they are inextricably mixed with the world's most perplexing social and economic problems."[2]

Peter Raven, the director of the Missouri Botanical Garden, wrote in a paper titled "We Are Killing Our World" with a similar sense of urgency regarding the magnitude of the environmental crisis: "The world that provides our evolutionary and ecological context is in serious trouble, trouble of a kind that demands our urgent attention. By formulating adequate plans for dealing with these large-scale problems, we will be laying the foundation for peace and prosperity in the future; by ignoring them, drifting passively while attending to what may seem more urgent, personal priorities, we are courting disaster."

Rethinking Worldviews and Ethics

For many people an environmental crisis of this complexity and scope is not only the result of certain economic, political, and social factors. It is also a moral and spiritual crisis which, in order to be addressed, will require broader philosophical and religious understandings of ourselves as creatures of nature, embedded in life cycles and dependent on ecosystems. Religions, thus, need to be reexamined in light of the current environmental crisis. This is because religions help to shape our attitudes toward nature in both conscious and unconscious ways. Religions provide basic interpretive stories of who we are, what nature is, where we have come from, and where we are going. This comprises a worldview of a society. Religions also suggest how we should treat other humans and how we should relate to nature. These values make up the ethical orientation of a society. Religions thus generate worldviews and ethics which underlie fundamental attitudes and values of different cultures and societies. As the historian Lynn White observed, "What people do about their ecology depends on what they think about themselves in relation to things around them. Human ecology is deeply conditioned by beliefs about our nature and destiny—that is, by religion."[3]

In trying to reorient ourselves in relation to the earth, it has become apparent that we have lost our appreciation for the intricate nature of matter and materiality. Our feeling of alienation in the modern period has extended beyond the human community and its patterns of material exchanges to our interaction with nature itself. Especially in technologically sophisticated urban societies, we have become removed from the recognition of our dependence on nature. We no longer know who we are as earthlings; we no longer see the earth as sacred.

Thomas Berry suggests that we have become autistic in our interactions with the natural world. In other words, we are unable to value the life and beauty of nature because we are locked in our own egocentric perspectives and shortsighted needs. He suggests that we need a new cosmology, cultural coding, and motivating energy to overcome this deprivation.[4] He observes that the magnitude of destructive industrial processes is so great that we must initiate a radical rethinking of the myth of progress and of humanity's role in the evolutionary process. Indeed, he speaks of evolution as a new story of the universe, namely, as a vast cosmological perspective that will resituate human meaning and direction in the context of four and a half billion years of earth history.[5]

For Berry and for many others an important component of the current environmental crisis is spiritual and ethical. It is here that the religions of the world may have a role to play in cooperation with other individuals, institutions, and initiatives that have been engaged with environmental issues for a considerable period of time. Despite their lateness in addressing the crisis, religions are beginning to respond in remarkably creative ways. They are not only rethinking their theologies but are also reorienting their sustainable practices and long-term environmental commitments. In so doing, the very nature of religion and of ethics is being challenged and changed. This is true because the reexamination of other worldviews created by religious beliefs and practices may be critical to our recovery of sufficiently comprehensive cosmologies, broad conceptual frameworks, and effective environmental ethics for the twenty-first century.

While in the past none of the religions of the world have had to face an environmental crisis such as we are now confronting, they remain key instruments in shaping attitudes toward nature. The unintended consequences of the modern industrial drive for unlimited economic growth and resource development have led us to an impasse regarding the survival of many life-forms and appropriate management of varied ecosystems. The religious traditions may indeed be critical in helping to reimagine the viable conditions and long-range strategies for fostering mutually enhancing human-earth relations.[6]

Indeed, as E. N. Anderson has documented with impressive detail, "All traditional societies that have succeeded in managing resources well, over time, have done it in part through religious or ritual representation of resource management."[7]

It is in this context that a series of conferences and publications exploring the various religions of the world and their relation to ecology was initiated by the Center for the Study of World Religions at Harvard. Coordinated by Mary Evelyn Tucker and John Grim, the conferences involved some six hundred scholars, graduate students, religious leaders, and environmental activists over a period of three years. The collaborative nature of the project is intentional. Such collaboration maximizes the opportunity for dialogical reflection on this issue of enormous complexity and accentuates the diversity of local manifestations of ecologically sustainable alternatives.

This series is intended to serve as initial explorations of the emerging field of religion and ecology while pointing toward areas for further research. We are not unaware of the difficulties of engaging in such a task, yet we have been encouraged by the enthusiastic response to the conferences within the academic community, by the larger interest they have generated beyond academia, and by the probing examinations gathered in the volumes. We trust that this series and these volumes will be useful not only for scholars of religion but also for those shaping seminary education and institutional religious practices, as well as for those involved in public policy on environmental issues.

We see such conferences and publications as expanding the growing dialogue regarding the role of the world's religions as moral forces in stemming the environmental crisis. While, clearly, there are major methodological issues involved in utilizing traditional philosophical and religious ideas for contemporary concerns, there are also compelling reasons to support such efforts, however modest they may be. The world's religions in all their complexity and variety remain one of the principal resources for symbolic ideas, spiritual inspiration, and ethical principles. Indeed, despite their limitations, historically they have provided comprehensive cosmologies for interpretive direction, moral foundations for social cohesion, spiritual guidance for cultural expression, and ritual celebrations for meaningful life. In our search for more comprehensive ecological worldviews and more effective environmental ethics, it is inevitable that we will draw from the symbolic and conceptual resources of the religious traditions of the world. The effort to do this is not without precedent or problems, some of which will be signaled below. With this volume and with this series we hope the field of reflection and discussion regarding religion and ecology will begin to broaden, deepen, and complexify.

Qualifications and Goals

The Problems and Promise of Religions

These volumes, then, are built on the premise that the religions of the world may be instrumental in addressing the moral dilemmas created by the environmental crisis. At the same time we recognize the limitations of such efforts on the part of religions. We also acknowledge that the complexity of the problem requires interlocking approaches from such fields as science, economics, politics, health, and public policy. As the human community struggles to formulate different attitudes toward nature and to articulate broader conceptions of ethics embracing species and ecosystems, religions may thus be a necessary, though only contributing, part of this multidisciplinary approach.

It is becoming increasingly evident that abundant scientific knowledge of the crisis is available and numerous political and economic statements have been formulated. Yet we seem to lack the political, economic, and scientific leadership to make necessary changes. Moreover, what is still lacking is the religious commitment, moral imagination, and ethical engagement to transform the environmental crisis from an issue on paper to one of effective policy, from rhetoric in print to realism in action. Why, nearly fifty years after Fairfield Osborne's warning in *Our Plundered Planet* and more than thirty years since Rachel Carson's *Silent Spring,* are we still wondering, is it too late?[8]

It is important to ask where the religions have been on these issues and why they themselves have been so late in their involvement. Have issues of personal salvation superseded all others? Have divine-human relations been primary? Have anthropocentric ethics been all-consuming? Has the material world of nature been devalued by religion? Does the search for otherworldly rewards override commitment to this world? Did the religions simply surrender their natural theologies and concerns with exploring purpose in nature to positivistic scientific cosmologies? In beginning to address these questions, we still have not exhausted all the reasons for religions' lack of attention to the environmental crisis. The reasons may not be readily apparent, but clearly they require further exploration and explanation.

In discussing the involvement of religions in this issue, it is also appropriate to acknowledge the dark side of religion in both its institutional expressions and dogmatic forms. In addition to their oversight with regard to the environment, religions have been the source of enormous manipulation of power in fostering wars, in ignoring racial and social injustice, and in promoting unequal gender relations, to name only a few abuses. One does not want to underplay this shadow side or to claim too much for religions' potential for ethical persuasiveness. The problems are too vast and complex for unqualified optimism. Yet there is a growing consensus that religions may now have a significant role to play, just as in the past they have sustained individuals and cultures in the face of internal and external threats.

A final caveat is the inevitable gap that arises between theories and practices in religions. As has been noted, even societies with religious traditions which appear sympathetic to the environment have in the past often misused resources. While it is clear that religions may have some disjunction between the ideal and the real, this should not lessen our endeavor to identify resources from within the world's religions for a more ecologically sound cosmology and environmentally supportive ethics. This disjunction of theory and practice is present within all philosophies and religions and is frequently the source of disillusionment, skepticism, and cynicism. A more realistic observation might be made, however, that this disjunction should not automatically invalidate the complex worldviews and rich cosmologies embedded in traditional religions. Rather, it is our task to explore these conceptual resources so as to broaden and expand our own perspectives in challenging and fruitful ways.

In summary, we recognize that religions have elements which are both prophetic and transformative as well as conservative and constraining. These elements are continually in tension, a condition which creates the great variety of thought and interpretation within religious traditions. To recognize these various tensions and limits, however, is not to lessen the urgency of the overall goals of this project. Rather, it is to circumscribe our efforts with healthy skepticism, cautious optimism, and modest ambitions. It is to suggest that this is a beginning in a new field of study which will affect both religion and ecology. On the one hand, this process of reflection will inevitably change how religions conceive of their own roles, missions, and identities, for such reflections demand a new sense of the sacred as not divorced from the earth itself. On the other hand, environmental studies can recognize that religions have helped to shape attitudes toward nature. Thus, as religions themselves evolve they may be indispensable in fostering a more expansive appreciation for the complexity and beauty of the natural world. At the same time as religions foster awe and reverence for nature, they may provide the transforming energies for ethical practices to protect endangered ecosystems, threatened species, and diminishing resources.

Methodological Concerns

It is important to acknowledge that there are, inevitably, challenging methodological issues involved in such a project as we are undertaking in this emerging field of religion and ecology.[9] Some of the key interpretive challenges we face in this project concern issues of time, place, space, and positionality. With regard to time, it is necessary to recognize the vast historical complexity of each religious tradition, which cannot be easily condensed in these conferences or volumes. With respect to place, we need to signal the diverse cultural contexts in which these religions have developed. With regard to space, we recognize the varied frameworks of institutions and traditions in which these religions unfold. Finally, with respect to positionality, we acknowledge our own historical situatedness at the end of the twentieth century with distinctive contemporary concerns.

Not only is each religious tradition historically complex and culturally diverse, but its beliefs, scriptures, and institutions have themselves been subject to vast commentaries and revisions over time. Thus, we recognize the radical diversity that exists within and among religious traditions which cannot be encompassed in any single volume. We acknowledge also that distortions may arise as we examine earlier historical traditions in light of contemporary issues.

Nonetheless, the environmental ethics philosopher J. Baird Callicott has suggested that scholars and others "mine the conceptual resources" of the religious traditions as a means of creating a more inclusive global environmental ethics.[10] As Callicott himself notes, however, the notion of "mining" is problematic, for it conjures up images of exploitation which may cause apprehension among certain religious communities, especially those of indigenous peoples. Moreover, we cannot simply expect to borrow or adopt ideas and place them from one tradition directly into another. Even efforts to formulate global environmental ethics need to be sensitive to cultural particularity and diversity. We do not aim at creating a simple bricolage or bland fusion of perspectives. Rather, these conferences and volumes are an attempt to display before us a multiperspectival cross section of the symbolic richness regarding attitudes toward nature within the religions of the world. To do so will help to reveal certain commonalities among traditions, as well as limitations within traditions, as they begin to converge around this challenge presented by the environmental crisis.

We need to identify our concerns, then, as embedded in the constraints of our own perspectival limits at the same time as we seek common ground. In describing various attitudes toward nature historically, we are aiming at *critical understanding* of the complexity, contexts, and frameworks in which these religions articulate such views. In addition, we are striving for *empathetic appreciation* for the traditions without idealizing their ecological potential or ignoring their environmental oversights. Finally, we are aiming at the *creative revisioning* of mutually enhancing human-earth relations. This revisioning may be assisted by highlighting the multiperspectival attitudes toward nature which these traditions disclose. The prismatic effect of examining such attitudes and relationships may provide some necessary clarification and symbolic resources for reimagining our own situation and shared concerns at the end of the twentieth century. It will also be sharpened by identifying the multilayered symbol systems in world religions which have traditionally oriented humans in establishing relational resonances between the microcosm of the self and the macrocosm of the social and natural orders. In short, religious traditions may help to supply both creative resources of symbols, rituals, and texts as well as inspiring visions for reimagining ourselves as part of, not apart from, the natural world.

Aims

The methodological issues outlined above were implied in the overall goals of the conferences, which were described as follows:

1. To identify and evaluate the *distinctive ecological attitudes,* values, and practices of diverse religious traditions, making clear their links to intellectual, political, and other resources associated with these distinctive traditions.

2. To describe and analyze the *commonalities* that exist within and among religious traditions with respect to ecology.

3. To identify the *minimum common ground* on which to base constructive understanding, motivating discussion, and concerted action in diverse locations across the globe; and to highlight the specific religious resources that comprise such fertile ecological ground: within scripture, ritual, myth, symbol, cosmology, sacrament, and so on.

4. To articulate in clear and moving terms *a desirable mode of human presence with the earth;* in short, to highlight means of respecting and valuing nature, to note what has already been actualized, and to indicate how best to achieve what is desirable beyond these examples.

5. To outline the most significant areas, with regard to religion and ecology, in need of *further study;* to enumerate questions of highest priority within those areas and propose possible approaches to use in addressing them.

In this series, then, we do not intend to obliterate difference or ignore diversity. The aim is to celebrate plurality by raising to conscious awareness multiple perspectives regarding nature and human-earth relations as articulated in the religions of the world. The spectrum of cosmologies, myths, symbols, and rituals within the religious traditions will be instructive in resituating us within the rhythms and limits of nature.

We are not looking for a unified worldview or a single global ethic. We are, however, deeply sympathetic with the efforts toward formulating a global ethic made by individuals, such as the theologian Hans Küng or the environmental philosopher J. Baird Callicott, and groups, such as Global Education Associates and United Religions. A minimum content of environmental ethics needs to be seriously considered. We are, then, keenly interested in the contribution this series might make to discussions of environmental policy in national and international arenas. Important intersections may be made with work in the field of development ethics.[11] In addition, the findings of the conferences have bearing on the ethical formulation of the Earth Charter that is to be presented to the United Nations for adoption within the next few years. Thus, we are seeking both the grounds for common concern and the constructive conceptual basis for rethinking our current situation of estrangement from the earth. In so doing we will be able to reconceive a means of creating the basis not just for sustainable development, but also for sustainable life on the planet.

As scientist Brian Swimme has suggested, we are currently making macrophase changes to the life systems of the planet with microphase wisdom. Clearly, we need to expand and deepen the wisdom base for human intervention with nature and other humans. This is particularly true as issues of genetic alteration of natural processes are already available and in use. If religions have traditionally concentrated on divine-human and human-human relations, the challenge is that they now explore more fully divine-human-earth relations. Without such further exploration, adequate environmental ethics may not emerge in a comprehensive context.

Resources: Environmental Ethics Found in the World's Religions

For many people, when challenges such as the environmental crisis are raised in relation to religion in the contemporary world, there frequently arises a sense of loss or a nostalgia for earlier, seemingly less complicated eras when the constant questioning of religious beliefs and practices was not so apparent. This is, no doubt, something of a reified reading of history. There is, however, a decidedly anxious tone to the questioning and soul-searching that appears to haunt many contemporary religious groups as they seek to find their particular role in the midst of rapid technological change and dominant secular values.

One of the greatest challenges, however, to contemporary religions remains how to respond to the environmental crisis, which many believe has been perpetuated because of the enormous inroads made by unrestrained materialism, secularization, and industrialization in contemporary societies, especially those societies arising in or influenced by the modern West. Indeed, some suggest that the very division of religion from secular life may be a major cause of the crisis.

Others, such as the medieval historian Lynn White, have cited religion's negative role in the crisis. White has suggested that the emphasis in Judaism and Christianity on the transcendence of God above nature and the dominion of humans over nature has led to a devaluing of the natural world and a subsequent destruction of its resources for utilitarian ends.[12] While the particulars of this argument have been vehemently debated, it is increasingly clear that the environmental crisis and its perpetuation due to industrialization, secularization, and ethical indifference present a serious challenge to the world's religions. This is especially true because many of these religions have traditionally been concerned with the path of personal salvation, which frequently emphasized otherworldly goals and rejected this world as corrupting. Thus, as we have noted, how to adapt religious teachings to this task of revaluing nature so as to prevent its destruction marks a significant new phase in religious thought. Indeed, as Thomas Berry has so aptly pointed out, what is necessary is a comprehensive re-evaluation of human-earth relations if the human is to continue as a viable species on an increasingly degraded planet. This will require, in addition to major economic and political changes, examining worldviews and ethics among the world's religions that differ from those that have captured the imagination of contemporary industrialized societies which regard nature primarily as a commodity to be utilized. It should be noted that when we are searching for effective resources for formulating environmental ethics, each of the religious traditions have both positive and negative features.

For the most part, the worldviews associated with the Western Abrahamic traditions of Judaism, Christianity, and Islam have created a dominantly human-focused morality. Because these worldviews are largely anthropocentric, nature is viewed as being of secondary importance. This is reinforced by a strong sense of the transcendence of God above nature. On the other hand, there are rich resources for rethinking views of nature in the covenantal tradition of the Hebrew Bible, in sacramental theology, in incarnational Christology, and in the vice-regency (*khalifa Allah*) concept of the Qur'an. The covenantal tradition draws on the legal agreements of biblical thought which are extended to all of creation. Sacramental theology in Christianity underscores the sacred dimension of material reality, especially for ritual purposes.[13] Incarnational Christology proposes that because God became flesh in the person of Christ, the entire natural order can be viewed as sacred. The concept of humans as vice-regents of Allah on earth suggests that humans have particular privileges, responsibilities, and obligations to creation.[14]

In Hinduism, although there is a significant emphasis on performing one's *dharma,* or duty, in the world, there is also a strong pull toward *moksa,* or liberation, from the world of suffering, or *samsāra.* To heal this kind of suffering and alienation through spiritual discipline and meditation, one turns away from the world (*prakṛti*) to a timeless world of spirit (*puruṣa*). Yet at the same time there are numerous traditions in Hinduism which affirm particular rivers, mountains, or forests as sacred. Moreover, in the concept of *līlā,* the creative play of the gods, Hindu theology engages the world as a creative manifestation of the divine. This same tension between withdrawal from the world and affirmation of it is present in Buddhism. Certain Theravāda schools of Buddhism emphasize withdrawing in meditation from the transient world of suffering (*saṃsāra*) to seek release in *nirvāṇa*. On the other hand, later Mahāyāna schools of Buddhism, such as Hua-yen, underscore the remarkable interconnection of reality in such images as the jeweled net of Indra, where each jewel reflects all the others in the universe. Likewise, the Zen gardens in East Asia express the fullness of the Buddha-nature (*tathāgatagarbha*) in the natural world. In recent years, socially engaged Buddhism has been active in protecting the environment in both Asia and the United States.

The East Asian traditions of Confucianism and Taoism remain, in certain ways, some of the most life-affirming in the spectrum of world religions.[15] The seamless interconnection between the divine, human, and natural worlds that characterizes these traditions has been described as an anthropocosmic worldview.[16] There is no emphasis on radical transcendence as there is in the Western traditions. Rather, there is a cosmology of a continuity of creation stressing the dynamic movements of nature through the seasons and the agricultural cycles. This organic cosmology is grounded in the philosophy of *ch'i* (material force), which provides a basis for appreciating the profound interconnection of matter and spirit. To be in harmony with nature and with other humans while being attentive to the movements of the *Tao* (Way) is the aim of personal cultivation in both Confucianism and Taoism. It should be noted, however, that this positive worldview has not prevented environmental degradation (such as deforestation) in parts of East Asia in both the premodern and modern period.

In a similar vein, indigenous peoples, while having ecological cosmologies have, in some instances, caused damage to local environments through such practices as slash-and-burn agriculture. Nonetheless, most indigenous peoples have environmental ethics embedded in their worldviews. This is evident in the complex reciprocal obligations surrounding life-taking and resource-gathering which mark a community's relations with the local bioregion. The religious views at the basis of indigenous lifeways involve respect for the sources of food, clothing, and shelter that nature provides. Gratitude to the creator and to the spiritual forces in creation is at the heart of most indigenous traditions. The ritual calendars of many indigenous peoples are carefully coordinated with seasonal events such as the sound of returning birds, the blooming of certain plants, the movements of the sun, and the changes of the moon.

The difficulty at present is that for the most part we have developed in the world's religions certain ethical prohibitions regarding homicide and restraints concerning genocide and suicide, but none for biocide or geocide. We are clearly in need of exploring such comprehensive cosmological perspectives and communitarian environmental ethics as the most compelling context for motivating change regarding the destruction of the natural world.

Responses of Religions to the Environmental Crisis

How to chart possible paths toward mutually enhancing human-earth relations remains, thus, one of the greatest challenges to the world's religions. It is with some encouragement, however, that we note the growing calls for the world's religions to participate in these efforts toward a more sustainable planetary future. There have been various appeals from environmental groups and from scientists and parliamentarians for religious leaders to respond to the environmental crisis. For example, in 1990 the Joint Appeal in Religion and Science was released highlighting the urgency of collaboration around the issue of the destruction of the environment. In 1992 the Union of Concerned Scientists issued the statement "Warning to Humanity," signed by over 1,000 scientists from 70 countries, including 105 Nobel laureates, regarding the gravity of the environmental crisis. They specifically cited the need for a new ethic toward the earth.

Numerous national and international conferences have also been held on this subject and collaborative efforts have been established. Environmental groups such as World Wildlife Fund have sponsored interreligious meetings such as the one in Assisi in 1986. The Center for Respect of Life and Environment of the Humane Society of the United States has also held a series of conferences in Assisi on

Spirituality and Sustainability and has helped to organize one at the World Bank. The United Nations Environmental Programme in North America has established an Environmental Sabbath, each year distributing thousands of packets of materials for use in congregations throughout North America. Similarly, the National Religious Partnership on the Environment at the Cathedral of St. John the Divine in New York City has promoted dialogue, distributed materials, and created a remarkable alliance of the various Jewish and Christian denominations in the United States around the issue of the environment. The Parliament of World Religions held in 1993 in Chicago and attended by some 8,000 people from all over the globe issued a statement of Global Ethics of Cooperation of Religions on Human and Environmental Issues. International meetings on the environment have been organized. One example of these, the Global Forum of Spiritual and Parliamentary Leaders held in Oxford in 1988, Moscow in 1990, Rio in 1992, and Kyoto in 1993, included world religious leaders, such as the Dalai Lama, and diplomats and heads of state, such as Mikhail Gorbachev. Indeed, Gorbachev hosted the Moscow conference and attended the Kyoto conference to set up a Green Cross International for environmental emergencies.

Since the United Nations Conference on Environment and Development (the Earth Summit) held in Rio in 1992, there have been concerted efforts intended to lead toward the adoption of an *Earth Charter* by the year 2000. This *Earth Charter* initiative is under way with the leadership of the Earth Council and Green Cross International, with support from the government of the Netherlands. Maurice Strong, Mikhail Gorbachev, Steven Rockefeller, and other members of the Earth Charter Project have been instrumental in this process. At the March 1997 Rio + 5 Conference a benchmark draft of the *Earth Charter* was issued. The time is thus propitious for further investigation of the potential contributions of particular religions toward mitigating the environmental crisis, especially by developing more comprehensive environmental ethics for the earth community.

Expanding the Dialogue of Religion and Ecology

More than two decades ago Thomas Berry anticipated such an exploration when he called for "creating a new consciousness of the multiform religious traditions of humankind" as a means toward renewal of the human spirit in addressing the urgent problems of contemporary society.[17] Tu Weiming has written of the need to go "Beyond the Enlightenment Mentality" in exploring the spiritual resources of the global community to meet the challenge of the ecological crisis.[18] While this exploration has also been the intention of both the conferences and these volumes, other significant efforts have preceded our current endeavor.[19] Our discussion here highlights only the last decade.

In 1986 Eugene Hargrove edited a volume titled *Religion and Environmental Crisis.*[20] In 1991 Charlene Spretnak explored this topic in her book *States of Grace: The Recovery of Meaning in the Post-Modern Age.*[21] Her subtitle states her constructivist project clearly: "Reclaiming the Core Teachings and Practices of the Great Wisdom Traditions for the Well-Being of the Earth Community." In 1992 Steven Rockefeller and John Elder edited a book based on a conference at Middlebury College titled *Spirit and Nature: Why the Environment Is a Religious Issue.*[22] In the same year Peter Marshall published *Nature's Web: Rethinking Our Place on Earth,*[23] drawing on the resources of the world's traditions. An edited volume titled *Worldviews and Ecology,* compiled in 1993, contains articles reflecting on views of nature from the world's religions and from contemporary philosophies, such as process thought and deep ecology.[24] In this same vein, in 1994 J. Baird Callicott published *Earth's Insights,* which examines the intellectual resources of the world's religions for a more comprehensive global environmental ethics.[25] This expands on his 1989 volumes, *Nature in Asian Traditions of Thought* and *In Defense of the Land Ethic.*[26] In 1995 David Kinsley issued a book titled *Ecology and Religion: Ecological Spirituality in a Cross-Cultural Perspective,*[27] which draws on traditional religions and contemporary movements, such as deep ecology and ecospirituality. Seyyed Hossein Nasr wrote his comprehensive study *Religion and the Order of Nature* in 1996.[28] Several volumes of religious responses to a particular topic or theme have also been published. For example, J. Ronald Engel and Joan Gibb Engel compiled a monograph in 1990 titled *Ethics of Environment and Development: Global Challenge, International Response*[29] and in 1995 Harold Coward edited the volume *Population, Consumption and the Environment: Religious and Secular Responses.*[30] Roger Gottlieb edited a useful source book, *This Sacred Earth: Religion, Nature, Environment.*[31] Single volumes on the world's religions and ecology were published by the Worldwide Fund for Nature.[32]

The series Religions of the World and Ecology is thus intended to expand the discussion already under way in certain circles and to invite further collaboration on a topic of common concern—the fate of the earth as a religious responsibility. To broaden and deepen the reflective basis for mutual collaboration was an underlying aim of the conferences themselves. While some might see this as a diversion from pressing scientific or policy issues, it was with a sense of humility and yet conviction that we entered into the arena of reflection and debate on this issue. In the field of the study of world religions, we have seen this as a timely challenge for scholars of religion to respond as engaged intellectuals with deepening creative reflection. We hope that these volumes will be simply a beginning of further study of conceptual and symbolic resources, methodological concerns, and practical directions for meeting this environmental crisis.

Notes

1. He goes on to say, "And that is qualitatively and epochally true. If religion does not speak to [this], it is an obsolete distraction." Daniel Maguire, *The Moral Core of Judaism and Christianity: Reclaiming the Revolution* (Philadelphia: Fortress Press, 1993), 13.

2. Gerald Barney, *Global 2000 Report to the President of the United States* (Washington, D.C.: Supt. of Docs. U.S. Government Printing Office, 1980–1981), 40.

3. Lynn White, Jr., "The Historical Roots of Our Ecologic Crisis," *Science* 155 (March 1967):1204.

4. Thomas Berry, *The Dream of the Earth* (San Francisco: Sierra Club Books, 1988).

5. Brian Swimme and Thomas Berry, *The Universe Story* (San Francisco: Harper San Francisco, 1992).

6. At the same time we recognize the limits to such a project, especially because ideas and action, theory and practice do not always occur in conjunction.

7. E. N. Anderson, Ecologies of the Heart: Emotion, Belief, and the Environment (New York and Oxford: Oxford University Press, 1996), 166. He qualifies this statement by saying, "The key point is not religion per se, but the use of emotionally powerful symbols to sell particular moral codes and management systems" (166). He notes, however, in various case studies how ecological wisdom is embedded in myths, symbols, and cosmologies of traditional societies.

8. *Is It Too Late?* is also the title of a book by John Cobb, first published in 1972 by Bruce and reissued in 1995 by Environmental Ethics Books.

9. Because we cannot identify here all of the methodological issues that need to be addressed, we invite further discussion by other engaged scholars.

10. See J. Baird Callicott, *Earth's Insights: A Survey of Ecological Ethics from the Mediterranean Basin to the Australian Outback* (Berkeley: University of California Press, 1994).

11. See, for example, The Quality of Life, ed. Martha C. Nussbaum and Amartya Sen, WIDER Studies in Development Economics (Oxford: Oxford University Press, 1993).

12. White, "The Historical Roots of Our Ecologic Crisis," 1203–7.

13. Process theology, creation-centered spirituality, and ecotheology have done much to promote these kinds of holistic perspectives within Christianity.

14. These are resources already being explored by theologians and biblical scholars.

15. While this is true theoretically, it should be noted that, like all ideologies, these traditions have at times been used for purposes of political power and social control. Moreover, they have not been able to prevent certain kinds of environmental destruction, such as deforestation in China.

16. The term "anthropocosmic" has been used by Tu Weiming in *Centrality and Commonality* (Albany: State University of New York Press, 1989).

17. Thomas Berry, "Religious Studies and the Global Human Community," unpublished manuscript.

18. Tu Weiming, "Beyond the Enlightenment Mentality," in *Worldviews and Ecology,* ed. Mary Evelyn Tucker and John Grim (Lewisburg, Pa.: Bucknell University Press, 1993; reissued, Maryknoll, N.Y.: Orbis Books, 1994).

19. This history has been described more fully by Roderick Nash in his chapter entitled "The Greening of Religion," in The Rights of Nature: A History of Environmental Ethics (Madison: University of Wisconsin Press, 1989).

20. *Religion and Environmental Crisis,* ed. Eugene Hargrove (Athens: University of Georgia Press, 1986).

21. Charlene Spretnak, *States of Grace: The Recovery of Meaning in the Post-Modern Age* (San Francisco: Harper San Francisco, 1991).

22. *Spirit and Nature: Why the Environment Is a Religious Issue,* ed. Steven Rockefeller and John Elder (Boston: Beacon Press, 1992).

23. Peter Marshall, *Nature's Web: Rethinking Our Place on Earth* (Armonk, N.Y.: M. E. Sharpe, 1992).

24. *Worldviews and Ecology,* ed. Mary Evelyn Tucker and John Grim (Lewisburg, Pa.: Bucknell University Press, 1993; reissued, Maryknoll, N.Y.: Orbis Books, 1994).

25. Callicott, *Earth's Insights.*

26. Both are State University of New York Press publications.

27. David Kinsley, *Ecology and Religion: Ecological Spirituality in a Cross-Cultural Perspective* (Englewood Cliffs, N.J.: Prentice Hall, 1995).

28. Seyyed Hossein Nasr, *Religion and the Order of Nature* (Oxford: Oxford University Press, 1996).

29. *Ethics of Environment and Development: Global Challenge, International Response,* ed. J. Ronald Engel and Joan Gibb Engel (Tucson: University of Arizona Press, 1990).

30. *Population, Consumption, and the Environment: Religious and Secular Responses,* ed. Harold Coward (Albany: State University of New York Press, 1995).

31. This Sacred Earth: Religion, Nature, Environment, ed. Roger S. Gottlieb (New York and London: Routledge, 1996).

32. These include volumes on Hinduism, Buddhism, Judaism, Christianity, and Islam.

Note on the Romanization of Chinese Terms: Dao or Tao?

This book employs the Pinyin system for transcribing, or "romanizing," Chinese words into alphabetical form, rather than the older Wade-Giles system. Hence, in these pages the Chinese character for the "Way" is rendered "Dao" rather than the more familiar "Tao." So also will the reader encounter such terminology as the *Daode jing*, in place of *Tao Te Ching* (for the ancient *Book of the Way and Its Power*), and Laozi and Zhuangzi for the sages more commonly known in Western sources as Lao Tzu and Chuang Tzu. The reasons for this alteration are largely technical and phonological and reflect current scholarly conventions. One other reason for insisting on a system that is relatively unfamiliar to a general audience is that its apparent strangeness (given the "traditional" usage of the nineteenth-century Wade-Giles system for English transliteration) calls attention to the important point that earlier discussions of the Daoist tradition were often distorted and misleading—especially in terms of the special Western fascination with the "classical" or "philosophical" *Daode jing* and the denigration and neglect of the later sectarian traditions. The simple fact is that older ways of understanding and naming the tradition do not reflect recent revisionary developments in Daoist scholarship.

Guide to the Pronunciation of Pinyin

Adapted from *Compton's Living Encyclopedia* (AOL 8/16/1995) and Paul Halsall's web page (http://acc6.its.brooklyn.cuny.edu/~phalsall/texts/chinlng1.html).

Pinyin	*Wade-Giles*	*Pronounced as*
b	p	b as in "be," aspirated
c	ts'	ts as in "its"
ch	ch'	as in "church"
d	t	d as in "do"
g	k	g as in "go"
ian	ien	as in "yen"
j	ch	j as in "jeep"
k	k'	k as in "kind," aspirated
p	p'	p as in "par," aspirated
q	ch'	ch as in "cheek"
r	j	approximately like the "j" in French "je"
s	s, ss, sz	s as in "sister"
sh	sh	sh as in "shore"
si	szu	ts plus oo as in "took"
t	t'	t as in top
x	hs	sh as in "she," thinly sounded
yi	i	approximately like the "ee" in "meet"
you	yu	y plus o as in "owl"
z	ts	z as in "zero"
zh	ch	j as in "jump"
zi	tzu	dz plus oo as in "took"

Acknowledgments

This volume grows out of the conference "Daoism and Ecology" held at the Center for the Study of World Religions, Harvard University, 5–8 June 1998. This was one of the final sessions in a groundbreaking series of conferences entitled "Religions of the World and Ecology" and organized by Mary Evelyn Tucker and John Grim, two pioneering and perspicacious scholars who believe that academics really do have a significant role to play in the larger scheme of things. Lawrence Sullivan, the director of the Center, must be thanked for his support, as must his incredibly efficient and hospitable staff at the Center (especially Malgorzata Radziszewska-Hedderick and Don Kunkel). Stephanie Snyder at Bucknell University was also extraordinarily proficient at maintaining lines of communication. We owe all of the above a special debt of gratitude. At the present time, the important work initiated at these conferences continues in the Forum on Religion and Ecology based at the Yenching Institute at Harvard, directed by Tu Weiming, and the Harvard Committee on the Environment, headed by Michael McElroy.

Organized by Norman Girardot and Livia Kohn, the conference was a rare gathering of Daoist scholars and Daoist practitioners from around the world that included a number of colleagues not directly represented in this volume. We want to extend special thanks, therefore, to Tu Weiming of Harvard, Toshiaki Yamada of Tokyo University, Franciscus Verellen of the École Française d'Extrême-Orient, Robert Weller of Boston University, Chu Ron Guey of the Academica Sinica-Taipei, Paul Kjellberg of Whittier College, Michael Puett of

Harvard University, Peter Nickerson of Duke University, and J. P. Seaton of the University of North Carolina, Chapel Hill (see the endnotes to Ursula Le Guin's "Dao Song" for Seaton's response at the conference). We also want to recognize the encouragement and support of several other scholars and two prominent Daoists who, for various reasons, were unable to participate in the conference—Anthony Yu of the University of Chicago, Julian Pas of the University of Saskatchewan, Ellen Marie Chen of St. John's University, John Lagerwey of the Chinese University of Hong Kong, Eva Wong of the Fung Loy Kok Taoist Temple, Colorado, and Mantak Chia of the Healing Tao Center. Furthermore, we want to register the spirited participation of the sizable audience of scholars, students, and other interested observers—all were "honorary Daoists" for the duration of the conference. Finally, we are grateful for the special help and participation of Ma Xiaohong of Temple University and three undergraduate students from Lehigh University: Christopher Spinney, Stephanie Nelson, and Annette Gavigan.

We want to note that while this volume largely depends on the papers presented at the conference, it is augmented by various specially solicited contributions (both new articles and several commissioned translations). These include the contributions by Stephen Field of Trinity University in Texas and Terry Kleeman of the University of Colorado. Of particular importance is the article by Zhang Jiyu (a sixty-fifth generation Celestial Master in China) and Li Yuanguo ("'Mutual Stealing among the Three Powers' in the *Scripture of Unconscious Unification*"), translated by Yam Kah Kean and Chee Boon Heng in Singapore, with scholarly annotations by Louis Komjathy at Boston University, and Zhang's "Declaration of the Chinese Daoist Association on Global Ecology," translated by David C. Yu. Livia Kohn also deserves special credit for her overall support and her deft compilation of the roundtable discussion of Daoist practitioners ("Change Starts Small: Daoist Practice and the Ecology of Individual Lives"). Moreover, James Miller's heroic work on the annotated bibliography (aided and abetted by Jorge Highland, Liu Xiaogan, Zhong Hongzhi, and Belle B. L. Tan) and sectional discussions (digesting and developing the comments of Richard G. Wang of the Chinese University of Hong Kong, Edward Davis of the University of Hawai'i, John Patterson of Massey University, New Zealand, and Russell Goodman of the University of New Mexico) deserves special commendation. Liu Xiaogan's exceptional help with the Chinese glossary, the arrangements for the translations, and the contributor biographies also requires particular acknowledgment. The triumvirate of editors were involved in all aspects of the preparation of this volume, but Norman Girardot was primarily responsible for drafting the introduction. However, in keeping with the typical ritual disclaimer on such occasions as this, all three editors collectively share the blame for any blunders and oversights perpetrated in the pages that follow.

The preparation of this book was greatly facilitated by the gracious professionalism of Kathryn Dodgson, the publications coordinator at the Center for the Study of World Religions. Marian Gaumer, Religion Studies Department Coordinator at Lehigh University, also greatly aided with various aspects of the production. We thank them for their patience and good cheer during a sometimes tedious and protracted process.

Lastly, let it be said that in reference to both the conference and this volume, the symbol and spirit of the "calabash-gourd" was an ever-present talisman of the fruitful potential of a Daoist approach to today's ecological problems. It is, after all, a pregnant emptiness that gives birth to all that exists—discursively, organically, and morally.

The Editors

Introduction

N. J. GIRARDOT, JAMES MILLER, and LIU XIAOGAN

> As for the Dao, the Way that can be spoken of is not the
> constant Way;
> As for names, the name that can be named is not the con-
> stant name.
> The nameless is the beginning of the ten thousand things;
> The named is the mother of the ten thousand things.
> Therefore, those constantly without desires, by this means
> will perceive its subtlety.
> Those constantly with desires, by this means will see only
> that which they yearn for and seek.
> These two together emerge;
> They have different names yet they're called the same;
> That which is even more profound than the profound—
> The gateway of all subtleties.
> —*Daode jing/Dedao jing* (Mawangdui B),
> chap. 1 (amended Henricks)

The Named and the Nameless

Daoism and ecology are often invoked as natural partners in contemporary discussions of environmental issues in the West. When looking to the religious and intellectual resources provided by various "world religions," it has therefore been a commonplace assumption that the Chinese tradition conventionally known as "Daoism/Taoism" reveals an obvious and particularly compelling affinity with global ecological concerns.[1] For most Western commentators until recently, Daoism primarily referred to the "mystical wisdom" found in several ancient "classical" texts (especially the *Daode jing* and *Zhuangzi*) and was seen to be fundamentally in tune with heightened contemporary fears about the increasingly fractured relations between humanity and the natural world. Popular testimony would even whimsically suggest that Pooh Bear and Piglet affirmed the profound ecological sensibility of the ancient Chinese Daoists.[2]

Unfortunately there has been very little serious discussion of this beguiling equation of Daoism and ecology. Too much has been simply, and sometimes fantastically, taken for granted about what is finally quite elusive and problematic—both concerning the wonderfully "mysterious" tradition known as Daoism and, in this case, the "natural" confluence of Daoism and contemporary ecological concerns. Among the shelves of Western books and articles written in the past twenty-five years about the religious, ethical, and philosophical implications of a worldwide "environmental crisis," there have been many passing allusions to a kind of Daoist ecological wisdom (often associated with Native American and other tribal-aboriginal perspectives, as well as with Pooh-like themes and the free-floating and universalized "Suzuki-Zen" of an earlier generation).[3] However, there is still no single work that is grounded in a scholarly understanding of the real complexities of the Daoist tradition and is also devoted to a critical exploration of the tradition's potential for informing current ecological issues.

Even in works generally well informed about various religions and ecological issues, a certain kind of romantic infatuation with a "classically pure" and timelessly essential Daoism (embedded within one or two ancient texts and connected with a few key themes) has tended to shape the overall discussion of how this tradition can be "applied" to the problems of the contemporary world. The question remains whether there is anything to be learned beyond various vague appeals to Laozi's enigmatic little treatise "The Way and Its Ecological Power," to Zhuangzi's playfully insightful parables about "useless" trees and gourds, or to popular visions of a Yoda-like Chinese sage wandering amidst a mist-laden cosmic landscape of craggy mountains, swaying bamboo, and lofty waterfalls. Despite these ongoing reveries, Daoism is increasingly being recognized as an exceedingly rich religious tradition with an immense textual and historical lore that defies any attempt to reduce its meaning to a few ancient texts or Forrest Gump platitudes. It is clear that many popular assumptions about Daoism say less about the real significance of the tradition for ecological concerns than they say about the desire and dominion of Western regimes of both scholarly and popular understanding which, in the words of the *Daode jing*, tend to "see only that which they yearn for and seek."[4] The difficult truth is that there is much that has not been named or known either about Daoism itself or about its possible contribution to recent environmental problems.

Since popular stereotypes are not easily dispelled, it is worth underscoring some of the more pervasive Western distortions about Daoism at the beginning of the twenty-first century. Thus it has been said—both seriously and flippantly—that Daoism is as Daoism does.[5] Daoism in the West at times seems to be a sitcom religion about "nothing" at all, a situation compounded by its resolute reliance on "non-action" or *wuwei*. As we know in this age of global MTV and the World Wide Web, such pop fabulations are often more mesmerizing and influential than the revisionary constructions by scholarly specialists, sinologists, and historians of religion in Paris, Kyoto, Beijing, or Boston. Daoism is, however, about nothing *and* something; and it takes the silly and the serious to tell the fullness of the Daoist story in China and in its contemporary manifestations throughout the world. At the very least, a heightened awareness of these difficulties will help to establish some imaginative footing for slowly walking a path back to the actual historical and cultural complexities of the tradition. Returning to these as yet unnamed aspects of Daoism at the same time provides the crucial pretext and context for naming some of the tradition's implications for ecological thought and practice.

While both popular misconceptions and scholarly "perplexities" abound concerning Daoism,[6] similar difficulties can be found in contemporary Western discussions of ecology, especially those harboring various apocalyptic emotions. The salvational urge for a definitive bio-spiritual reformation of life on earth to some extent represents a discursive artifact of an Enlightenment and liberal Protestant "post-millennialist" missionary agenda hidden within the authoritative structures of knowledge in the West.[7] In this sense, things are decidedly deep and ominously foreboding during these days of millennial passage. There is, consequently, much overly portentous talk about the special spiritual *gravitas* of both Daoism and ecology—that is, the mystical ecoprofundities of the *Daode jing* along with deep ecology, a deeper socioecology and ecofeminism, and an even deeper bio-religiosity of Gaia-Earth.[8] Needless to say, the real life-and-death issues of environmental concern are not well served by too quickly conflating a romantic fantasy about Daoism with a certain kind of evangelical passion for ecological damnation and salvation.

Finally, we need to remember that throughout the long Chinese (and now Western) history of the tradition, individual Daoists have often resisted overly hasty and sentimentalized presumptions about the Ways taken and not taken. Buddhists and Buddhologists, for example, have been considerably more "engaged" with contemporary ecological issues than Daoist practitioners and scholars in either Asia or the West.[9] This situation no doubt reflects the more developed nature of Buddhist scholarship and the presence of a substantial tradition of acculturated self-reflection on the part of Western Buddhists, but it also hints at a *wuwei*-inspired caution among Daoists in the past and present regarding interventionist forms of crisis management and overly assertive forms of social engagement. While today some living Daoist masters are recommending the need for concerted social action to combat the accelerated destruction of China's sacred mountains (see Zhang Jiyu's "Declaration of the Chinese Daoist Association on Global Ecology" in this volume), a few contemporary Daoists still seem to prefer a more muddled and less meddlesome methodology (e.g., in this volume, the comments of the American Daoist Liu Ming in "Change Starts Small").[10] This kind of instinctive wariness, though sometimes simply contrarian and polemical, has both a historical and an ethical rationale in Daoist tradition. Ever since the time of the *Zhuangzi*, some Daoists have avoided "huffing and puffing" after an overly instrumental form of virtue—not an unimportant consideration during morally ambiguous periods such as our own, when "charity" has often become a corporate commodity.

The Ecological Landscape in China

Regardless of a *wuwei*-ish prudence among some Daoists and the evangelical simplifications in certain aspects of the Western rhetoric of immediacy and profundity, there are real and pressing ecological problems affecting the world today.[11] Moreover, the complex synergistic issues of life on this fragile biosphere, issues which are always relational and ecological in nature, certainly have important scientific, moral, and religious implications for every nation on earth. The truth is that China, the ancestral homeland of Daoism, constitutes a dramatically disturbing case of ecological neglect.[12] Indeed, a balanced appraisal of the ecological condition in China today is difficult and often discouraging. While the destruction of the natural environment, especially involving deforestation and desertification, has a long and sad history, it cannot be denied that there has been an accelerated deterioration of the ecological situation in China during the last half of the twentieth century, particularly following the rapid economic expansion since the early 1980s.

According to Zhang Kunmin, secretary-general of the Chinese Council for International Cooperation on the Environment and Development, there are five major problems concerning the protection and conservation of the environment in contemporary China. The first of these issues involves the immense Chinese population which, even with stringent birth policies, has a net yearly growth of more than thirteen million people (a number nearly equal to half of the Canadian population). Second, there is the incredible rate of urbanization in China where the population in the cities has increased by 180 million from 1978 to 1995, plus an additional 50 million or so of a kind of "floating" population. This is a growth which is accompanied by an exponential escalation of pollution, waste, sewage, and transportation problems. A third major difficulty is the rapid, and often unbalanced and uncontrolled, economic expansion since the 1980s. There has been, for example, a disproportionate development of heavy and chemical industries, which produce a tremendous amount of pollutants. Moreover, burning coal is the major source of energy in China, a situation which seriously aggravates the overall quality of air. A fourth and related consideration involves the inadequate Chinese investment in organizations and equipment concerned with environmental protection and improvement. Finally, and this is where the traditional religions have a clear role, there has been a general lack of public consciousness regarding ecological problems and a failure to develop comprehensive national policies of environmental control.[13]

The facts concerning ecological deterioration in contemporary China are grim and have obvious global implications. A study by the Washington-based World Resources Institute has concluded that nine of the ten worst air-polluted cities in the world are found in China.[14] Other statistics concerning deforestation, desertification, and water pollution are equally alarming.[15] There are, however, some promising signs that go back to 1972 when the Chinese government sent a delegation to the First United Nations Conference on the Human Environment in Stockholm. Then, in 1979, China promulgated the Environmental Protection Law for "trial implementation," finally adopting it as law in 1989. The upgrading of the Chinese National Environmental Protection Agency to the State Environmental Protection Administration in March 1998 represents an important recent development and an attempt to put more managerial authority into environmental planning.

Nevertheless, it must be said that there is still a low level of governmental action on environmental issues in China.[16] More encouraging than these sporadic and halfhearted official efforts is the emergence and growth of an environmental consciousness among the general public in China. In the past few decades, an increasing number of journalists, writers, scholars, religious leaders, workers, and farmers have begun to speak out on the ecological situation. It has been in this grassroots context that modern Daoists have started, in a small way, to contribute to the protection and renewal of natural resources, especially to projects concerning reforestation. Thus, under the leadership of the abbot Yang Chenquan, the Daoist priests of Mt. Wudang temple in Qinghai Province have since 1982 grown 1.73 million trees and have recultivated many acres of grassland. Likewise Fan Gaode, the old abbot of the Huashiyan Daoist temple in Gansu Province, is said to have made the "barren hills" green again.[17]

These actions by Daoists in China are noteworthy, but at the same time they are quite modest and were influenced directly or indirectly by various Western environmental movements. Furthermore, if Daoism somehow has a special ecological wisdom going back to the very foundations of the tradition, why has there been such a woeful record of environmental concern throughout Chinese history, and why, for that matter, have the actions of contemporary Daoists been so meager and relatively restricted? While these questions are tentatively addressed by several of the papers in this book, they remain problems that touch upon the general history of Chinese civilization and have no easy answers. In terms of China's immediate problems, it must also be specifically asked how and in what way Daoism, or any of the other traditional religions and philosophies, can make a greater and more systemic contribution to the environmental situation. Part of the answer no doubt involves various Western-influenced, short-term "techno-fix" methods for tempering and recycling aspects of rampant economic development.[18] However, it would seem that the long-term regeneration and sustainable care of the overall environment in China will even more depend on a broad national "sino-ecological" commitment that draws upon traditional values creatively reinterpreted and reappropriated by contemporary religious leaders and scholars both in China and in the West.[19] In the intensely pluralistic context of the postmodern world, effective environmental efforts in particular countries will require a global consciousness and cooperative methodologies informed by the distinctive cultural insights of individual traditions (such as Daoism within the Chinese context).

The creative application of traditional Daoist values (values that, as this volume shows, cannot be restricted to a few classical texts) to contemporary problems will perhaps only be determined in relation to a hermeneutical strategy that understands the whole environmental problem in its specific and practical interrelationship with each of the "ten thousand things" making up the natural world. Such an awareness may be called a kind of latter-day Daoist perspective if we keep in mind Roger Ames's distinction between the "local and focal" in ecological questions and David Hall's observation that the ancient Lao-Zhuang texts "celebrate the insistent particularity of items comprising the totality of things."[20] So also does James Miller, in the spirit of the Highest Clarity texts, call for an imaginative realization of a "Daoist ecotheology" that fosters "respect" for the totality of the cosmic environment.[21] What is needed to conjoin an emergent "Eco-Daoism" with the meaningful passions of "Deep Ecology" is a more insistent concern for the reciprocal interrelationship of all the constituent parts of the Dao as a cosmic body or landscape.

Remembering the *Zhuangzi*'s meditation on the relativity of understanding and behavior, the Dao is always to be found in the large *and* the small, in the gigantic peng bird *and* the lowly "piss and shit" of the world, in the snow leopard *and* the snail darter. The nameless is known only in and through the named, in the recalcitrant details of all the myriad life-forms that are subject to both regeneration and degradation. This perspective applies theoretically and pragmatically both to our own appreciation of the full historical complexity and cultural intertextuality of the Daoist tradition and to a contemporary "Daoist" response to any disruptions in the delicate balance and incredible biodiversity of things. In a way that interestingly (and sometimes esoterically) expands on the *Laozi* and the *Zhuangzi*, this concern for the dynamic interaction of all forms of life is also envisioned by the loosely organized Daoist religious tradition, which was ritually concerned with how particular human persons, concrete local communities, and regional natural environments comprise the corporate and constantly transforming Body of the Dao. In this regard, it is noteworthy that practicing Daoist masters in China have recently emphasized the "symbiotic mutuality" of "Heaven, Earth, and humankind" and have issued an ecological statement stressing Daoism's "unique sense of value," which judges human affluence in terms of the preservation of the many different species of life (see in this volume the articles "Stealing among the Three Powers" and "A Declaration of the Chinese Daoist Association," both involving Zhang Jiyu; Zhang is a sixty-fifth generation descendant of Zhang Daoling, the founder of the Celestial Masters tradition, and a vice president of the Chinese Daoist Association).[22]

Particular Parts of the Way

The particular contribution this book makes to an embryonic Daoist perspective on contemporary ecological problems is its concern for the fullness of the Daoist tradition—that is, the incredible corpus of "revealed" texts, the complex ritual and meditational practices, composite sociological forms, practical eclectic ethics, and soaring cosmic vision associated with the eighteen-hundred-year history of the living Daoist religion. It is this exceptionally luxuriant but little understood tradition comprising thousands of scriptures and dozens of sectarian movements that is still sorely neglected in popular Western discussions about Daoism. Rather than what was often called only a vulgar degeneration of the "pure" philosophy enunciated in the classic texts, the Daoist religion "names" an amorphous amalgamation of cultural phenomena that equals the sociological and intellectual complexity of medieval and reformation Christianity in Europe. Fortunately, given advances in recent Daoist scholarship—including the increasing availability of accurate translations of significant Daoist religious scriptures, various new interpretive perspectives coming from the comparative history of religions and other disciplines, and a revived scholarship by native scholars and practicing Daoists in China and the West—a significant revision of our understanding of Daoism is now possible.[23] In like manner, the study of the organized Daoist religion in the past and present is also leading to new insights concerning the meaning and use of the *Daode jing* and *Zhuangzi* (see especially the papers in section four of this volume).

What is found in the pages that follow offers no simple or straightforward conclusions regarding a Daoist approach (or approaches) to current environmental issues. This book, nonetheless, does constitute the only collection of articles discussing the ecological implications of both the earliest "classical" texts and the fascinating yet often bewildering Daoist religious scriptures. This is a work that not only challenges many popular assumptions about the earliest Daoist texts (especially the difficulties of too quickly reading a Western-style ecological consciousness into the early "philosophical" writings associated with Laozi and Zhuangzi; see, for example, the positions argued by Russell Kirkland and Lisa Raphals in this volume), but also embraces a contextualized approach to the complex cultural significance of Daoist religious thought and social practice in the Chinese past and in the more pluralistic present. To some degree, therefore, this book marks a new stage within the evolution of Daoist studies because it shows that Daoist scholars (both in the West and in Asia) have reached a stage of confluent hermeneutical sophistication that for the first time allows for a project of contemporary global discourse. Buddhologists could have done this fifty years ago, but it would not have been possible in Daoist studies even five to ten years ago.

The new perspectives coming from recent scholarship on the "real" religious Daoism of the Chinese people do not necessarily invalidate everything we thought we knew about the sage sayings in the *Laozi* and *Zhuangzi* (the ancient texts have, after all, always inspired and influenced the organized religious tradition—as, for example, Kristofer Schipper's "Study of the Precepts of the Early Daoist Ecclesia" and Zhang's "Declaration" demonstrate). They do, however, strongly suggest that we will have to expand our horizons concerning Daoism's "philosophical," "religious," "theological," and "ethical" understandings of the dynamic interconnectedness of human and cosmic life. In the most basic sense, Daoism—whether associated with the early

texts or the later organized religion—does have something important
to say regarding many ecological questions. What it suggests, however, is almost always more contradictory and provocative than we
could ever have imagined when constrained by the neatly polarized
categories of an early "mystical" philosophy (*daojia*) and a corruptly
superstitious and ritualistic later religion (*daojiao*). In the best sense
of the postmodernist critique of Western scholarship, essentializing
definitions of Daoism must be replaced by the messy particularity of
various "Daoisms" interacting with all aspects of Chinese tradition.

The Daoist religious tradition consists of numerous schools and
syncretistic sectarian movements that cannot be easily categorized or
summarized.[24] Nevertheless, it may be helpful to indicate that, as distinct from the discursive "protohistory" of the tradition associated
with ancient texts like the *Laozi* and *Zhuangzi*, the history of Daoism
as a self-consciously organized religion goes back to movements at
the breakup of the Han dynasty (second and third centuries C.E.), especially the Tianshi, or Celestial Masters, tradition affiliated with the
revelations to Zhang Daoling (traditional dates, 34–156 C.E.). Two other
important revelatory textual traditions followed the Celestial Masters
movement in the fourth and fifth centuries. One of these was known
as the Shangqing, or Highest Clarity, tradition, which stressed visionary experience and practice; the other came to be called the Lingbao,
or Numinous Treasure, tradition and emphasized ritual practices.
Both of these amorphous traditions not only drew upon indigenous
aspects of Chinese religious tradition but also incorporated significant
aspects of Buddhism. All subsequent movements were influenced by
these early forms of revealed Daoism. From the fifth to the tenth century, the various Daoist sectarian religious groups were loosely organized, and their scriptures were systematized in an open-ended
"canon" that came to be known as the *Daozang*, or Treasury of the
Dao. New reformist types of Daoist religion emerged from the tenth
through the fourteenth century—among which were the schools of
"internal alchemy" (*neidan*), new liturgical traditions, and several
syncretistic schools that accented a morality combining Daoist, Buddhist, and Confucian values. Of the Daoist movements developing after the eleventh century, two traditions continue to the present day.
The first of these is the Southern, or Zhengyi (Orthodox Unity), form
of Daoism, which was traditionally centered at Mt. Longhu in south

China and claims to continue the ritualistic and priestly traditions of the ancient Celestial Masters.[25] The second tradition is the Northern, or Quanzhen (Complete Perfection), Daoism, which is today nominally based at the White Cloud Abbey in Beijing. It continues the meditation tradition of "inner alchemy" and shows strong affinities with Chan Buddhism and Neo-Confucianism.[26] Finally, it must be noted that in recent years various forms of transplanted and acculturated Daoism have sprung up in North America and Europe.[27]

Lastly, we want to emphasize that, for all of the advances in Daoist scholarship, there are still many aspects of the tradition that have not yet been adequately studied. Related to these historical and textual gaps in our understanding are also the larger methodological issues having to do with the definition of the tradition, the nature and significance of the ancient "classical" texts, the complex dynamics of Han and Six Dynasties religious history as it relates to the origins of the organized religious tradition, the important interaction with Buddhism and the imperial state, and so on. It was never our intention to resolve all of these scholarly difficulties in this book. Suffice it to say, therefore, that we have deliberately operated with a broadly inclusive understanding of "Daoism," one that honors the sociological and religious distinctiveness of the organized Daoist sectarian movements, but one that also allows for the inspirational role played by the ancient "classical" texts and for the diffuse interaction of Daoist movements (however defined) with all sorts of eclectic ideas and practices associated with the traditional yin-yang/ *wu-xing* cosmology (e.g., as seen in this volume in sections three and four, such things as geomancy [fengshui], traditional medical practices, *qigong*, and martial arts—none of which are specifically "Daoist"). Given the current state of our knowledge, it has seemed best to proceed with an assemblage that incorporates a broad range of historical and cultural phenomena that in some fashion were "named" in Chinese sources as having "Daoist" affinities. Such a strategy largely begs the definitional problem of "Daoism," but at the same time it does honestly reflect the real confusion surrounding many of these issues. In this way, also, we purposely insisted on an approach that would encompass some interesting Western literary redactions of certain Daoist themes—most prominently, in this case, the work of Ursula K. Le Guin.

The Ecological Landscape of Religious Daoism

The Daoist religion that emerged during the third through fifth centuries is profoundly ecological in its theoretical disposition, but in practice does not conform easily to Western notions of what this should entail.[28] This is because some of the most prominent forms of Daoist religious cosmology recommend the transformation of the individual as a celestial being who is fully translucent to the cosmic environment in which he or she is situated. While some Daoist schools emphasize the collective and institutionalized ritual regeneration of the society and the cosmos, many forms of the Daoist religion, especially those influenced by the Highest Clarity scriptures, are typically and ideally concerned with "perfected persons" (*zhenren*). Such "immortals" or "transcendent beings" (*xian*) are able to penetrate beyond the gross physicality of ordinary existence to achieve an attentive harmony with the subtle and mysterious ("alchemical") transformations of the Dao (the ever-changing flow of cosmic processes) at its root, primordial level.[29] A dynamic ecological system that transparently links the "lower-outer-physical," or earthly, and the "higher-inner-spiritual," or cosmic, levels of human life is therefore the presupposition behind much of Daoist religious thought and practice. From this kind of cosmological perspective, the interpenetrating "bodies" of individuals, society, the natural world, and the infernal and celestial spheres truly constitute a cosmic landscape pulsating with life.

Cosmic Ecology

The Daoist universe is one and nameless, but infinitely diverse and particular. Its unity is implied by the fact that all dimensions of existence, from the budding of a flower to the orbit of the stars, may be denominated in terms of qi, the fundamental energy-matter of the universe whose dynamic pattern is a cosmic heartbeat of expansion (yang) and contraction (yin). Its diversity is a function of the complex interaction of the myriad cosmic processes, both light and fluid and heavy and dense. The universe is a single, vital organism, not created according to some fixed principle, but spontaneously regenerating itself from the primal empty-potency lodged within all organic forms of life.

It is not quite correct, therefore, to speak theoretically of an "ecology," as though there were an intellectual principle (*logos*) for comprehending one's cosmic environment (*oikos*). The *Daode jing* warns that if we speak of the Dao, such speaking must be in-constant, un-usual, or extra-ordinary. This has led, on the one hand, to an intense skepticism about the ability of human rationality to grasp properly its situation within the universe, and, on the other hand, to the flowering of a religious tradition dependent upon revelations from supreme celestial beings, those most attentive to the subtle workings of the primordial Dao. In the former case, human institutions (including this scholarly compendium) must bear the rhetorical brunt of criticism: the transformations of the universe are especially beyond the grasp of those who rely upon their long years of learning. In the latter case, it is only by being initiated into the sacred texts and proper lineages of transmission that one is able to comprehend and thereby transcend the ordinary dimensions of human existence. Knowing in the Daoist sense is always alchemical and ecological in nature since it depends on the revelatory experience and practice that comes in and through the transformation of the human body in corporate relation with all other particular bodies.

The Ecology of the Body

The hermeneutical principle on which much Daoist religious practice rests is that of the mutual interpenetration of all dimensions of being (all of which represent various gravid, liquid, and ethereal manifestations of qi), with the body as the most important field for the interaction of cosmic forces. Properly visualized within the body, gods (i.e., personified or psychologized nodal centers of spiritualized energy) dwell in their palaces, the constellations of the heavens are made manifest, and a pure and refined qi comes to flow. From this mysterious energy the alchemical embryo or immortal body is generated, and the adept is eventually reborn as a celestial immortal. This "bio-spiritual" practice is dependent upon traditional Chinese medical theory, which views the body as a complex system of interacting energy circuits. Illness, broadly speaking, is symptomatic of some defect of circulation, perhaps a blockage, a seepage, excess, or desiccation. "Religion" therefore is not the denial or the overcoming of physical existence, but its gradual refinement to an infinitesimal point of astral translucence. The idea of "salvation" that is suggested in the

Daoist religion is fundamentally medicinal—that is, concerned with the "healing" regeneration and rejuvenation of the organic matrix of life.

The Ecology of Time

Time is a function of the calendar: days and years are not numbered but named according to the interaction of two zodiacal cycles of twelve and ten. The *Jiuzhen zhongjing* (Central Scripture of the Nine Perfections), an important text of the Highest Clarity revelations, for example, details the correlation of cycles of colors, bodily organs, and divinities with days of the year and times of day. When all the cycles mesh, the possibility for radical transformation reaches its zenith. On a much larger scale, the Buddhist-influenced *Zuigen pin* (The Roots of Sin) speaks in terms of millions of cycles of kalpa revolutions, and outlines the degeneration of human culture from a simple organic community to complex civilizations based on law codes where corruption and vice are prevalent. Each kalpa cycle ends with the total destruction of the cosmos and then begins again. In either case Daoism encourages us to take a radical perspective on our temporal situation. Time is not something that passes and is then irretrievably lost. There is no kairos-moment that requires a decision of apocalyptic consequence. Human civilization and all life is inscribed within cycles far greater than can be comprehended.

Local Ecology

Because of the vast comprehensiveness of the Daoist cosmic ecology, and not in spite of it, the arena for all human action is the immediate environment. Only by paying attention to the minute details of one's local context is one able to penetrate to the deep roots of the Dao. Popular Chinese culture is full of ways for human beings to micromanage their particular environment, from fengshui (the strategy of arranging one's immediate area to take full advantage of its natural environment) to *taijiquan* (*t'ai-chi-ch'üan*; the embodiment of cosmic patterns to properly attune the self in the world). Daoism has particularly emphasized the importance of small beginnings and local perspectives not as ends in themselves, but as a strategy. The advice of the *Daode jing* is to be low, soft, weak, and nonassertive. The *Zhuangzi* praises the spontaneous skillfulness of craftspeople that cannot be easily taught in words, but is achieved only by the repeated practice of an individual in a highly particular context. Religious practices begin with the purification of mind and body and take for granted the respect for all living beings in one's immediate environment. Religious communities enshrine such attitudes in precepts that are the precondition for more proscriptive methods (see in this volume Kristofer Schipper's discussion of the "one hundred and eighty precepts" associated with the Celestial Masters tradition). Caring about the extinction of the snow leopard or panda is like the concern among contemporary Daoist masters today for the pollution and gross commercialization of pilgrimage mountains in China; the whole is effected only by means of a profound respect for all the particular manifestations of life.

Reversion and Spontaneity

The ecoreligious goal of Daoist meditational and ritual practice is to mirror unobtrusively the dynamic spontaneity of one's environment, to become imperceptible and transparent as though one were not at all. This goal is made all the more remote by the complex web of social and intellectual structures layered throughout history that form the cultural flux in which human life is trapped. The path toward pure spontaneity thus consists always in a "healing" reversion or undoing. This reversion can occur mentally through "sitting in oblivion" (*zuowang*), physically through the generation of an immortal embryo, collectively through communal ritual, and even cosmogonically through alchemical practices founded on the principle that degenerative natural processes can be reversed and restored to their original pristine state (*hundun*).

Constructing Nature

Daoism proposes a comprehensive and radical restructuring of the way in which we conceive of our relationship to nature and our cosmic environment. This imaginative act does not readily lend itself to the solution of the problems of modern society except inasmuch as it challenges the very foundations of our economic, political, scientific, and intellectual structures. At the same time, however, as Daoism becomes more influential in the West, even as it is misunderstood, it surely exerts a positive influence with respect to understanding what it means to be embedded in a cosmic landscape. In such an understanding, "nature" is not something outside of us to be dealt with after the fashion of a mechanic repairing a car, but is both a mental attitude to be carefully cultivated and the true condition of one's body, which contains the infinite dimensions of cosmic reality within itself. Ultimately, nature is to be constructed and visualized time and again. The terrain of our most authentic ecological concern, therefore, is first and foremost the landscape of the religious imagination.[30]

Imagining Daoism Today

Having set out a preliminary sketch of a particular biospiritual worldview of traditional religious Daoism, we are still left with questions about the relevance and creative application of such perspectives to contemporary ecological problems in China and the world. Perhaps, however, these questions should be framed in another way. Thus, it might more fruitfully be asked, "Who speaks for Daoism today?" The answer is not as obvious as it may seem, since exactly who or where the Daoists are today is no easy matter, except to say that there are various fragmentary traditions that continue in China and in the Chinese diaspora, as well as a rudimentary and acculturated Western or American-style Daoism and several related "Daoist" practices. Given the disjointed and sometimes dispirited world of modern-day Daoist practitioners, perhaps it is more properly the "cultured elite," the scholars, who speak authoritatively for Daoism. Certainly, when it comes to a historical and textual understanding of the tradition, the scholarly community has a lot to say that is important and salutary. In fact, what has been called the partial "resurrection of the Daoist body," after the disastrous vicissitudes of modern Chinese history, owes much to the labor and influence of scholars during the past quarter century.[31] Finally, it may be asked whether even popular commentators have something to offer to the contemporary appropriation of a kind of global and ecologically aware Daoism. As we have already indicated, on the one hand there is much that is simply silly and simpering about many contemporary Western popularizations of the Dao.

On the other hand, there is a world of difference between the Pooh Bear perspectives offered by Benjamin Hoff and those much more rigorous and unsentimental literary fabulations envisioned by Ursula Le Guin (see in the this volume the chapter by Jonathan Herman and the epilogue). This is a difference that finally has to do with the hard alchemical work of the human imagination (*solve et coagula*)—that is, the creative deconstructive reinterpretation and ritual transformation that gives new meaning and ongoing life to any human tradition.[32]

When it comes to who legitimately speaks for Daoism today, we are too often left with a kind of Dao Wars. The popularizers ignore the scholars; the scholars mock the popularizers; and the practicing Daoists, whether in China or the West, remain mostly quiet (as maybe they should). There is still much to be learned about the history of Daoism, but let us be wary of blithely replacing the "purely" philosophical and mystical Daoism of an earlier generation of scholarship with the "real" religious and scriptural Daoism known today only by a few scholarly experts. Neither the trope of the "spiritually 'pure'"or the "historically 'real'" completely captures the imaginative "truth" of Daoism in the past and present. Moreover, the ongoing life of the tradition in both China and the West today confronts a public crossroads of ecological concern that requires a reinterpretation of the past in relation to the contemporary situation. This calls for a creative reappropriation in the present of the earlier Daoist tradition that is both deferential and differential.

During this chaotic period of millennial turning, when virtual worlds are replacing the natural world, the time seems ripe for some Daoist perspectives on the ecological problems of our current situation. Assuredly, these perspectives will be neither definitive nor redemptive, but they may contribute to the gradual and periodic ritual renewal of life on this planet. Furthermore, in a post-Tiananmen Chinese world of Coca-Cola communism, the Daoist tradition, in both its past configurations in China and its contemporary global transformations, has something important to say about the ecological role of the religious imagination for a young generation of Chinese studying at Beijing University and working at McDonalds. It is unlikely that such young urban Chinese will be perusing the canonical Daoist scriptures. But the danger is, perhaps, that the *Tao of Pooh* will be read in Chinese translation before Le Guin's *The Dispossessed*. Finally, it may be said that all of us—urban Chinese and global citizens of the twenty-first century—need the important repository of Daoist efforts to envision the embeddedness of human life within a cosmic landscape. We require a Daoist perspective on these matters if we are to have the creative resources necessary for imagining and realizing a new, and more "translucent," world of ecological harmony.

The Way Taken

In this volume we have tried to create a flexible structure that respects the current difficulties in the discussion of Daoism and ecology and yet moves toward a productive engagement of the issues. The sectional groupings are somewhat artificial, but, in keeping with the multifaceted nature of the tradition and our inclusive concerns, they serve to organize a rather diverse assortment of papers. There is, however, some logic to our arrangement. After setting forth the mythic landscape of the traditional Daoist vision of organic life as generated from a bipartite cosmic gourd (Stephen Field's epic poem in the prologue, "The Calabash Scrolls"—a work that evokes much of the agrarian rootedness of Chinese tradition, especially in the metaphorical sense wherein all the "ten thousand things" are but the offspring of a cosmic wonton or primordial man known fondly by Daoists as Hundun or Pangu),[33] we proceed from a consideration of the general problems compromising any discussion of Daoism and ecology (section one) to an analysis of perspectives found in Daoist religious texts (section two) and within the larger Chinese cultural context (section three). The papers in section four build on the earlier papers by delineating some of the key issues found in the "classical" texts. These papers then lead to a set of ecological observations on the applicability of modern-day Daoist thought and practice in China and the West (section five and the epilogue). As a coda to each of the major sections, we have appended some synoptic discussion of the themes and questions raised by the individual papers. These short concluding statements on each of the sectional groupings reflect both our own editorial concerns and also some of the commentary provided by respondents at the Harvard conference. At the very end, we have included an annotated bibliography of works on Daoism and ecology.

The first sectional grouping of papers ("Framing the Issues") specifically takes up the theoretical and historical complications associated with a Daoist approach to the environment. Jordan Paper's presentation ("'Daoism' and 'Deep Ecology': Fantasy and Potentiality") gives us a provocative overview of these difficulties while at the same time suggesting some corrective strategies. Joanne Birdwhistell's contribution ("Ecological Questions for Daoist Thought: Contemporary Issues and Ancient Texts") critically addresses some important ecological themes as problematically related to the earliest texts and pointedly raises further questions from a feminist perspective. Michael LaFargue's paper ("'Nature' as Part of Human Culture in Daoism") extends Birdwhistell's discussion with an insightful and "confrontational" hermeneutical appraisal of the meaning of "nature" as seen in the *Zhuangzi* and *Laozi*. Closing out this section and expanding the discussion beyond the ancient "proto-Daoist" texts are Terry Kleeman's suggestive reflections on cosmic "order" as found in the Daoist religion ("Daoism and the Quest for Order").

Following this section is a series of important papers ("Ecological Readings of Daoist Texts") devoted to the analysis of Daoist religious scriptures. The discussions in this section by Chi-tim Lai (on the *Taiping jing,* or *Scripture of Great Peace*), Robert Campany (on Ge Hong), and Zhang Jiyu and Li Yuanguo (on the *Yinfu jing,* or *Scripture of Unconscious Unification*) are all pioneering explications of particular religious texts, but it can be said that Kristofer Schipper's paper on some early Daoist ecological "precepts" (found in the text known as the *Yibaibashi,* or *The One Hundred and Eighty Precepts*) has special historical significance and contemporary resonance. Speaking both as an initiated Daoist priest and a renowned academic scholar, Schipper affirms the proposition that religious Daoism traditionally "did not only think about the natural environment and the place of human beings within it, but took consequential action toward the realization of its ideas."

The papers in section three ("Daoism and Ecology in a Cultural Context") constitute an especially eclectic grouping inasmuch as they deal generally and comparatively with various cultural themes and folk practices that have some traditional "Daoist" affinity or significance. Thus, Thomas Hahn ("An Introductory Study on Daoist Notions of Wilderness") interestingly lays out some of the crucial historical and cultural context for understanding the ideas of "nature" and "wilderness" in Chinese tradition and Stephen Field ("In Search of Dragons: The Folk Ecology of Fengshui") discusses some of the

origins of fengshui as one of the "longest lived traditions of environmental planning in the world." From a broad cultural perspective, E. N.
Anderson ("Flowering Apricot: Environmental Practice, Folk Religion, and Daoism") gives us a perceptive anthropological meditation
on agricultural tradition, aspects of Daoist practice, and Chinese folk
religion as related to both the past and present. Finally, Jeffrey Meyer's
paper ("Salvation in the Garden: Daoism and Ecology") evocatively
suggests the relevance of Chinese "gardening" as a creatively "inven-
tive" metaphor for a modern Daoist approach to ecology that stresses
a collaborative relationship between the natural and the human.

Building on some of the insights brought forth by the earlier papers, the next section ("Toward a Daoist Environmental Philosophy")
includes a series of speculative reflections on the significance (or lack
thereof) of the "classical" texts for a contemporary ecological philosophy. David Hall's and Roger Ames's papers (respectively, "From
Reference to Deference: Daoism and the Natural World" and "The
Local and the Focal in Realizing a Daoist World") are especially intriguing postmodernist reinterpretations of the *Daodejing* and *Zhuangzi*.
These papers (by authors who are frequent philosophical collabora-
tors) are powerfully illustrative of how ancient Daoist texts can lend
themselves to creative philosophical appropriation. Russell Kirkland
("'Responsible Non-Action' in a Natural World: Perspectives from
the *Neiye, Zhuangzi,* and *Daode jing*") and Lisa Raphals ("Metic Intelligence or Responsible Non-Action? Further Reflections on the
Zhuangzi, Daode jing, and *Neiye*") more argumentatively take up the
contested discourse surrounding the ancient meaning and contemporary moral relevance of *wuwei* ("non-action"). Kirkland's hard position concerning the radical non-interventionist implications of *wuwei*,
though contrary to what some would say is the "scholarly consensus,"
is nevertheless an important reminder of the difficult "otherness" of
ancient texts. In keeping with LaFargue's perspective on these matters, Kirkland provides us with a "confrontational hermeneutics" that
resists too easy (and gravely anachronistic) appropriations of ancient
Daoist texts and ideas. Raphals effectively supplements and extends
Kirkland's argument by discussing various forms of "non-interventionist" or "indirect" action in the early Daoist texts and in ancient
Greek tradition. On the other hand, Liu Xiaogan ("Non-Action and
the Environment Today: A Conceptual and Applied Study of Laozi's
Philosophy," a paper interestingly augmented by Zhang Jiyu's Daoist

"declaration" in the following section) not only finds a more activist ethic present in the ancient texts, but also provides us with his own interpretive application of *ziran* ("spontaneity" or "self-so") and *wuwei* to modern ecological problems.

The final section ("Practical Ecological Concerns in Contemporary Daoism") includes papers that theoretically and practically "apply" various aspects of the Daoist tradition to the contemporary ecological situation. Thus, James Miller articulates the ecological implications of Daoist visionary experience as seen in the Highest Clarity tradition ("Respecting the Environment, or Visualizing Highest Clarity"), and Jonathan Herman cogently argues for the significance of the American novelist Ursula Le Guin's imaginative redaction of Daoism. From a more pragmatic perspective are Zhang Jiyu's "Declaration of the Chinese Daoist Association on Global Ecology" and the fascinating roundtable discussion by contemporary Western practitioners of various Daoist and quasi-Daoist arts ("Change Starts Small: Daoist Practice and the Ecology of Individual Lives," a discussion with Liu Ming, René Navarro, Linda Varone, Vincent Chu, Daniel Seitz, and Weidong Lu).

The volume concludes with an epilogue made up of Ursula K. Le Guin's haunting remarks on her life as a self-styled American Daoist and literary ecologist. This is followed by Le Guin's plaintive "Tao Song," a short poetic refrain that picks up and extends Stephen Field's initial cosmogonic epic about gourds, organic life, and the Dao. In Le Guin's trenchant sense of things, we are left with a dark yet hopeful song of organic life—verses which tersely and wisely capture much of the Daoist roughhewn celebration of nature.

Ways within a Cosmic Landscape

To conclude these introductory comments, we return to the thematic metaphor of the "landscape" of life, especially as embodied in traditional landscape paintings, gardening, and the cultivation of miniature gardens (*penjing*) in China. Typically, a Chinese landscape painting (or the microcosm of a garden within a basin) is expressive of the dynamic interrelatedness of the cosmic (the celestial "frame" or "space" of the painting or container), natural (mountainous forms, vegetation, and water), and human (both individual wayfarers and expressions of social life, such as roads and buildings) spheres of life—particularized manifestations of the biosphere that often transparently merge into an organic whole by virtue of an all-pervasive cloudy mist or vaporish qi. Important in these "small worlds" or multiperspectival tableaus of the unity and particularity of life (the manifest or named Dao) is a kind of double irony. Thus, what is "natural" is always in relation to the constructed, imagined, or artificial presence of humanity. At the same time, the natural artificiality of the "landscape" of life, unlike Greek artistic tradition, primarily refers to the profoundly humbled significance of humans in relation to the greater whole. The craft of Chinese landscape art and miniature gardens achieves its "natural" effect and "humanistic" significance by being conspicuously artificial *and* nonanthropocentric.[34]

It is this necessary but subdued role of humanity in cooperative relation with nonhuman nature and the cosmos (the "gardening" theme brought out so effectively by Jeffrey Meyer) that hints at the mythological story of creation associated with the cosmic giant known as Pangu (or Pon Ghu in Field's poem) born of the primordial egg, wonton, or gourd (see the prologue to this volume). The human world is in fact the dismembered body of Pangu from whose body lice are spawned human beings.[35] From the very beginning, therefore, humans have infested the greater landscape of life and are cooperatively responsible for the overall health or disease of cosmic life. The question becomes, then, whether this relationship will evolve parasitically and destructively, or symbiotically and productively. What comes to the fore when reflecting on these images is the ubiquity of organic, agricultural, and medicinal metaphors that valorize an intimate cooperation of the human and natural worlds. In some ways, these ideas (as with the overall traditions of landscape painting and gardening) are more pan-Chinese than specifically Daoist.[36] Nevertheless, it can be said that Daoists—more so perhaps than either courtly Confucian bureaucrats or sophisticated Buddhist monks—tended to remember ancient mythic themes and ritual practices as ways to rearticulate, temporarily, imaginatively, and artificially, the original unbroken wholeness of individual bodies, particular social worlds, and the infinite cosmos.

The collaborative or participatory relationship with nature generally promoted by the tradition of landscape painting and gardening in China is not a prescription for passivity. As in the broad Daoist spirit of *wei-wu-wei*, or effective nonegotistical action, humans should respond actively and creatively to the sinuous and often degenerative turnings of life. Thus, a landscape painting commonly depicts the humbled, yet responsive, wayfarer who is consciously striving to find an ascending path up (and into) the mountain of life. Both the destination and the journey have significance in landscape painting. And as Schipper has reminded us, Daoists may even provide us with precepts, signs, and talismans along the way—passports back to an interconnected cosmos. Here again is suggested a kind of generalized Daoist lesson about negotiating the byways of contemporary ecological concern. In many ways, the brokenness and dis-ease of bodies and spirits, as well as the devious bypaths of the mountainous body of life, must be accepted. But this means that it is incumbent upon all of us who inhabit this increasingly fragmented cosmic landscape to walk (together with other wayfarers) a path that cherishes and cultivates the healing interrelatedness of all the "ten thousand things." We embrace the unnamed Dao of the cosmos only through the myriad speciated *de*'s of our own local environment—our own patch and parchment of garden.

Daoists may not always be the first to act in times of crisis, nor are they likely to work out elaborate theories of engaged social action, but they have always known that it is imperative to take up a way of life that responds in a timely and imaginative fashion to the dangers of neglect, imbalance, distortion, and degradation that inevitably affect human relations with the natural and cosmic worlds. What is needed is a bodily and spiritual resurrection of what Tuan Yi-fu calls a "topophilia"—that is, an aesthetic respect and a practical love for one's particular life-scape, a love that has general ecological import because of its rootedness in the specific topography of a lived body and local environment.[37] Coming to the end of our journey within the confusing realms of Daoism and ecology is, then, only to be in a position to begin the work of knowing and healing again. In time and because of time, all things—including the natural world itself—require attentive cultivation and responsive care. This, after all, is the "natural" way of things. It is one of the ways—which might be called a "Daoist" or transformative way—to live gracefully, reciprocally, and responsibly within the cosmic landscape of life.

Notes

1. Concerning the "easy" and "natural" assumption of a special affinity between Daoist tradition and ecological concerns, see, among other examples, J. Baird Callicott, *Earth's Insights* (Berkeley and Los Angeles: University of California Press, 1994), 67–75. As Callicott says (p. 67), "contemporary Western environmental ethicists scouring Eastern traditions of thought for ecologically resonant ideas and environmentally oriented philosophies of living have been drawn chiefly to Taoism."

2. See Benjamin Hoff's two best-selling "new age" commentaries on Daoism, *The Tao of Pooh* (New York: Penguin Books, 1982) and *The Te of Piglet* (New York: Dutton, 1992). On the whole fascinating topic of Americanized "pop" Daoism or Dao-Lite, see N. J. Girardot, "My Way: Teaching the *Tao Te Ching* and Taoism at the End of the Millennium," forthcoming in *Teaching the Tao Te Ching,* ed. Warren Frisinia (New York: Oxford University Press).

3. On the experiential "Suzuki-Zen" see Robert H. Sharf, "The Zen of Japanese Nationalism," in *Curators of the Buddha: The Study of Buddhism under Colonialism,* ed. Donald S. Lopez, Jr. (Chicago: University of Chicago Press, 1995), 107–60.

4. Concerning the checkered history of Western regimes of knowledge concerning Chinese tradition, see J. J. Clarke, *Oriental Enlightenment: The Encounter Between Asian and Western Thought* (New York: Routledge, 1997), 37–53; and N. J. Girardot, *The Victorian Translation of China: James Legge's Oriental Pilgrimage* (Berkeley and Los Angeles: University of California Press, forthcoming). Especially important is J. J. Clarke, *The Tao of the West: Western Transformations of Taoist Thought* (London and New York: Routledge, 2000).

5. See also the versions of a particularly popular scatological "definition" of Daoism in relation to other religions; for example, "Shit Happens in Various World Religions" at www.ee.pdx.edu/ ~alf/html/shit-religions.html and the "Canonical List of Shit Happens" at www.humorspace.com/ humor/lists/lshit.htm.

6. See Nathan Sivin, "On the Word 'Taoist' as a Source of Perplexity," *History of Religions* 17 (1978): 303–30.

7. On the nineteenth-century cultural history of the "Protestant," "missionary," and "postmillennial" agenda inherent in much Orientalist discourse and comparative religions, see Girardot, *Victorian Translation of China.* Specifically with regard to Daoism, see N. J. Girardot, "'Finding the Way': James Legge and the Victorian Invention of Taoism," *Religion* 29 (1999): 107–21. An illustration of some of the difficulty and silliness inherent in the conflation of quasi-religious environmental apprehensions with an enlightened reform of "traditional" and "superstitious" Chinese religious practices is seen in the heavily Westernized Chinese community of Taiwan. Thus, an "environmentally friendly" governmental minister in Taipei recently urged people to stop the wasteful practice of burning wads of imitation spirit-money for the dead. Hsieh Chin-ting, head of the Department of Civil Affairs suggested that using a credit card system in temples would be more ecologically and religiously efficacious since the dead could charge as much as they desired in the afterworld without causing the living to pollute the earthly realm. Directly linking these environmental interests with the traditional "three teachings" of Confucianism, Daoism, and Buddhism, Minister Hsieh, as a kind of latter-day Confucian bureaucrat, said he was acutely con-

cerned that "the tons of imitation banknotes burned each year [were] a waste of natural resources." His solution to this problem was his strong recommendation that "Buddhist and Taoist temples take the lead in bringing about the change" to ghostly credit cards. This article appeared as a syndicated "News of the Weird" item and appeared in the *Allentown (PA) Morning Call*, 2 October 1993, under the heading of "Give Dead Credit; Save a Taiwan Tree."

8. On the "depth" of the contemporary ecological movement, see especially Michael E. Zimmerman, *Contesting Earth's Future, Radical Ecology, and Postmodernity* (Berkeley and Los Angeles: University of California Press, 1994), and *The Green Reader: Essays Towards a Sustainable Society*, ed. Andrew Dobson (San Francisco: Mercury Books, 1991). The best known of the deep ecologists is the Norwegian philosopher Arne Naess; see his "The Shallow and the Deep, Long-Range Ecology Movement: A Summary," in *The Deep Ecology Movement: An Introductory Anthology*, ed. Alan Drengson and Yuichi Inoue (Berkeley: North Atlantic Books, 1995). For an interesting discussion of the Daoist implications of Naess's deep ecology, see Vanessa Phillips, "The *Tao Te Ching* and Its Relation to Deep Ecology," *Lehigh Review* 7 (spring-fall 1999): 31–39.

9. Among other works, see especially *Buddhism and Ecology: The Interconnection of Dharma and Deeds*, ed. Mary Evelyn Tucker and Duncan Ryuken Williams (Cambridge, Mass.: Harvard University Center for the Study of World Religions, 1997).

10. On the contemporary Daoist concern for the destruction of China's holy mountains, see Martin Palmer, "Saving China's Holy Mountains," *People and the Planet* 5, no. 1; URL: www.oneworld.org/patp/vol5/feature.html.

11. Much of the material in this section was contributed by Liu Xiaogan.

12. On the situation involving the Three Gorges Dam, see Wu Ming, "A Disaster in the Making," *China Rights Forum*, spring 1998, 4–9. A recent discussion of the problem of air pollution in China is found in the Associated Press story on the Beijing "Blue Skies Project" by Elaine Kurtenback, printed in *The Morning Call* (Allentown, Pa.), 23 March 1999, D1, D6.

13. Zhang Kunmin, "China's Environmental Strategy and Environmental Literature" (paper presented at the International Conference on Humankind and Nature: Literature on the Environment, Singapore, 27–30 February 1999).

14. Richard Louis Edmonds, "The Environment in the People's Republic of China Fifty Years On," *China Quarterly* 159 (1999): 644.

15. According to an official report in 1997, desertification of land throughout China had increased to 27.3 percent. In the 1960s, desertification expanded at the yearly rate of 1,560 km^2; by the early 1980s, this had increaed to 2,100 km^2. See *Diqiu, Ren, Jingzhong* [The earth, humankind, and the alarm] (Beijing: China Environmental Science Press, 1997), 153. In the western provinces, the percentage of forested land has been greatly depleted—e.g., 0.35 percent in Qinghai, 0.79 percent in Xinjiang, 1.54 percent in Ningxia, 4.33 percent in Gansu, and 5.84 percent in Tibet. See the article in *Lianhe Zaobao*, 27 December 1999. From the 1950s to 1970s, deforestation to create new farmland caused the percentage of the forested land in Xishuangbanna to be reduced from 70 percent to 26 percent, and from 35 percent to 26 percent in Hainan. Similarly, because of the movement to reclaim farmland from

lakes, the area of the second large Dongting Lake shrank by 60 percent, and the first large Poyang Lake by 50 percent. (Fu Hongchun; "Hongxing Chuqian de Jingjixue," *Lianhe Zaobao,* 1999.) The seven major river systems were considered badly polluted or barely acceptable according to the test in 1997, and groundwater and coastal regions are polluted to various degrees. See Zhang Kunmin, "China's Environmental Strategy and Environmental Literature."

16. Edmonds, "The Environment in the People's Republic of China," 641.

17. *Dao Fa Ziran yu Huanjing Baohu,* ed. Zhang Jiyu (Beijing: Huaxia Press, 1998), 200–201.

18. Most recently during his trip to the United States in April 1999, the Chinese Prime Minister Zhu Rongji participated in a forum on the environment and "spoke frankly of 'the devastation of Mother Nature' in China as a result of soil erosion, deforestation, and emissions from factories, cars and coal-burning furnaces, the country's main source of heat"; Joseph Kahn,"Two Accords with China Billed as Icing Become Part of a Simpler Cake," *New York Times,* 10 April 1999; www.nytimes.com/library/world/asia/041099china-us.html. For additional discussions of the environmental situation in contemporary China, see the bibliography on energy and environment in China and East Asia, compiled by Timothy C. Weiskel, found on "East and Southeast Asia. An Annotated Directory of Internet Resources," http://newton.uor.edu/ departments&programs/asianstudiesdept/china-science.html.

19. On the international organization of "sino-ecologists," see the following URL: http://sevilleta.unm.edu/~yyang/sino-eco/about.html.

20. See, in this volume, Roger T. Ames, "The Local and Focal in Realizing a Daoist World," and David L. Hall, "From Reference to Deference: Daoism and the Natural World."

21. See, in this volume, James Miller, "Respecting the Environment/Visualizing Highest Clarity."

22. Also see Palmer's discussion of these developments in his "Saving China's Holy Mountains," p. 2.

23. Important new translations of Daoist religious scriptures are found in: *The Taoist Experience: An Anthology,* ed. Livia Kohn (Albany: State University of New York Press, 1993); Steven Bokenkamp, *Early Daoist Scriptures* (Berkeley and Los Angeles: University of California Press, 1997); and Eva Wong, *Teachings of the Tao* (Boston: Shambhala, 1997).

24. See Isabelle Robinet, *Taoism: Growth of a Religion,* trans. Phyllis Brooks (Stanford, Calif.: Stanford University Press, 1997).

25. On Orthodox Unity Daoism practiced in Taiwan today, see especially Kristofer Schipper, *The Taoist Body,* trans. Karen Duval (Berkeley and Los Angeles: University of California Press, 1993).

26. For an engaging fictional portrait of the founders of the Complete Perfection school, see Eva Wong's translation *The Seven Taoist Masters* (Boston: Shambhala, 1990).

27. There is still no reliable discussion of these Westernized forms of Daoism, but see Solala Towler, *A Gathering of Cranes: Bringing the Tao to the West* (Eugene, Oreg.: Abode of the Eternal Tao, 1996). See also the *Frost Bell,* the interesting newsletter of "Orthodox Daoism in America" published in Santa Cruz, California. Liu

Ming (= Charles Belyea) is the leader of this organization. On Liu Ming, see, in this volume, "Change Starts Small: Daoist Practice and the Ecology of Individual Lives."

28. Most of this section ("The Ecological Landscape of Religious Daoism") was originally published by James Miller as "Daoism and Ecology," in *Earth Ethics* 10, no. 1 (fall 1998): 26–27.

29. On the phenomenology of Daoist "immortals," see especially Isabelle Robinet, "The Taoist Immortal: Jesters of Light and Shadow, Heaven and Earth," in *Myth and Symbol in Chinese Tradition*, ed. N. J. Girardot and John S. Major, symposium issue of the *Journal of Chinese Religions* 13-14 (1985–1986): 87–106. Concerning the origins of the Highest Clarity tradition see also Isabelle Robinet, *Taoist Meditation*, trans. Julian Pas and N. J. Girardot (Albany: State Universtiy of New York Press, 1992).

30. From a comparative perspective see, for example, Simon Schama, *Landscape and Memory* (New York: Alfred A. Knopf, 1995).

31. For a discussion of some of these issues, see N. J. Girardot, "Kristofer Schipper and the Resurrection of the Taoist Body," in Kristofer Schipper, *The Taoist Body*, trans. Karen Duval (Berkeley and Los Angeles: University of California Press, 1993), ix–xviii.

32. For an interesting discussion of how the "bad scholarship" of popular interpretations of Daoism may sometimes result in "good religion," see Julia M. Hardy, "Influential Western Interpretations of the *Tao-te-ching*," in *Lao-tzu and the Tao-te-ching*, ed. Livia Kohn and Michael LaFargue (Albany: State University of New York Press, 1998), 165–88.

33. On these mythic themes, see N. J. Girardot, *Myth and Meaning in Early Taoism: The Theme of Hun-tun* (Berkeley and Los Angeles: University of California Press, 1988).

34. See, among other works, Mai-mai Sze, *The Tao of Painting: A Study of the Ritual Disposition of Chinese Painting* (New York: Pantheon Books, 1956); Dusan Pajin, "Environmental Aesthetics and Chinese Gardens," http://dekart.f.bg.ac.yu/~dpajin/gardens/; Lothar Ledderose, "The Earthly Paradise: Religious Elements in Chinese Landscape Art," in *Theories of the Arts in China*, ed. Susan Bush and Christian Murk (Princeton, N. J.: Princeton University Press, 1983), 165–183; and Kiyohiko Munakata, "Mysterious Heavens and Chinese Classical Gardens," *RES* 15 (1988): 61–88. On miniature gardens, see the classic work by Rolf Stein, *Le Monde en petit: Jardins en miniature et habitations dans la pensée religieuse d'Extrême-Orient* (Paris: Flammarion, 1987), translated by Phyllis Brooks as *The World in Miniature* (Stanford: Stanford University Press, 1990).

35. For some discussion of the Pangu mythology, see Stephen Field, "In a Calabash: A Chinese Myth of Origins," *Talus* 9/10 (1997), particularly pp. 52–55. See also Yuan Ke, *Dragons and Dynasties: An Introduction to Chinese Mythology*, trans. Kim Echlin and Nie Zhixiong (London: Penguin Books, 1993); and Anne Birrell, *Chinese Mythology: An Introduction* (Baltimore: Johns Hopkins University Press, 1993).

36. See Miranda Shaw, "Buddhist and Taoist Influences on Chinese Landscape Painting," *Journal of the History of Ideas* 49 (1988): 183–206; and Shen Shan-hong, "The Influence of Tao in the Development of Chinese Painting" (Ph.D. diss., New York University, 1978).

37. On the theme of "topophilia," see Tuan Yi-fu, *Topophilia: A Study of Environmental Perception, Attitudes, and Values* (New York: Columbia University Press, 1974). It should be noted that the geographer Tuan is not at all sanguine about organized religions (including Daoism) contributing to a revived topophilia in the contemporary world. For a statement from the "sociobiological" perspective concerning the interconnectedness of all life forms, see Edward O. Wilson, *Biophilia: The Human Bond with Other Species* (Cambridge, Mass.: Harvard University Press, 1984).

Prologue: The Calabash Scrolls

STEPHEN L. FIELD

Chinese mythology is one of the least known of the major world traditions, due mainly to the lack of a great Chinese epic in the style of the *Iliad*, which is an unfortunate omission when one considers the profusion of epical gods and heroes in the preclassical Chinese tradition. My poem "In a Calabash," from which the following excerpt is taken, is an attempt to remedy that literary void by providing the English-speaking world with a plausible Chinese cosmogony.[1] From the various early myths dealing with origins, I selected three and joined them in linear fashion to narrate a tale of the origin of the cosmos and creation of the Chinese race: Space-time is generated spontaneously by the accidental annihilation of Hun Dun in a Daoist "big bang," and the subsequent emergence and demise of the cosmic giant Pon Ghu forged the macrocosmos and its counterpart, the primal egg, from which the siblings Shewah and Fuh He are hatched into paradise.[2] Shewah then fashions human beings from yellow clay before repairing the devastated world for her human descendants. The creation of "black-haired Man" brings the Daoist cosmogonic mythos full circle, from the prelapsarian cosmic source back to a new beginning—in this case, two human children escaping in the primordial gourd-egg. The calabash, or bottle gourd, is an ancient and pervasive symbol of salvation in the Chinese tradition.[3] As such, it may also be an appropriate symbol of a Daoist ecology, an understanding that human beings, at least because of their chthonic origins, are an integral part of the environment that gave birth to them.

The tear in the eye of the galactic storm is Hnn Tnn, the Hollow,
Interminable orb.
Now the seed of Hhn Thn grows in girth in the midst of the fathomless void,
Until his bulk contains the whole of naught,
A bloated sack, a calabash, a wineskin bulging at the seams.
Ten thousand aeons for his macrocosmic body, bag of space,
To turn itself completely inside-out,
Height to revolve into cavernous depth,
Sun into a vanished organ.

Hwn Twn swallows the seas.
Lightning dragons too are sucked into that maw.
Shu, Shamanka of the North, and Hu, the Shaman of the South,
Are sealed by the depths of ocean and perceive on recovering,
Not a hole, not a crack, not a sound anywhere.

> "We have seven openings," they count,
> "For seeing, hearing, eating, breathing.
> He," they grope at Hun Tun's head, "has none.
> We shall cut, we shall core. . . ."

> > They fasten on the breath of thought,
> > > Begin the rites of penetration
> In the search for buried light:
> Their tool of breath now sows the seed of hearing,
> > Now sews the thread of sight—
> And just when Hurn Turn thinks to blink—
> Out shoots awhirl a needle or a root,
> To tap the sack and puncture
> Inner space. Look!
> That trickle—
> Silver, subtle as a star—
> > Seeping through a
> > Thousand
> Leagues of
> Sea!

Then with further penetration comes the incipient yet brilliant fascination—
    Mind—
Sharpened by the pair of smiths to complex recollections of desire.

Horn Torn suffers more with each long burst of light.
Light is ambivalent wealth, the arrow of love, the flash of anger,
That he no longer can extinguish.
Light slays him.
His sides are rent.

A new, sudden, unpredictable cosmos takes the dragons by storm.
They are knocked unconscious by the tides of unleashed emotion.
Hon Ton implodes, sucking Shu and Hu into a spinning vortex.
Won Don explodes, revealing the earth's uncertain origin.
Seven blows or portals in a breath-bag,
Seven horns or orifices in the microcosmic head of Hun Dun.

Pon Ghu emerges, creature of paradox, dwarf yet giant.
His tinctured body seems all silver and gold.
The crown of light he wears like the motif of a star
Is a beacon to the surf at his boatlike feet,
A witness to the embroidered tapestry of sky at his fingertips.
Bundled in his titan arms, an egg encases double embryos,
A swollen, swirling crucible of yin and yang.

He tries his eyes. . .
Their dazzling glare reveals the pod of space that is his gaol.
Yet the Pon Ghu knows no bounds:
He bellows with a thundering rage, then charges.
For 18,000 years he batters with that bullhide ram,
But every inch he gains his body grows.
Undaunted, he crouches, he leaps,
And with his sword of light he severs the sack.
But the cataclysm kills him.

The shock of worlds dividing 90,000 miles asunder
Knocks Pon Ghu off his feet.
He falls full length and backwards, crashing, spread-eagled.
Promptly is his breath the wind that shapes the clouds
Into reflections of mouth, flaring nostrils, lightning rage, thunder.
One eye glares.
Darkness falls in a curtain of blood
Revealing myopic other eye wobbling in the night sky.

                    Pon Ghu's
              Body now floats in space.
           Hands and feet have taken root
          In opposite directions.
                              Arms and legs like wings
           Stretch behind him backwards
                    To encompass a new earth and ocean.
          Knees are drawn upwards
                              Into mountain peaks.
                Elbows and hands
                        Seem to sweep downwards
              From squares of trunk
                    Or globe of body.

Flesh of soil glimmers.   Bamboo hair vaguely shines.
Sweat of   Jade.   Precious ores.
Veins.   Mountain waves.
Arterial cracks.   Rivers.

Soon will come swarming droves of fleas, teeming populations,
The myriad clans and the hundred names of two-legged black-haired Man,
To inherit the yellow earth, Pon Ghu's plague.

Lying liberated in the crook of an arm, the egg of plague and bicameral yolk
Is nourished in the nest of Earth, in the caress of Heaven,
For another ten thousand years.
Creeping mist is the aura of Sky; branching wood is the aura of Earth.
Fog and leaf feed the transmutation, and the egg is quickened.
It begins to shake.
A scratching noise betrays a dragon claw, five-toed,
But a manly hand appears instead, five-fingered,
Then a head and brawny torso.
Another head emerges with sable locks
To rival the snow-clad shoulder of the mountaintop.
With a last concerted heave serpent tails untwine
And siblings burst asunder out the fractured calabash.

Fructified by the summer wind, human eyes shine like pools of pearl.
Perhaps in answer to some silent command,

Human heads bow past saurian scales toward the four quarters of Earth,
Affording the newborn their first feast of sight,
Sun-spangled and sonorously baritone the day of primal awakening.

To climb and cling like apes is second nature to their limbs,
So the hatchlings forage lichee nuts, grub for the ginseng root,
Search the trees for the monkey peach, and drink the dew.
Companions are the Qhi-lin and Phoenix, Bearcat and Tiger;
They follow the giant Turtle.
At sunrise they romp and roam about the pristine forests, vales and meadows.
At sunset they follow the flame of the Fire Star
In the flight of the Cerulean Dragon.
At night they sleep in mountain caves the death-like slumber of ancient trees.

Ever in sync with the natural regularities
Twice each year are the playful twins aware of an impending change.
Today the sun reversed its course and began its annual journey to the south.
The dragons dance a solstitial rite to the confluence of Heaven and Earth.
Looking east to the Heart of the Dragon and south to the calling Phoenix,
They bow to the four quarters of the ecliptic.
With tails twining in the co-mingling vapors,
They resume for the bat of a mammalian eye
The reptilian brain of ancestral Shu and Hu.

> I awaken to lie on a bank of yellow clay with pieces clinging to my limbs.
> A spring breeze laps waves upon the shore; the sun is beginning to rise.
> > What's this?
> I reach down to the oozing clay, scoop up a handful, squeezing, kneading,
> Divide the lump, sculpt a long, gainly limb, then another bough or leg
> And lean the two together—windward, leeward,
> Affix a third piece for the trunk, two tawny arms, a handsome head,
> > Voila!

The pattern finally reveals itself,
When Shewah fashions the bust in her own reflected form.
Next she cradles the sculpture and blows into its mouth
As if she played her bamboo flute.
It abruptly jerks its legs and arms and comes to life with a snarl.
Then it springs from her embrace and bumps and bounds up the riverbank.

So the woman without stopping builds another and another.
Some she licks to life as if tasting placenta.
Some she bathes in the river water.
Some she fondles and strokes until the petting wakes the spirit of the clay.
But when the dark-haired kit first tastes salt of the lady's palms,
First gazes into those lizard eyes,
It fears for its miserable life and escapes from the clutch of the creatrix.
Soon a score of them are perched in the trees.
The next figure she forms in yellow clay
Bites her as she swaths it in her sweaty arms,
So she hurls it to the center of the river by a leg.
     Ai, yeow!
Resolved no longer to knead clay like dough
She picks up instead a stray piece of cord,
Drags it like a fishnet through the mud,
Then snaps and sends a glob of amber flying.
     Ei?
It lands with a flap and changes to a boy
Who runs in circles screaming like a piglet.
Delighted, she claps her hands, and dabs and lobs again.
     Hai!
The splotch becomes a girl turning somersaults.
Soon the daubs have populated hills and vales with a hundred of the tiny
    humanoids.
As the sun revolves into the lake, the dragon woman rests from her labors.
Her tail curls around her slumbering form,
As the chatter in the trees becomes a roar.

The hoard of flea-like, hairy striplings tumbles down
And scampers, lopes and crawls about the peopled hills,
Grazing on the pullulating fare.
By now the local fauna feels the malodorous presence of the new bipedal
    beast,
And each carnivore, and omnivore of near and far
Has winged or hoofed or snaked its way to this valley of ordure.
When tooth and claw and fang is bared,
Satiated fleshbags bound up their brushy haunts
Roosting up above the carnage—breast and bone.
Naked in their ignorance, dumb of mouth and cold,
Scared of death and always defecating,
This is black-haired Man.

Meanwhile, Shewah lies coiled on the loam,
When out of nowhere bursts the murk of hated Gongong, Fiend of the Flood,
Intent on ravishing the matriarch of Man.
But the lady rises on his lewd advance and strikes him with her lashing tail.
The demon reels, then charges with his massive head,
Batters at the base of Bujou Shan, pillar of the sky.
Soon Asymmetric Mountain falls, causing Earth to rise and Heaven slip
At the northwest quadrant of the land.
The rising earth sends rushing waters flooding ever eastward—
The stars since Gongong's fit of rage have flown toward the north.
Pleased with the enormity of this his first calamity,
Gongong calls the rains down to drown that heartless worm.
Laughs loud and long does dragon dam before retrieving her bamboo flute,
Whereupon she blows a note to rival thunder,
Shake the roots of mountains, and fire the forge of molten rock.
For ages do they battle, ranging wide as flood and fire consume the land.

But what of quaking Man, barefoot and prisoner of bush?
From the vantage point of crunching bone,
From the sound of canine lapping blood,
From the smell of brimstone and the freeze of arctic rain
Something snaps in the lemming brain.
Crazed, the mass of naked beasts leaps to certain death into the mounting
    flood,
But for two, a boy and girl, pubescent,
Who take refuge in a hollow bottle gourd.
The rising tide tears at the mooring vine until it breaks,
And the calabash is launched.
It floats there among the flotsam until the one remaining raven-sun
Flies fourscore circuits of the sky.
Embraced by shell like seeds they feed by day on melon rind
And dream by night of lizard primogenitor,
Little knowing that the hands that shaped their yellow limbs
Are just then locked in mortal combat.

At long last with the help of Turtle,
The demon Gongong weakens and is buried deep beneath the arctic ice and
    snow.
But now no longer separate is the Earth from Heaven.
The sky has fallen to the ground and ocean mingled with the moon.

Shewah kneels:
    I bid you, Turtle, sacrifice your sturdy limbs.
And these she stands at Earth's four corners, propping up the sky.
Then she dredges from the galactic alluvium gemstones of every hue
To mix a mortar fit to mend the broken sky.
Her duty done, she sighs, and joins the Turtle beneath the placid waves.

                    t
                    h
                    e
                calabash
               bangs   on
             solid ground
              and breaks
                  open
                 spilling
             melon   seeds
             and   boy   and
             girl,   her   belly
              bulging like a
               bottle gourd.

Notes

1. For the entire poem, see "In a Calabash: A Chinese Myth of Origins," *Talus* 9/10 (1997): 52–97. This excerpt collects and revises books 1–3, 12, and the epilogue from the original twelve books, prologue and epilogue, plus scholarly introduction.

2. At this point my original epic describes a complex "fall" from this paradise, and the subsequent destruction of earth by the flood fiend, Gongong. See N. J. Girardot's *Myth and Meaning in Early Taoism* (Berkeley and Los Angeles: University of California Press, 1983) for a discussion of the Daoist concept of "paradise lost."

3. Zhuangzi recounts the story of a man who could float across a river in a giant calabash. In Daoist iconography the bottle gourd (often seen hanging from his gnarled staff) is the immortal's receptacle for the elixir of long life. It is also the emblem of the Chinese apothecary.

I. Framing the Issues

"Daoism" and "Deep Ecology": Fantasy and Potentiality

JORDAN PAPER

Prologue

Since the inception of a doctoral program in the Faculty of Environmental Studies at York University a half decade ago, I have served either as the thesis supervisor or on the dissertation committee of those graduate students interested in religion. This introduced me to "deep ecology" and its use of Asian and Native American traditions, or, more precisely, what were assumed to be such traditions. In this essay, I will critically review the most frequently quoted deep ecology formulations pertinent to Daoism, compare them with the most often cited relevant studies by philosophers familiar with the Chinese materials, relate both sets of studies to skeptical ones, and conclude with a brief exposition of reasonable possibilities.[1]

Having participated in two previous, related conferences,[2] however, I am well aware that a sanguine response is preferred: from a religious tradition, one is expected to advance a solution to the environmental crisis. But if any aspect of Chinese tradition has had a major effect on me, it is that aspect which is the antithesis of Western intellectualizing. Chinese theorizing proceeds from real problems by developing means for resolving those specific problems. To the contrary, Western theorizing tends to proceed from ad hoc premises to create theories only tangentially related to the problems to which they may be applied. Hence, Western intellectuals and Western-influenced Chinese intellectuals tend to derive theories from Chinese texts that have little if any relationship to Chinese modes of thinking. Similarly,

Chinese religion, as virtually all non-Western religions, focuses on religious experience, whether this be personal ecstatic experience, experience from participating in rituals, or experience of superhuman beings via possessed spirit mediums or the work of shamans. Again in contrast, Western religions tend to proceed from belief in religious principles, dogma, theological formulations, and so on, although there are always subgroups which focus on religious experience.

A second pattern of behavior I have learned from Chinese culture is to criticize perceived errors in understanding and reacting to problems. Traditional Chinese government early institutionalized such criticism in a Board of Censors, and my doctoral thesis, from so long ago that it now seems to me a previous incarnation, focused on the literary remnants of such an official.[3] Accordingly, I hope that if not forgiven, there will at least be a degree of understanding in my taking more of a critical than an enthusiastic note with regard to this volume. I consider the imminent destruction of this planet for human habitation through the predation of humans so immediate and momentous that I have become rather immoderate in my impatience toward the assumption that donning the rose-colored glasses of romantic interpretations of other cultures will save the world.

The very topic of this volume raises serious concerns. Most of those involved cannot speak for or from within Daoism as a religion, in that the only Daoists per se, from a Chinese perspective, are initiated priests or monks-nuns.[4] With regard to Daoism as a philosophical orientation (*daojia*), we are essentially speaking of two enigmatic texts: *Zhuangzi* and *Laozi/ Daode jing*. The earliest strata of the former proceeds from those of the aristocracy who had no need to be productive members of society and focused on individual ecstatic experience. The latter, in the received version (i.e., excluding recently excavated variant versions), seems to have initially functioned as an ideological justification for a ruthless, totalitarian government. Both then came to be perceived as relevant to the search for the extension of life by the elite, and the latter written by a divinity. Only with the developed civil service system, as part of the elite ideology complimentary to the more pragmatically oriented aspects of *rujia* thought, do we find modes of thought that might be realistically relevant to the environmental crisis. Again, except for the nuns and monks of the Quanzhen order, Daoist religious (*daojioa*) rituals function together with the normative religious rituals and concerns of Chinese religion.

It is difficult to understand how artificially separating out either actual Daoist thought or Daoist religious practices from their ideological or religious context, according to the Western rather than the Chinese understanding, can contribute anything meaningful to resolving our urgent environmental problems.

Recent usages of the term "Daoism" have further confused the situation. There are now sectarian attempts in Taiwan to claim the term for normative Chinese religion in general.[5] This in turn is leading some new religions to claim the term for themselves, to the consternation if not indignation of the Daoist priests. On the mainland, the Chinese government is attempting to enfold all of indigenous Chinese religion within the Quanzhen order of Daoism as a means to control an otherwise amorphous, noninstitutional religion, since the order itself is under a government ministry. Unfortunately, a half millennium of the Western insistence on a trinitarian concept of distinctly separate religions in China, while ignoring normative Chinese religion, has led to this new use of the term "Daoism" for indigenous Chinese religion as a whole. This convention all but replaces the unfortunate mistranslation of "Confucianism" for *rujia*, the term for the dominant ideology of the former Chinese civil service system, as a term for normative Chinese religious practices, practices which long preceded the time of Kungfuzi, let alone *daojiao*. Finally, I have noticed, in the Toronto area at least, that Westerners who run commercial teaching institutions based on Chinese practices, such as *taiji*, or who are involved with various forms of healing practices based on Chinese modes tend to preface these institutions or practices with the term "Daoist," while those from China with similar involvements do not.

Although several scholars of Chinese religion have long attempted to point to Chinese religion as a singular, complex religious construct,[6] this has not as yet caught on, although we comfortably use the term Judaism to refer to the religion of the Jews or Hinduism to refer to the religion of Hindus, the Western term for the majority of South Asians who do not identify themselves with a specific term. It would now be most confusing to use one term of the ersatz trinity of Chinese religions for the whole. For this volume to remain meaningful, it is preferable that the term "Daoist" not be used without specific cultural referents and that we do not refer to a panoply of practices that may or may not be practiced specifically by Daoists in the Chinese sense, such as fengshui, divination, spirit possession, and *qigung*, or if we do, that we understand these practices to be part and parcel of Chinese religion in general, not specific to either *daojiao* or *daojia*.

   In spite of these concerns, I do believe this volume can make at least two positive contributions. First, these articles can assist in bringing to an end the romanticism of deep ecology with regard to its construction of "Daoism." Deep ecology could then proceed in more viable directions, particularly by trying to work within and modifying Western traditions through being informed by a more realistic understanding of Daoism. Second, by looking at Chinese ideology holistically from a Chinese perspective, while fully acknowledging that environmental degradation is already mentioned in virtually the earliest extant texts and that contemporary China is an environmental basket case, I do believe we can bring forward articulated attitudes of long standing, beginning with the *Mengzi* rather than the *Zhuangzi*, and also reflected in *jiao* (renewal) rituals, that may, indeed, be useful in reconceptualizing attitudes toward the environment both here and in China.

Deep Ecology

Deep ecology can be traced at least as far back as a seminal article, published three decades ago, by Lynn White, Jr., "The Historical Roots of Our Ecological Crisis," in which he pointed to Western philosophical and religious traditions as the basis of our present plight: "Human ecology is deeply conditioned by beliefs about our nature and destiny. . . ."[7] But White was also dubious about the application of non-Western traditions to the Western experience, in contrast to the later deep ecologists. The term itself seems first to have been used by the Norwegian philosopher Arne Naess in 1973.[8] The term was intended to distinguish between "environmentalism," the practical side of environmental concerns, and "ecosophy," or philosophical ecology, which brings in not only Western thinkers but also Native American thought, as well as Daoism and Zen Buddhism, to address fundamental ways of understanding the environment.

   According to my graduate students, the most influential works with regard to deep ecology's use of Daoism, verified by the deteriorated condition of these books in my university library, are John Clark, *The Anarchist Moment: Reflections on Culture, Nature and Power* (1984), although it does not address the term directly; Bill Devall and George Sessions, *Deep Ecology* (1985); and Dolores LaChapelle, *Sacred Land, Sacred Sex—Rapture of the Deep: Concerning Deep Ecology and Celebrating Life* (1988).[9] These works find the Western religious paradigms to be the root of the ecological crisis and virtually irremedial, and they turn to non-Western religious traditions—of which Daoism, as they understand it, is preeminent—for a new foundation for understanding life and the world, in order to solve environmental ills. Before turning to these and related depictions of Daoism, for purposes of comparison, we might quickly review writings on these topics by those who have spent their scholarly lives studying Chinese texts in the original language.

Sinologist Philosophers and Their Friends

The best-known writings on Daoism and ecology by philosophers who are not environmental specialists are found in two volumes, both published in the late 1980s; these scholars are associated with either the University of Hawaii or each other. Since two of these authors are involved with this volume, my comments will be brief, as they will be providing essays of their own.

Volume 8 (1986) of *Environmental Ethics* contains articles by Roger T. Ames—"Taoism and the Nature of Nature"—and Chung-ying Cheng—"On the Environmental Ethics of the *Tao* and the *Ch'i*."[10] Ames provides a thorough and reasonable analysis of the topic and provides a crucial admonition, one that tends to be ignored by those environmentalists espousing deep ecology who have read the article:

> Taoism's [Daoism's] concreteness returns us to our own particularity as the beginning point of the natural order. We cannot play the theoretician and derive an environmental ethic by appealing to universal principles, but must apply ourselves to the aesthetic task of cultivating an environmental *ethos* in our own place and time, and recommending this project to others by our participation in the environment.[11]

My sole caveat to this point, relevant to the concluding discussion, is that I find nothing specifically Daoist here, but rather that this is in the nature of traditional Chinese thought as a whole.[12]

With an erudite discourse on Chinese terms that are not in and of themselves exclusively Daoist, Cheng formulates a position that could "eventually lead to a maturity of the environmental ethics of the *Tao* and *Ch'i*, and it is in terms of this mature form . . . that we will be able to speak of the embodiment of a new reason in the mind of man."[13] Of course, by then humans may already have rendered themselves extinct due to environmental degradation.

Nature in Asian Traditions of Thought: Essays in Environmental Philosophy (1989), edited by J. Baird Callicott and Roger T. Ames, contains three articles relating to Daoism: the article by Ames previously mentioned, retitled "Putting the *Te* back into Taoism"; an article by Graham Parkes, "Human/Nature in Nietzsche and Taoism"; and David L. Hall, "On Seeking a Change of Environment."[14] From the introduction, subtitled "The Asian Traditions as a Conceptual Resource for Environmental Philosophy," it is apparent that the editors intended the anthology for an audience of deep ecologists. They note that recasting "environmental philosophy may not go far enough. A deeper break with traditional Western philosophical commitments may be required."[15] But they also note that

> There has been a general assumption that Eastern traditions of thought could provide important conceptual resources for this project, and there has been a lot of loose talk about how they might. But with the exception of a handful of essays, no direct and extensive work by experts in Eastern thought has been undertaken of the environmental philosophy problematic.[16]

This was the inspiration for the volume.

Parkes finds that

> Since there is no evidence that Nietzsche knew anything about Taoism, the number of his ideas about nature which correspond to Taoist views suggests that there may be further, hitherto unexplored philosophical (re)sources in the Western tradition which may prove helpful to consider in our current predicament.[17]

Hall elaborates aspects of "the Taoism of classical China" in its "cultural matrix" pertinent to developing an "altered sense of human nature and of nature per se." He seems not to concern himself as to why it had not done so within its own "cultural matrix."

Deep Ecology and Fantasies of Daoism

For want of some order, the more popular deep ecology texts will be addressed by date of publication. John Clark's *The Anarchist Moment* (1984) has but a single chapter on non-Western phenomena. That chapter is exclusively on Daoism, and begins with the sentence: "The *Lao Tzu* is one of the great anarchist classics."[18] Indeed, he understands it to be the preeminent work on the topic "East or West." Clark is aware of the major translations of the work at the time his book was written and quotes from the Wing-tsit Chan version, refers to relevant articles by Roger Ames and David Hall, and points to the works of Ursula Le Guin, whom he considers "perhaps the most widely read contemporary anarchist writer, and also a Taoist."[19] Among the reasons for Clark's conclusion, I will list but the most dubious ones, for in the text he finds both compassion and the rejection of coercion. Perhaps he missed such key passages as ". . . the sage governs the people by emptying their minds and filling their bellies, and by weakening their aspirations and strengthening their bones," and he dismisses the reference to treating people as straw-dogs.[20] Perhaps Clark is reading the text from a Mahāyāna Buddhist perspective. He also writes of "a Taoist community" as if it actually exists (aside from religious communities) and of "the Taoist ruler-sage" as more than a hypothetical figure.[21] Certainly, this is a *Laozi* that radically diverges from the sinological perspective, at least that common interpretation that finds the extant text, reinforced by the library excavated at Mawangdui, originally compiled to provide an ideological basis to *fajia* totalitarianism.[22]

Devall and Sessions, in *Deep Ecology* (1985), find, among "Eastern sources," the *Laozi* and the writings of Dōgen particularly inspiring for their purpose.[23] They are familiar with Clark's work, for they have a virtually identical, unattributed sentence: ". . . the Taoist way of life is based on compassion, respect, and love for all things."[24] There is oblique reference to *ziran* (self-actualization/ spontaneity/ nature-natural), which I would agree is most apropos:

> One metaphor for what we are talking about is found in the Eastern Taoist image, the *organic self*. Taoism tells us there is a way of unfolding which is inherent in all things.[25]

Deep Ecology contains no sustained argument, but rather bits and pieces of poetry, rituals, and the like, considered relevant to the theme of the work. There are several poem-translations from the *Laozi* by Tom Early[26] and a detailed plan for an "Autumn Equinox Taoist Celebration," seemingly held at Dolores LaChapelle's center in Silverton, Colorado.

Dolores LaChapelle's extended rhapsody *Sacred Land, Sacred Sex—Rapture of the Deep* (1988) centers on Daoism; the chapter entitled "Taoism" "constitutes the deep core of this work."[27] Her interest began several decades ago, on reading volume two of Joseph Needham's *Science and Civilization in China*, and developed through studying *taiji* for decades and subsequently teaching the subject:

> Now after all these years of gradual, deepening understanding of the Taoist way, I can state categorically that all these frantic last-minute efforts of our Western world to latch on to some "new idea" for saving the earth are unnecessary. It's been done for us already—thousands of years ago—by the Taoists. We can drop all that frantic effort and begin following the way of Lao Tzu [Laozi] and Chuang Tzu [Zhuangzi].[28]

LaChapelle understands that Western thinkers dismissed Daoism because they confused it with the practices of "superstitious village priests."[29] Certainly, she holds no truck with religious Daoism, although she readily creates "Taoist" rituals of her own. Written in a New Age style, the book presents an exposition of a Daoism based on personal living and selected, extensive reading, ready to be adopted by those of her readers who choose to do so.

The above three works are the primary sources on Daoism for environmental studies students, as well as for others who come to the topic through writings by environmentalists. There is another work, from the same period and written from the same romantic standpoint, that is frequently referred to in the environmental literature. The article by Po-keung Ip, "Taoism and the Foundations of Environmental Ethics," is often quoted as representative of the indigenous Chinese scholarly perspective—according to the statement provided in the anthology where the article is found, the author is a secretary at Lingnan College in Hong Kong. Depicting Western philosophy as "patently anti-environmentalistic," Ip argues that Daoist philosophy is able to fulfill all that is needed to transcend the human predicament and, at the same time, is "compatible with science and is thus capable of providing a minimally coherent ethics." By "Taoism," Ip means the *Laozi* and excerpts from the *Zhuangzi* found in Wing-tsit Chan's source book on Chinese philosophy, as well as Joseph Needham's references to the *Huainanzi*—there are no references to or mention of the texts in Chinese. Ip understands Daoism to be uncontaminated by religion. For example, Ip notes that *Tien* (Sky) "was a highly naturalistic notion which has no strong religious connotation." Apparently, he was unaware that *Tien* was so sacred that, for anyone other than the imperial couple, to sacrifice to *Tien* and *Ti* (Earth) was, ipso facto, treason.[30]

The most recent work in this grouping is Peter Marshall's *Nature's Web: Rethinking Our Place on Earth* (1992), the first chapter being "Taoism: The Way of Nature." Confused with regard to Chinese literary chronology, Marshall, referring to the *Zhuangzi*, finds that

> The first clear expression of ecological thinking appears in ancient China from about the sixth century BC. . . . The Taoists . . . offered the most profound and eloquent philosophy of nature ever elaborated and the first stirrings of ecological sensibility.[31]

His references are to the most common translations utilized by those in environmental studies: the Gia-Fu Feng and Jane English free rendering of the *Laozi*, the dated Herbert Giles translation of the *Zhuangzi*, and the Victorian, James Legge translation of the *Yi*. As are many of these authors, Marshall is pleased that Daoism has but a minimal relation to religion, terming it a pure philosophy, and finds that it models the most desirable form of government:

> Although containing elements of mysticism, the Taoists' receptive approach to nature encouraged an experimental method and a democratic attitude . . . [Taoism] provides the philosophical foundations for a genuinely ecological society and a way to resolve the ancient antagonism between humanity and nature which continues to bedevil the world.[32]

While most sympathetic with many of the views expressed by the writings of deep ecology and tangential works, I find the Daoism expressed simplistic to the point of absurdity: it is ahistorical; the authors have no understanding that the interpretation of the texts continually changed over time, as did Chinese culture, according to changing needs and values; they assume a philosophy can be based on two enigmatic, early texts, one containing contradictory parts developed over several centuries, and the second probably edited from diverse sources; they presume that there were single authors for both works; they provide contemporary Western interpretations of the texts, which is acceptable only if they recognized what they were doing; and, strangest of all—since these authors critique the coldness and distance of Western post-enlightenment philosophy—they celebrate an assumed antimystical, antireligious orientation in texts clearly related to both mysticism and religion in China. Given that the deep ecologists are seeking a fundamental transformation of culture, their attitude toward religion is puzzling: philosophies do not transform cultures, religions do.[33]

The romanticized translations tend to be read literally, with little understanding of metaphor. The descriptions of an ideal rustic life of small villages separated by virgin forests already reflected an idealized antiquity in these last Zhou texts. In any case, the texts were written for the literate aristocracy, not peasants. Even the epitome of the aristocratic Daoist "drop-out," Tao Qian, over half a millennium later had servants[34] and played at farming when he was not drinking with his wealthy, aristocratic friends or writing poetry. To assume that reversion to a simple rusticity could be applied to the billion and a quarter people of grossly overpopulated China, when it could not at a time when the population was less than 10 percent of the present-day, is asinine.

In summary, the Daoism of the deep ecologists is utterly a modern Western one. That a Western Daoism can solve a crisis assumed to be brought on by and unredeemable through Western thinking implies a logical contradiction.

The Skeptics

As early as 1968, Yi-Fu Tuan, a geographer, published his "Discrepancies between Environmental Attitude and Behaviour: Examples from Europe and China," which remains as cogent and definitive today as when it was written.[35] He exemplifies the pragmatism of Chinese thinking in pointing out the obvious:

> The history of environmental ideas, however, has been pursued as an academic discipline largely in detachment from the question of how— if at all—these ideas guide the course of action, or how they arise out of it.[36]

Tuan notes the environmentalists' references to the *Laozi,* but points to actually stated environmental concerns in the *Mengzi* (the most important *rujia* [Confucian] text for the last one thousand years), early offices for inspecting forests, and memorials to the throne. Tuan briefly lays out the evidence for early environmental destruction in China. He ends by comparing the European formal garden and the Chinese naturalistic garden, suggesting "that these human achievements probably required *comparable* amounts of nature modification."[37]

Anyone who has traveled through China other than in tour groups would have, I trust, a difficult time in understanding how Chinese traditions can be lauded for ideal environmental attitudes. While environmental destruction has enormously accelerated in the twentieth century, the degradation is long-standing. Vast areas of forests were denuded and many waters became highly polluted well over a millennium ago. For those who are but aware of China vicariously, Vaclav Smil's *China's Environmental Crisis* should serve as a necessary corrective.[38]

More recently, Holmes Rolston, III, presented an argument similar to Tuan, but from the perspective of philosophy, in "Can the East Help the West to Value Nature?" Of all the articles researched for this essay, it was the only one razored out of its library binding (and replaced by a photocopy), indicating it is a popular article among students and leading one to wonder, as university teachers often do, if students read to the end of an article. Rolston begins with the question of his title. Pointing to the test of the ability of these traditions to resolve environmental problems in Eastern nations (he incorrectly assumes that these problems did not exist in premodern times), he concludes: "My own judgement is that the East needs considerable reformulation of its sources before it can preach much to the West."[39]

Possibilities

Aside from the fanciful, simplistic interpretations of "Daoism" by the deep ecologists, all of the previously discussed interpretations, including some by sinologists, disturb me in two regards. First is the separating out of Daoist thought, let alone two particular texts, from Chinese thought as a whole. This ignores the intellectual context, as well as the actual applications, of this aspect of Chinese thought, especially on the subject of environmental concerns. Yi-Fu Tuan, who is aware of the wealth of relevant material points this out, but so, indirectly, does J. Baird Callicott, the religionist who has probably published more on religion and ecology than anyone else.

In a chapter entitled "Traditional East Asian Deep Ecology" in *Earth's Insights*, Callicott has sections on "Taoism," where he notes that Western environmental ethicists "have been drawn chiefly to Taoism," and "Confucianism," where he posits that "By transposing the Confucian social model from the human to the biotic community, the ecological holist is encouraged to add a fourth dimension to the web of life."[40] Actually, this already took place centuries ago.

As Chinese religion evolved, *xiao* (filial piety), the fundamental aspect of social relationships of the *rujia* tradition based on natural, nuclear-family ties, was stretched to the realm beyond humans. Wang Yangming, the early sixteenth-century *rujia* theorist—a *rujia* that incorporated aspects of *daojia* as well as a sinified Buddhism—who so influenced literati ideology, wrote:

> Everything from ruler, minister, husband, wife, and friends to mountains, rivers, spiritual beings, birds, animals, and plants should be truly loved in order to realize my humanity that forms one body with them, and then my clear character will be completely manifested, and I will really form one body with Heaven [Sky], Earth, and the myriad things.[41]

For Wang Yangming, family not only included the state, it included the natural world, a realization undoubtedly arising from the ecstatic religious experience of union with the entire environment. What blocks this understanding is simply greed, the contemporary global ethic rapidly spreading around the planet from its Western, capitalistic roots:

> Thus the learning of the great man [person] consists entirely in getting rid of the obscuration of selfish desires in order by his[/her] own efforts to make manifest his[/her] clear character, so as to restore the condition of forming one body with Heaven [Sky], Earth, and the myriad things, a condition that is originally so, that is all.[42]

In the concluding section, however, Callicott finds value not in separating the strands of Chinese thought, but in their actual gestalt:

The potential for the development of an explicit indigenous Chinese environmental ethic based on classical Chinese thought is tremendous. And the potential contribution of classical Chinese thought to deep ecology, ecofeminism,[43] and, more generally, to a global consciousness and conscience is equally great.[44]

Hence, the question arises as to why, in the series of Harvard University conferences leading to these volumes, "Confucianism" and "Daoism" were not only split apart, but whether this separation was inadvertently emphasized by holding one of the conferences near the beginning of the series and one near the end. In the foreword to this series of volumes, the series editors Mary Evelyn Tucker and John Grim consistently do not separate the two: "The East Asian traditions of Confucianism and Taoism remain, in certain ways, some of the most life-affirming in the spectrum of world religions" (with the caveat in a note that China is not immune to environmental degradation).[45] They then go on to detail this evaluation, never splitting Chinese thought into distinctly different intellectual traditions. It is only when the Western deconstruction of Chinese thought is disregarded—and when the focus on the "classics," while ignoring the history of their interpretation and development, is also disregarded—that we can find much of relevant value in the tradition as a whole.

My second concern in the above discussed interpretations of Daoism is the continuation of the Christian missionary dismissal of Chinese religion as ignorant superstition.[46] I have not yet been able to wrap my mind around the conundrum of negatively viewing Chinese religion while positively valuing the writings of those who were not only a part of it—particularly since Chinese religion and family are coterminous—but who, as government officials, were also priests in the state rituals as well as in their clan sacrifices. Certainly, the two texts that are at the basis of deep ecology, the *Laozi* and the *Zhuangzi*, are essentially religious texts, and it is their religious understanding, as well as the developments that later took place in Daoism as a religion (*daojiao*)—ignored by all the deep ecologists—that offers the most useful possibilities for saving the planet.[47]

Ruminations on the mystic experience (the ecstasy of self-loss/ union) are transmitted by but a few cultures: those that are socially stratified (for nonproductive ecstacies to be valued) and literate (for transmission to those of other cultures or other times)—the monotheistic, South Asian, and East Asian traditions. These ruminations may have existed in the Central American civilizations, but due to the thoroughness of the Spanish friars' destruction of their libraries, we may never know. The South Asian traditions tend to understand the experience of the void (*śūnyatā*) as the escape (*mokṣa, nirvāṇa*) from ceaseless existence (*saṃsāra*), the world understood as illusion (*māyā*). The monotheistic traditions, in positing a disjunction between a single creator and the created, tend to understand the experience of loss of self as merging with the creator or an aspect of it, of transcending the world. Only in China, where life was understood to be singular (hence, the search for longevity) and existence to be enjoyed rather than avoided or denied, did the experience enhance the importance of this world, at least until one died and one's body was recycled.

This experience, or rather the coming out of it, led to the understanding that we constantly create not only the world around us, but ourselves (*ziran*), based on the actual experience of going from nothingness (*wu*, "zero-experience") to somethingness (*yu*) to singularity (*dao*), and that singularity splitting into two (*tiendi* and/or *yinyang*), three, and quickly, the myriad things. This world that we create may in its essence be *wu*, but it is all we have, and we might as well enjoy it as it is, that is, naturally (*ziran*). By extension, we cannot enjoy it unless the environment for human life survives, requiring active involvement in environmental protection when the environment is threatened. For well over a millennium, this experience was articulated through the Three Incomparables (*sanjue*)—poetry, calligraphy and painting, especially of landscapes (*shanshui*: mountains-waters)—which specifically stand for numinous Earth. This mode of communication is perhaps epitomized in Su Shi's well-known "Red Cliff *Fu* (prose-poem) #1."[48]

Since we are social beings, we must enjoy the world with others, and it is in this regard that *daojia* thought works in conjunction with *rujia* thought. We can only enjoy it with others through understanding fundamental relationships that involve duties and responsibilities. These relationships, given an understanding of the essential falseness of distinctions, include not only humans, but everything that exists. Moreover, we cannot enjoy it through unnecessary, wasteful consumption, due to meaningless greed—the *bête noire* of many *rujia* theorists and practitioners for two and a half millennia. We must be exceedingly careful in dealing with problems arising from nature, however, or we may exacerbate, rather than ameliorate, the problems.

While such cogitation arose from a few and, more often than not, fell on deaf ears, others were creating ritual ceremonies that celebrated renewal and ritual behaviors that reinforced the connections between human and nonhuman, including Earth, Waters, and Sky. These rituals combined with those continuing from the inception of horticulture—the offering of gifts and gratitude to the soil itself for the food on which our lives depend, as well as to the dead of the family, on whom our family's well-being depends (from the perspective of Chinese and other civilizations arising from horticulture). And these rituals are not just of the distant past, for they continue today and are constantly being renewed, particularly in the most industrialized of Chinese cultures: Taiwan, Singapore, and Hong Kong. As well, in spite of a half-century of Marxist-influenced suppression as unscientific superstition, shrines to Grandmother and Grandfather Earth are reappearing in mainland China's agricultural fields. The rigid *daojiao jiao* rituals are, to a degree, being supplanted by the spontaneous (that is, based on inspiration from the deities) *fahui* (the *fojiao* [Buddhist] term being borrowed) renewal rituals of the contemporary society of mediums in Taiwan.[49] It is these rituals in particular that have the potential to articulate forcefully through the voices of the deities (via possessed mediums) the growing environmental awareness to be found there.

Since normative Chinese religion has not been a topic of the Harvard conferences on religion and ecology, a brief example may suggest the rich possibilities in these regards.[50] For example, Chinese religion, as virtually all nonmonotheistic traditions, has always celebrated and understood Earth and its produce as numinous beings. In traditional times, the imperial couple sacrificed to Sky and Earth; regional governors, as well as the emperor, sacrificed at altars to Soil and Grain; and farming families continue to sacrifice at small shrines in the farm fields to Grandmother and Grandfather Earth (the names varying according to local usages) and have their images on family altars in farmers' homes. Understanding the Earth to be not only numinous but also capable of denying humans food mitigates against deliberate abuse. To the contrary, both agribusinesses and the former massive agricultural communes in China and the former Soviet Union desacralized Earth, allowing it to be raped at will.

None of this can be directly meaningful in Western culture. But indirectly, it can suggest means for reinterpreting Western traditions to reverse rapid environmental degradation. As liberation theology, inspired by the continuation of Native American lifestyles and understandings in Central America, clashed with invested power, including that of the Vatican, with regard to destructive political domination, similar theologies could, and are starting to, speak to the global destruction of human habitation. Only awareness of the realities of Chinese thought—pragmatic thought—and rituals as a whole can suggest viable responses for Christians. Those seeking to create alternatives to the dominant Western traditions are not likely to find a firm basis in fantasies that never existed. They too need to understand the realities, both positive and negative, of Chinese culture for models that will stand some chance of surviving the rapidly changing fashions of New Age religion. This is what we as scholars of Chinese thought and religion can offer.

Notes

1. Appreciation is due to Adrian Ivankiv and Sherry Rowley for providing lists of deep ecology texts pertinent to Daoism.

2. Jordan Paper and Li Chuang Paper, "Chinese Religions, Population, and the Environment," in *Population, Consumption, and the Environment: Religious and Secular Responses,* ed. Harold Coward (Albany: State University of New York Press, 1995), 173–91; and Jordan Paper, "Chinese Religion and Ecology" (paper presented in "Spiritual Ecology," 1997 spring lecture series, Boston Research Center for the 21st Century).

3. The central part was published as *The* Fu-Tzu: *A Post-Han Confucian Text,* T'oung Pao Monograph 13 (Leiden: E. J. Brill, 1987).

4. This point is made by Kristofer Schipper in his *The Taoist Body,* trans. Karen Duval (Berkeley and Los Angeles: University of California Press, 1993), 3.

5. For example, the title of Michael Saso's excellent encapsulation of normative Chinese religious practices, especially in Taiwan: *Blue Dragon, White Tiger: Taoist Rites of Passage* (Washington, D.C.: The Taoist Center, 1990); as well as the thrust of Schipper, *The Taoist Body.*

6. E.g., Lawrence Thompson's *Chinese Religion: An Introduction,* from its first edition in 1973 to the most recent, fifth edition (Belmont, Calif.: Wadsworth, 1996); and my own elaboration and arguments in Jordan Paper, *The Spirits Are Drunk: Comparative Approaches to Chinese Religion* (Albany: State University of New York Press, 1995).

7. Lynn White, Jr., "The Historical Roots of Our Ecological Crisis," *Science* 155 (1967): 1205.

8. Peter Marshall, *Nature's Web: Rethinking Our Place on Earth* (New York: Paragon House, 1994 [1992]), 413. See Arne Naess, "Identification as a Source of Deep Ecological Attitudes," in *Deep Ecology,* ed. Michael Tobias (San Diego: Avant Books, 1985).

9. John Clark, *The Anarchist Moment: Reflections on Culture, Nature and Power* (Montreal: Black Rose Books, 1984); *Deep Ecology: Living As If Nature Mattered,* ed. Bill Devall and George Sessions (Salt Lake City: Peregrine Smith Books, 1985); Dolores LaChapelle, *Sacred Land, Sacred Sex—Rapture of the Deep: Concerning Deep Ecology and Celebrating Life* (Silverton, Colo.: Finn Hill Arts, 1988).

10. Roger T. Ames, "Taoism and the Nature of Nature," *Environmental Ethics* 8 (1986): 317–50; Chung-ying Cheng, "On the Environmental Ethics of the *Tao* and the *Ch'i,*" *Environmental Ethics* 8 (1986): 351–70.

11. Ames, "Taoism and the Nature of Nature," 348.

12. See Hajime Nakamura, *Ways of Thinking of Eastern Peoples,* trans. Philip P. Wiener (Honolulu: East-West Center Press, 1964).

13. Cheng, "On the Environmental Ethics of the *Tao* and the *Chi,*" 370.

14. Roger T. Ames, "Putting the *Te* back into Taoism," in *Nature in Asian Traditions of Thought: Essays in Environmental Philosophy,* ed. J. Baird Callicott and Roger T. Ames (Albany: State University of New York Press, 1989), 113–44; Graham Parkes, "Human/Nature in Nietzsche and Taoism," in Callicott and Ames, *Nature in*

Asian Traditions of Thought, 79–98; David L. Hall, "On Seeking a Change of Environment," in Callicott and Ames, *Nature in Asian Traditions of Thought*, 99–112.

15. "The Asian Traditions as a Conceptual Resource for Environmental Philosophy," in ibid., 3.

16. Ibid., 11–12.

17. Parkes, "Human/Nature in Nietzsche and Taoism," 80.

18. Clark, *The Anarchist Movement*, 165.

19. Ibid., 188.

20. *Laozi* 3 and 5.

21. Clark, *The Anarchist Movement*, 180 and 185.

22. Recently excavated fragmentary texts suggest that early versions that did not continue may have had different foci.

23. Devall and Sessions, *Deep Ecology*, 100.

24. Ibid., 11; compare to Clark, *The Anarchist Movement*, 176.

25. Devall and Sessions, *Deep Ecology*, 100.

26. I could not find a bibliographic reference.

27. LaChapelle, *Sacred Land, Sacred Sex*, 340.

28. Ibid., 90.

29. Ibid., 91.

30. Po-keung Ip, "Taoism and the Foundation of Environmental Ethics," in Eugene C. Hargrove, *Religion and the Environmental Crisis* (Athens, Ga.: University of Georgia Press, 1986), 102–5.

31. Marshall, *Nature's Web,* 9.

32. Ibid., 23.

33. "Philosophy" is understood to refer to theorizing about the nature of knowledge by intellectuals for intellectuals. "Religion" does not necessarily mean formal institutions but concerns the expression of that which is central to both a cultural and an individual understanding of life itself, including the notions of cosmos, time, continuity, disruption, identity, and reality, and particularly all those activities which pertain to life, that is, those behaviors which create and connote significance and meaning among those who take part.

34. The fragment of a letter by Tao Qian retained in his official biography mentions his sending one of his servants to care for his son.

35. Yi-Fu Tuan, "Discrepancies between Environmental Attitude and Behaviour: Examples from Europe and China," *The Canadian Geographer* 12 (1968): 176–91. References are to the reprint in *Ecology and Religion in History*, ed. David and Eileen Spring (New York: Harper Torchbooks 1974), 91–113. (The seminal article by Lynn White, Jr., referred to earlier in the paper, may also be found in this volume.)

36. Tuan, "Discrepancies between Environmental Attitude and Behaviour," 91.

37. Ibid.,111–12; emphasis added.

38. Vaclav Smil, *China's Environmental Crisis: An Inquiry into the Limits of National Development* (Armonk, N.Y.: M. E. Sharpe, 1993). This book is a follow-up to Smil's *The Bad Earth* (1983), which was reviewed in many major journals, was translated into Chinese (1988), and was available to the formative deep ecology writers.

39. Holmes Rolston, III, "Can the East Help the West to Value Nature?" *Philosophy East and West* 37 (1987): 172, 189.

40. J. Baird Callicott, *Earth's Insights: A Survey of Ecological Ethics from the Mediterranean Basin to the Australian Outback* (Berkeley and Los Angeles: University of California Press, 1994): 67, 84–85.

41. Wang Yangming, "Inquiry on the *Great Learning,*" in Wing-tsit Chan, trans., *Instructions for Practical Living and Other Neo-Confucian Writings by Wang Yangming* (New York: Columbia University Press, 1963), 273.

42. Ibid.

43. There are more possibilities in this regard than generally realized. See the chapters pertaining to China in my *Through the Earth Darkly: Female Spirituality in Comparative Perspective* (New York: Continuum, 1997).

44. Callicott, *Earth's Insights,* 85.

45. Mary Evelyn Tucker and John Grim, series foreword in this volume, xxiv.

46. For a fuller development of this and related points, see Jordan Paper, *The Spirits Are Drunk.*

47. Saving China itself requires a reconceptualization of family from a patrilineal basis to a bilateral one in order for the policies stabilizing the population to succeed, given that the primary obligation in Chinese religion is to continue the family. For an exposition of this point, see Paper and Paper, "Chinese Religions, Population, and the Environment."

48. For a translation, see Jordan Paper and Lawrence G. Thompson, *The Chinese Way in Religion* (Belmont, Calif.: Wadsworth Publishing, 1998), 194–96.

49. See Jordan Paper, "Mediums and Modernity: The Institutionalization of Ecstatic Religious Functionaries in Taiwan," *Journal of Chinese Religion* 24 (1996): 105–30.

50. Negative aspects have already been treated in Paper and Paper, "Chinese Religions, Population, and the Environment."

Ecological Questions for Daoist Thought: Contemporary Issues and Ancient Texts

JOANNE D. BIRDWHISTELL

Introduction

The term Daoism applies to many kinds of beliefs and practices, including those that now might be differentiated as religious, philosophical, and proto-scientific. My concern here is to examine how some of Daoism's ideas and values may be used to respond to issues of contemporary ecological discussions, especially with a focus on the environmental implications of the ideas. Despite the ancient and uncertain origins of many of its texts, certain aspects of Daoism appear to be still viable in Chinese culture—although clothed in new forms and identities. This discussion thus aims to make explicit some of the relevant beliefs both of Daoism and Chinese culture. My analysis joins previous efforts, mostly Western until recently, to probe cultural views for their environmental ramifications. Although many aspects of Daoism are recognized as favorable to the environment, others are problematic. This latter kind of idea must not be ignored, however, for it helps demonstrate how cultural beliefs that engender destructive activities remain powerful by "hiding" in other ideas.

As Western thinkers and activists in the twentieth century have become increasingly concerned with ecology, they have developed a variety of positions, many of which are still evolving. The views articulated have depended on how conditions and problems in all areas of life have been analyzed, not only in those areas specifically relating to the environment. Reflecting different emphases, some of the important stances consist of deep ecology, social ecology, ecofeminism, ecotheology, spiritual deep ecology, Earth First!, Greenpeace, environmentalism, and (comparative) environmental ethics.[1] These positions have disagreements, but most accept certain assumptions—for example, that human beings are not separate from the natural world, that all aspects and areas of the earth are interconnected, that environmentally destructive activities are related to, and supported by, a wide range of beliefs, and that human beings now face an ecological crisis.

Two primary aims have motivated their varied analyses: to analyze cultural and philosophical ideas in order to show how ideas about nature and the environment are intertwined with other cultural ideas and with often-unrecognized assumptions, and to propose new, less destructive, ways to conceive of the world. It is assumed that people need to change their views if they are to change their behavior, but people don't always know what they believe! Karen J. Warren, an ecofeminist, has suggested, for instance, eight kinds of links that have existed, and still do exist, between the domination of nature and the domination of women.[2] Others, such as Aldo Leopold, one of the first ecologists, and Arne Naess, a deep ecologist, have proposed new conceptions of the world that aim to overcome the deficiencies commonly attributed to the (Western) mainstream traditional philosophical position.[3] Many have pointed out that we need not only better knowledge about the earth, but better knowledge about our knowledge itself, including the behavioral implications of our assumptions, categories, and values.

Religious and philosophical thinkers have discussed ecology in many ways, often choosing topics and methods that have previously been fruitful in other fields of concern. Here I have selected four issues that are particularly important, and I examine how some early Daoist texts might respond to them. All involving matters of perspective and the implications of particular positions, these four issues consist of: 1) grand narratives and little narratives; 2) centrism and domination; 3) the particular or local and the general or universal; 4) the human story and the cosmic story. Since there are considerable differences among the texts, it is risky to make generalizations about them as a whole, and thus I have limited my comments to specific texts, particularly those whose ideas are most easily translatable to modern times.[4] I have focused especially, but not exclusively, on the *Zhuangzi* because it has been one of the most influential texts in Chinese culture. My conclusions find that this text and early Daoist thought (pre-

Han and Han dynasties) leave a mixed legacy in regard to ecological thinking. This conclusion should be helpful, however, for it can serve to remind us of the difficulty in identifying problematic assumptions and their wide-ranging ecological effects.

Grand Narratives and Little Narratives

The issue of whose story gets told and accepted by a society and whose stories are suppressed or simply not told, is important, for those who "tell the story" are able to shape people's actions, ideas, values, and identities in all sorts of significant ways. A culture's grand, or master, narrative does not simply present an account of the historical development of that culture; it justifies why things are as they are. Containing implicit and powerful values in the very categories it uses, a grand narrative represents the perspective of the dominant groups (although others usually accept it too), and it generally eliminates, without acknowledgment, other views.

Whether larger or smaller, narratives are important in that they contribute to forming an identity, they help give meaning to life, and, even more, they appear to be critical to the existence of a viable society. Here, "identity" should not be understood in the sense of an eternal, unchanging essence, but rather as something that is constructed as part of a historical process involving many kinds of relationships. This conception of identity entails seeing one's self or one's group as part of a specific history and specific place, it recognizes that the self or group is socially and linguistically constructed, and it accepts that people have some capacity for acting intentionally.

Realizing the ecological importance of "who tells the story," ecologists have analyzed how the grand narrative of the modern West has supported the ecological destruction associated with capitalism and state socialism.[5] Although most now reject this grand narrative, they do not agree on what should take its place. One major issue is whether the construction of a new grand narrative should be attempted, or whether little narratives might be more effective in changing people's views and actions.

Western thinkers have not seriously integrated the experiences of other major civilizations into their analyses, however. For instance, massive ecological destruction has also occurred in China, and long before the nineteenth and twentieth centuries, but for different reasons and in different ways from ecological destruction in the West.[6] Consideration of the narratives of other cultures may enhance current analyses, including those that ask what values are dominant in a culture, what views help to provide its identity, and what concerns serve to motivate the actions of the leadership group and ordinary people.

Early Daoist texts contain both grand narratives and little narratives. Those of the former type are mostly "countercultural" narratives, in that they reject the grand narrative of the Confucian view. Here I shall use as examples two selections from the *Zhuangzi*, although there are other similar stories, both in the *Zhuangzi* and in other texts.[7] By attacking the ostensibly harmful results of the Confucian way and by offering their own alternative vision, these stories attempt to establish a nondestructive human relationship with the surrounding environment. Constructing a different cultural identity, they emphasize close relations between humans and nature, here conceived in terms of the earth, living things, and the Way (*Dao*) and its Virtue (*De*).

My first example is a story from chapter 9, "Horses' Hoofs." It describes an ancient, utopian time of "Perfect Virtue," a period when people were not rushed or agitated and when mountains and lakes were not marred by such human constructions as trails and bridges. All living things lived with their own kind, developing to their full potentials, and the various species lived peacefully in close proximity to each other. This was a time characterized by the Way and its Virtue, but it was destroyed when the (Confucian) sage appeared, with his concepts of morality. Confucian notions of rightness and benevolence served to divide and destroy the ancient harmony.

My second example, a story from chapter 14, "The Turning of Heaven," tells about the rule of those early sage rulers the "Three August Ones" and the "Five Emperors." It is claimed that their methods of rule brought confusion instead of order, and their so-called wisdom led to environmental chaos. The air became such that the sun and moon no longer shone brightly, hills were stripped of their trees and streams became sluggish (with silt), and even the regularity of the four seasons was upset. "Not a living thing was allowed to rest in the true form of its nature and fate." Even worse, these rulers did not regard their actions as shameful, but "considered themselves sages!"

The Perfect Virtue story emphasizes a closeness that was believed to have once existed between human beings and the natural world. It especially draws attention to the similar "principles" or patterns that characterize life—of humans, animals, and plants. These patterns include both the social nature of life, that of animals as well as humans, and the compatibility of all types of life-forms. That all existence is social is an important assumption, for it implies a priority for the group on the basis of the inborn characteristics of living things. Indeed, this story does not even entertain the notion of an individual identity separate from a group.

The compatibility of all life-forms is a second, equally important, assumption, for it rejects the view that humans can live well only if they can control or destroy other life-forms or the natural world that they inhabit. Like many of the stories, this story criticizes Confucian actions and justifications, put forth in the name of such virtues as benevolence and rightness, on the grounds that Confucians in effect reject the "compatibility" assumption and transform the "social" into a narrow range of experience, specifically only that of human beings. The *Laozi* also echoes these ideas in many of its passages, such as: "When the great way falls into disuse, there are benevolence and rectitude"; and "Exterminate the sage, discard the wise, and the people will benefit a hundredfold."[8]

With a few changes in language, the story of the Three August Ones could be a description of the effects of modern industrialization, as it speaks critically of air and water pollution, deforestation, and the destruction of habitat. While it is possible that such language was primarily metaphorical, other texts and passages refer to similar kinds of environmental problems. For example, there is the anecdote about the prehistorical Kua Fu (Bragging Father), which appears in several kinds of eclectic texts from the Han period, including the *Huainanzi*, the *Classic of Mountains and Seas*, and the *Liezi*. Kua Fu's thirst was so great that even though he drank the Yellow and Wei Rivers dry, he still died of thirst. As Mark Elvin has suggested, rather than view the story of Kua Fu as being about someone who misjudged his own strength, it can be interpreted as representing the Chinese "destruction of the systems of the natural infrastructure that support life—here symbolized by the exhaustion of the great rivers, insufficient to satisfy Kua Fu's all-consuming thirst."[9]

Whether metaphorical or descriptive or both, the Three August Ones story assumes vital connections between human beings and the conditions of their habitat, which includes Heaven and Earth and all living things. Nonetheless, both in ancient China and the contemporary world, environmental destruction has often resulted incidentally from beliefs and actions concerned with some other specific problem. For instance, the Chinese often destroyed forests because they were viewed as "hideouts for bandits and rebels" and as places where uncivilized people lived.[10] Problems of environmental degradation were, and are, related to a constricted view of human beings, one that discounts, and is often blind to, the varied interactions among humans and other living things and their places of habitation. By the time the early Daoist texts were compiled, moreover, massive reshaping of the environment was already underway, especially through water control projects.

Narratives like these two offer alternatives to the grand narrative of Chinese culture accepted by Confucian thinkers and many historians. Daoist stories reject a cultural identity that apparently entailed considerable destruction of the natural world, a phenomenon which even the *Mencius*, a Confucian text, refers to in its famous story of Ox Mountain, which had become stripped of its original vegetation. An important ecological issue that appears in many of these narratives is the method of ruling. Daoist texts such as the *Zhuangzi*, *Laozi*, and *Liezi* reject rulers' efforts to shape and change people's hearts, for such efforts eventually lead people to see themselves as separate from one another and ultimately separate from other creatures and the environment. This destructive type of ruling (carried out by the mythological sage rulers and culture heroes Huang Di, Yao, Shun, and Yu) is concerned with manipulating and thwarting the natural inclinations and actions of people as a way to control the world.

The Daoist alternative found in its grand narrative described here emphasizes the view that processes of change are inherent in all things; thus, the way to flourish is to accept and go along with change—to accord oneself with Dao (the Way). As the *Laozi* claimed, if the rulers hold fast to the Way, "the people will be equitable, though no one so decrees."[11] The criticism of the *Zhuangzi*, *Laozi*, and other texts suggests that, because the Confucians have not recognized the many dimensions of the human habitat, their focus is too narrow and so their actions will eventually end in failure. The rejected Confucian grand narrative tends to discount the state's violations of Dao in the realm of the earth and the myriad things.

Centrism and Domination

The issue of centrism concerns situations of domination, and so it takes many forms, including anthropocentrism, androcentrism, and ethnocentrism. Rejecting the claim of a universal viewpoint, those concerned with this issue attempt to identify the implicit perspectives from which theories and concepts are formulated. Here, the notion of anthropocentrism in relation to ecological thinking will be examined.[12] While some ecologists claim that anthropocentrism is necessary and unavoidable, others see it as a major contributor to environmental destruction. These kinds of disagreements can be attributed both to differences in the definitions of anthropocentrism and to differences in fundamental assumptions about human knowledge and the world.

Postmodern thinkers of all varieties hold that it is impossible not to have a human-centered approach, because there is no view without a perspective, no view "from nowhere." All human knowledge is anthropocentric in some way. If anthropocentrism only entails this idea, it does not necessarily imply the human oppression of nature.[13] Some deep ecologists disagree, claiming that all anthropocentrism is oppressive toward the environment. There are also those who maintain that it is possible for humans to view things from the perspectives of other things, to have a nonhuman perspective.[14] However, a position that claims to disregard specific human interests and that presents itself as neutral or claims the ability to take another's view is certainly suspect, for past experience has shown that it may easily entail unrecognized assumptions that support a pattern of abuses called "custom."

Daoist texts offer a variety of comments on this issue, but only a few points can be considered here. The *Zhuangzi* in particular contains numerous conversations and stories about viewpoints and whether all things can agree on what is right. Although this concern may have been linked to other issues in the original historical context, such as the problem of making generalizations and thinking abstractly, later interpreters have regarded the issue of perspective as itself important. In order to discuss several different ideas, I shall cite three examples.[15]

In chapter 2, "Discussion on making all things equal," Wang Ni points out that men, loaches, and monkeys each live in a different kind of place, and so he asks, who knows the right place to live? Since men, deer, centipedes, and hawks each eat different kinds of things, which one knows the proper taste of food? The same question is also applied to standards of beauty, standards of right and wrong, and other such distinctions.

In chapter 7, "Fit for emperors and kings," Jie Yu, a madman (the type of person who is often considered a true sage in Daoism), compares the ruler's efforts to make people obey his "principles, standards, ceremonies, and regulations" to such feats as "trying to walk on the ocean, to drill through a river, or to make a mosquito shoulder a mountain!" In contrast to an emphasis on external behavior, the government of a true sage would involve, he claims, the self-cultivation of the ruler and then would aim to ensure that the proper conditions prevail so that all things, even a bird and a mouse, could develop and act according to their inborn capacities.

In chapter 17, "Autumn floods," Prince Mou narrates the conversation between a frog in a caved-in well and a great turtle of the Eastern Sea. The frog thought his world was the best, full of nicks and crannies, but the turtle got stuck before he could hardly get a foot into the well. The turtle then described the pleasures of the Eastern Sea, with its distance across of more than a thousand *li* and its depth of more than a thousand fathoms.

These conversations suggest several ways of considering anthropocentrism and ecology. The first dialogue supports the idea that living things all have their own requirements for life, and so there is no one standard for all. Thus, humans should not force their standards on other creatures if they wish other creatures to flourish according to their natural potentials. In addition, like all living things, humans themselves have their own standards and conditions of life that are based on their inborn nature. If rulers attempt to change these natural standards, humans will not flourish as they otherwise would. The conclusion is that humans must regard as important those concerns that are specifically human. Humans have no choice but to be anthropocentric to some extent, but ideally, that entails an appreciation of the variations in the requirements for life of all different creatures.

Also important for ecological thinking is the further assumption that things and their habitats go together; what a thing is cannot be separated from its habitat. The destruction of certain conditions would lead to the destruction of those life-forms intertwined with those conditions. The anthropocentric position presented here thus does not accept that humans should use their special abilities to destroy other living things and habitats simply because they have the power to do so. Such actions would eventually destroy the very world on which their lives depend. This point was clearly recognized by Chinese officials by the late imperial period, if not earlier, but they lacked sufficient power and incentive to act effectively.[16]

The second dialogue above reaffirms the idea that all things have, according to their "species," their inborn natures that condition their existence. In offering impossible situations as counterexamples, this conversation mocks that kind of anthropocentric position that wants to reshape the conditions of life of all creatures, including human beings. Such efforts will ultimately fail. The anthropocentric position supported here actually goes one step beyond a "control" or "neutral" position vis-à-vis living things, for it claims that the sage should act so as to help all creatures flourish. This imperative does not mean superficial types of actions, such as occasionally feeding wild creatures, but rather making sure that the conditions of their existence continue to prevail—including the conditions for human existence.

The third dialogue, between the well frog and the sea turtle, makes the point that all living things (not just humans) view the world from the perspectives of their own situations. The perspectives of all things, even humans, are conditioned and limited. In addition, humans are sufficiently similar to other creatures, so that analogies can be made between animals and humans, and dialogues between animals can be substituted for dialogues between humans. Not ridiculed or rejected, such dialogues are accepted as realistic.

While humans may think that their position is closer to that of the great sea turtle than that of the little frog, such may not be the case. The belief of people in their ability to control nature may not be so different from the certainty of the frog. Moreover, the very limitations of one's viewpoint, human or otherwise, prevent one from seeing how restricted one's viewpoint is. Ordinary life does not enable one to get outside of one's perspective and assumptions. Thus, anthropocentrism is as inevitable for humans as the "turtle-centric" and "frog-centric" positions are for the sea turtle and the well frog. What human beings can do, however, is recognize that there are limitations to the human perspective, even though what those limitations are may not be known, and they can start caring for other creatures and habitats. These passages thus seem to say that, unless one makes other, unstated assumptions, a "centric" position does not necessarily entail domination.

Other Daoist texts reveal that unstated assumptions were made in different contexts, however, and so the anthropocentric position expressed above was not shared by all texts labeled "Daoist." For example, sections in the *Huainanzi*, such as chapter 4, "The Treatise on Topography," represent a Han dynasty form of Daoism, generally termed Huang-Lao.[17] Taking a stronger anthropocentric perspective, a passage on the "beautiful things" of each region reveals a recognition and acceptance of the variation of the surrounding world, with each mountain or region producing products different from the others. However, the products of the outlying regions are regarded as exotic, and only those of the central region are seen as necessary for civilization. The standards for judging things, whether it concerns their beauty or their worth, are thus the standards of Chinese civilization and its ruling elite, not the standards of the things and regions themselves. A hierarchy of values is characterized by the allowance of only one legitimate perspective.

A passage describing the qi (matter-energy) of different kinds of places adds to the above position by providing a qi foundation, that is, an inborn relationship, between a region and the type of thing produced in that region. The sages are considered the best kind of thing and the central region produces many sages. Thus the central region, the site of Chinese civilization, has better qi than other regions.

While passages from the *Zhuangzi* suggest that an anthropocentric position is not necessarily destructive toward the natural world, many of those from the *Huainanzi* lay the groundwork and justification for the exploitation of the environment. Key assumptions in the *Zhuangzi*'s perspective are: that all knowledge is from a specific, interested point of view, and thus no viewpoint is neutral; that different types of things can exist harmoniously in the same world and that the success of one type does not require the failure of another; that there is a natural fit between places and creatures and so diversity is a fundamental characteristic of the flourishing of life; that human knowledge has limitations, even if not recognized, and thus the importance of keeping an open mind; and that there is an imperative for human responsibility in nurturing the well-being of things and habitats. However, the addition of just one belief from the *Huainanzi*, that the perspective of the Chinese state and civilization takes precedence over all other things, thoroughly changes the meaning of the *Zhuangzi*'s anthropocentrism.

The Particular or Local, and the General or Universal

A third issue that concerns ecologists is the relation between the focus of people's interests and their development of a moral stance. The question of focus is generally understood in terms of whether people pay attention to local and particular conditions, or to general and universal conditions, or to both. At issue is what kind of focus best contributes to developing a moral position. Both ecofeminists and deep ecologists have had much to say on this issue.

Ecofeminists often, but not exclusively, emphasize the importance of personal relationships and particular emotional attachments for the development of a moral position, and so they tend to pay attention to the particular and the local.[18] Deep ecologists generally emphasize the need to care for "the whole." Stressing the universal, they question how, and whether, a concern for human beings and for a local region can be developed into a concern for nonhuman beings and for the world as a whole. From an ecofeminist viewpoint, however, the deep ecologists' position continues, or is close to continuing, the abstract rationalism and the universalizing character of modern, Enlightenment thought. Still, some ecofeminists side with deep ecologists who fear the dangers of too much emphasis on the local, the regional, or "the tribal," without an accompanying larger view. Although there are different matters to resolve, the problem of developing a moral position seems to exist no matter the level at which one fixes one's interest.

The question has also been raised whether a focus on either the universal or the local inherently possesses a moral orientation, through the very construction of a story or formulation of a concern. Some, like Ynestra King, an ecofeminist, do not think that a local focus (or, bioregionalism) entails a moral stance.[19] And, even if it does, it is not necessarily one that is conducive to caring for the whole. On the other hand, deep ecology has been charged with moral failure by not having a genuine concern for all humans, and for a while some people saw possible ideological links between it and the Nazi movements of the mid-twentieth century.[20]

Before we consider some Daoist responses to the issue of focus, a few further comments about deep ecology and some comparisons and contrasts might be helpful. Deep ecology seems to offer a position that is compatible with many Daoist texts, for it focuses on the whole by assuming the interrelationship of all life. According to Jim Cheney, an ecofeminist, the ethical stance supported by deep ecology emphasizes "interdependence, relationship, and concern for the community in which we [humans] are imbedded," and it entails an opposition to ethical positions that emphasize "individual rights, independence, and the moral hierarchy implied in the rights view."[21] Michael Zimmerman has observed that deep ecology is opposed to "modernity's atomistic, hierarchical, dualistic, and abstract conceptual schemes."[22] It is especially known for its advocacy of the notion of an "expanded self," or "ecological self," which incorporates the entire cosmos and which, for some, has ties with the Hindu notion of *ātman* (the eternal, unchanging, real Self).

The views of deep ecology and the early Daoist texts actually differ in many ways, including the positions to which they are opposed. Early Daoism does not support any of the moral alternatives that deep ecology faces, either those that stress community relations or individual rights. Atomism was not an issue in the ancient Chinese world, and their dualisms involve relationships more akin to polarity than metaphysical otherness. However, abstract conceptual schemes did begin to develop about the time of the *Zhuangzi* and *Laozi* and became widespread slightly later, by Han times. Although Dao has many meanings, in the sense of a pattern of constant change inherent in all things, Dao does involve a type of abstract conceptual scheme, as do the yin-yang and Five Phases conceptions of the Han.

Hierarchy is also present in both deep ecology and early Daoism, although one must consider whether the oppression and environmental destruction that are often claimed to be a result of hierarchy are actually the result of hierarchy itself or of other beliefs linked to hierarchy. If the latter is the case, oppression and destruction would probably remain even if hierarchical relations were eliminated. Moreover, if we ask whether deep ecology and early Daoism both make the assumption that social domination and ecological destruction go together, the question is virtually a non-question for Daoism, since, for the Chinese, there is no world without social hierarchy and domination. Deep ecologists assume social domination and ecological destruction go together, but their solution may unwittingly continue the problem by "hiding" social domination in their new notion of the self, as some ecofeminists have suggested.

A troubling question raised about deep ecology is whether its holism involves a notion of the self that is a (Western) masculinist conception of self—that is, an atomistic conception of self, as opposed to a (Western) feminist, web-like, relational conception of self.[23] This difference is important because those who have formed one kind of self behave and think in different ways from those who have formed another kind. The concern is that deep ecology's ecological (or expanded) self could be a transformed version of the old, universal self, the male, which served as the standard against which others (lessers) measured themselves. Advocating the self of deep ecology could thus lead to the continued exploitation of women and others (including the nonhuman world) who might be swallowed up by this expanded self or for whom a notion of expanded self makes no sense, given that their sense of self is already permeable, pliable, relational, responsive, and without strict boundaries.[24] In the case of the latter notion of self, there is, in effect, nothing to expand.

Given this suspicion, which all ecologists do not share, about the expanded self, an important question is whether the early Daoists' conception of Dao ends up being functionally similar and so equally problematic ecologically. The concept of Dao in the *Zhuangzi* and *Laozi* often focuses on and emphasizes the universal (pattern of change) while depreciating particular, individual experiences. For instance, in chapter 13, "The Way of Heaven," the particulars of the sensory, natural world are not valued for themselves:[25]

> What you can look at and see are forms and colors; what you can listen to and hear are names and sounds. What a pity!—that the men of the world should suppose that form and color, name and sound are sufficient to convey the truth of a thing.

And from the *Laozi*:[26]

> I do my utmost to attain emptiness;
> I hold firmly to stillness.

The *Zhuangzi*, *Laozi*, and *Liezi* further take the paradoxical position of valuing the female and so seeming to value women, but by identifying the female with the source of life and of all things, they end up not treating women as full human beings and so actually depreciating them. For instance, in a passage referring to ideas from the *Laozi*, the *Zhuangzi* says:[27]

Lao Dan said, "Know the male but cling to the female; become the ravine of the world. Know the pure but cling to dishonor; become the valley of the world." Others all grasp what is in front; he alone grasped what is behind. He said, "Take to yourself the filth of the world." Others all grasp what is full; he alone grasped what is empty. . . .

From the *Liezi* there is this account, with a version also in the *Zhuangzi*:[28]

Only then did Liezi understand that he had never learned anything; he went home, and for three years did not leave his house.

He cooked meals for his wife,
Served food to his pigs as though they were human,
Treated all things as equally his kin,
From the carved jade he returned to the unhewn block,
Till his single shape stood forth, detached from all things.
He was free of all tangles
Once and for all, to the end of his life.

Depreciating women outright, the *Liezi* also says:[29]

I have very many joys. Of the myriad things which heaven begot mankind is the most noble, and I have the luck to be human; this is my first joy. Of the two sexes, men are ranked higher than women, therefore it is noble to be a man. I have the luck to be a man; this is my second joy. . . .

Depreciation of women as full human beings goes along with a depreciation of the particulars of nature (Heaven, Earth, and the myriad things) and of particular sensory and emotional experiences. The ultimate value is Dao, a primary understanding of which is the pattern of life and change. Although the early Daoist texts identify Dao with all things, large and small, they also identify Dao with what is inborn, as opposed to what is acquired; the inner, as opposed to the outer; and genuine qualities as opposed to superficial. In the *Zhuangzi* in particular, Dao is identified with Heaven. Heaven is above and in the male position; Earth is below and in the female position. From at least Han times on, moreover, Heaven was male and Earth was female. Thus, the set of associations that eventually developed linked the female to the group of acquired and superficial things, those not Dao.

Liezi's accomplishment of being able to forget social distinctions (the "carved jade") and return to Dao (the "unhewn block") is symbolized by taking over the position of his wife. The female role of staying at home and cooking meals was equivalent to his treating pigs as human and treating all things as equally his kin. These situations are all symbols of becoming one with Dao. As representatives of the unhewn block, Liezi's wife and the female role thus are part of the background for Liezi. Consequently, women have no way to achieve what Liezi did—to become "free of all tangles," for they already are at home and already cook the meals.

It is not claimed, moreover, that Liezi's wife has already achieved the state of the unhewn block by carrying out her assigned social role. To suggest that she could become free by a reversal in gender roles (as Liezi did)—that she take on the male role and find work in society—would make no sense, for the point is to escape from society's distinctions and rules. The path to achieving detachment from things and returning to Dao, by symbolically becoming the female, is not open to women.

A similar viewpoint appears in the *Laozi* reference from the *Zhuangzi* text, namely, that it is not a "human" perspective being represented, but a specifically elite male perspective. A woman already is "the female." There is no social instruction for her to "know the male" as there is for males. Despite the apparent honoring of the female, the associations with the female are such low things as the ravine of the world, dishonor, the valley of the world, what is behind, the filth of the world, and what is empty. Laozi can value these low things because he is not them, but he can choose a strategy of acting like them if it is beneficial to him to do so.

These ideas do not allow a woman the option of choosing alternative values and positions. Only someone in the elite male position has the possibility of acting deliberately or not following his inborn characteristics or his heart. Those in the female position, along with plants and animals, lack that freedom, for they are analogous to those aspects of the world that cannot act purposefully or initiate action. From an ecological viewpoint, the ideas associated with the surrounding natural world coincide with those associated with the female. They include such notions as lowness, dishonor, filth, and emptiness. Like the female, Earth's habitats and the myriad things are also not regarded as actors in themselves. Not assumed to have significant life courses in themselves, both habitats (such as rivers and mountains) and the myriad things are the conditions in which elite men act.

The concern, whether justified or not, that deep ecology's expanded self is covertly a version of the atomistic, masculinist self thus has a parallel in certain Daoist texts, in that the ideal of becoming "free" and "one with Dao" makes sense only for someone in the elite male position. By treating that position as the human and the female position as symbolic of Dao, early Daoism does not acknowledge women as fully human. Its stance tacitly entails social practices that are destructive to actual women as well as to actual parts of the natural world, for it is part of a view that sees particular things, events, and places as transitory and inconsequential moments in the larger process of constant change (Dao).

In other words, despite occasional comments in praise of the local and the particular, the ultimate focus of early Daoism is on the universal rather than on the particular. Although its androcentric value system favors inborn, natural capacities, the particulars of human experience do not have great worth. Things, places, feelings, and events in nature are not regarded as valuable "moments" in themselves. While the *Zhuangzi* and *Laozi* may not speak directly to the ecological issue of whether a local or universal focus contributes better to developing a moral position, their positions imply that only Dao, as a universal process of change, is moral—moral in the sense of being the ultimate value. The particulars of Earth and of the myriad things are left without genuine moral protection.

The Human Story and the Cosmic Story

A fourth issue for ecologists is whether a moral position can be developed that joins humans with the rest of the cosmos, or whether moral positions require, at least to some extent, a separation between the "human story" and the "story of the cosmos." The conceptions of both human beings and the cosmos are important in this issue. In regard to humans and going beyond biological aspects, a fundamental question is whether it is sufficient, for ecological purposes, to view humans as individual entities with social dimensions or whether it is also necessary to acknowledge the reality of social groups.[30] In accepting the latter position, social groups are not to be seen as either tangible or metaphysical entities, but as social entities, real in the sense that communities or ecosystems are real. The very formulation of a distinction between the cosmic story and the human story echoes the dichotomy between Heaven (or Dao) and human in the *Zhuangzi*, the uncarved block and vessels in the *Laozi*, and nature and culture in Western thought.[31]

In speaking to this issue, deep ecologists emphasize the interrelatedness of all things and so do not want to separate the human and the cosmic. Ecofeminists are somewhat suspicious of the implications of this position, however, and advocate some degree of separateness. They are concerned that deep ecology does not sufficiently acknowledge the difference between humans as biological organisms and as members of a social community and so could support positions that are immoral or amoral.[32]

Another difficulty focuses on the claim that one result of Enlightenment thought has been to disembody human beings, by separating mind and body and by valuing what stands for mind and denigrating what stands for body. This disembodiment has in turn led to separating human beings from nature. Charlene Spretnak, an ecofeminist, has further asserted that such foundational movements as Renaissance humanism, the scientific revolution, and the Enlightenment, as well as deconstructive postmodernism, have contributed to a diminished view of what is human and have "framed the human story apart from the larger unfolding story of the earth community."[33] The solution, she says, is not only to expose "the power dynamics inherent in the metanarratives of the modern world view," but also "to break out of the conceptual box that keeps modern society self-identified apart from nature and to reconnect with a fuller, richer awareness of the human as an integral and dynamic manifestation of the subjectivity of the universe."[34]

The question here is how the views of the human in the Daoist texts compare to these Western conceptions that are fundamentally biological, or disembodied, or social. While the texts recognize the role of language in shaping experience, there is no evidence that they promote or even recognize the idea that human experience is socially constructed. That is, despite their recognition of different viewpoints and the arbitrariness of names, they accept that there are "givens" in the world not subject to human control or social convention. One such given in the *Zhuangzi*, for instance, is the distinction between "what is of Heaven" (or Dao) and "what is of humans." Other givens include Heaven, Earth, and the myriad things; the realm or the world (*tianxia*, "under Heaven"); the world or this generation (*shi*); and the family (variously defined). Comparable ones are found in the other texts, such as the distinction between the named and the nameless, the constant and that which is not constant, found in the first chapter of the *Laozi*. Most texts, Daoist and non-Daoist, further accept certain social givens, such the ruler and the people (*min*).

Daoist texts contain accounts that may be compared to, although they are not the same as, the biological view, in that human beings and all things are seen as embedded, on the level of their bodies, in the processes of the natural world. The *Zhuangzi*, for instance, teaches people to accept the inevitable transformations of things (*wu*), even such transformations as becoming "all crookedy" or having an arm transformed into a rooster.[35] However, the particular transformations are not important; only Dao is. By valuing Dao as the creative, transforming cosmic process and by devaluing particular embodiments and physical characteristics, the *Zhuangzi* and *Laozi* put forth their versions of the disembodied view of the human.

Along with their (biological-type) views of the embeddedness of all things in the cosmic processes and their (disembodied-type) views that all things are in constant transformation and are subject to Dao, neither the *Zhuangzi* nor *Laozi* accept that social groups are a type of thing (*wu*). That is, there is no claim that society, or "the world," or any specific social group, is embedded in the cosmic process in the way that things are. The *Zhuangzi*'s teaching of constant transformation is not extended to society or to any social group, and the *Laozi* rejects changes in society by urging a return to primitive conditions of life, with no vehicles, weapons, or writing.[36]

The world (*tianxia* or *shi*), a human social collective, is "self-identified" apart from Dao. Rather than recognize that it had become characteristic of the world to build bridges and roads and to write and value books, the *Zhuangzi* and *Laozi* criticize these kinds of activities. Although not all of the world's changes are environmentally destructive, these texts value the ancient time of simplicity and harmony and do not accept that all the changes of the world are simply more transformations in the constant process of cosmic transformation.

By rejecting that transformations can apply to the world as well as to things, these texts ultimately view the human story as one concerning individual human beings who need to overcome their attachments to the social and sensory particulars of their lives. The human story is not one about the human collective as a whole or one about particular groups. Thus, like the Western foundational movements criticized by Spretnak, and despite their countercultural narratives noted above, the *Zhuangzi* and *Laozi* also frame "the human story apart from the larger unfolding story of the earth community."

Although "the world" is not recognized as a thing, the social dimension of being a human being is assumed. The reality of relationships like father and son, husband and wife, is not challenged. However, the *Zhuangzi* in particular does teach that if one develops to the utmost all of one's Heaven-given, inborn capabilities, including one's social and physical abilities, one can end up being chained or destroyed—by the world. For instance, one passage equates a man's studying the (Confucian) Dao and learning the principles of things with a monkey's developing its quickness or a dog's becoming a good hunter. They "end up chained." Another passage tells how delicious fruit trees end up being abused and suffering an early death if they produce their fruit, while the gnarled oak tree lives its full life.[37]

The position of the *Zhuangzi* and *Laozi* on this issue thus presents difficult problems from an ecological viewpoint. Their view "shrinks the human story" by opposing changes in the connections between the world—that is, humans acting together as a collective body—and the surrounding, natural environment and within social groups themselves. It opposes, but does not attempt to repair, activities that are judged destructive. In its rejection of both social activities associated with the Confucians and any ultimate value to sensory experience, these texts help support a position that actually denigrates both the particulars of the natural processes of the environment and many of the inborn capacities given to humans by Heaven.

In both the ancient Chinese and contemporary worlds, we see, however, that communal living and the formation of groups are as much a part of the human story as an individual person's experiences. By viewing the flourishing of society as entailing the enslavement of people and the flourishing of individual human beings as requiring that they cultivate only some of their natural capacities while repressing others, including those that are sensory and socially related, the texts in effect say that one's body and much that is associated with it are not important. Although their position links individual humans with the rest of the cosmos, it avoids addressing critical aspects of the human condition by separating the collective human story from the story of the cosmos. Change is valued for the latter but not for the former. Consequently, the *Zhuangzi* and *Laozi* lack a position on this issue that would be helpful for systematically addressing ecological issues.

Conclusion

Daoist ideas can be used to respond to contemporary ecological issues, such as those concerned with narratives, with centric positions, with tensions between the particular and the universal, and with the place of human beings in the cosmos. Although some ideas offer helpful perspectives on ways to conceive of the relationships between humans and the world, others are questionable and perhaps objectionable. The androcentric bias, for instance, must be given further analysis, so that the ecologically destructive effects of hidden associations and connections can be recognized.

Although more fully developed in texts other than the *Zhuangzi* and even the *Laozi*, such groups of associations as Earth, the female, stillness, completing, and lowly—as opposed to Heaven, the male, activity, initiating, lofty—implicitly support a view that sees the realm of the earth as the stage for elite male performances, a site where political activities reach completion. Thus, while many passages are critical of civilization's destruction of the earth, including its rivers and mountains, fundamental cultural associations remain, buried in the thinking, that do not enable the earth and those things associated with it to be conceived as worthwhile in themselves, or as actors able to make choices. Recognizing such problems in texts like the Daoist ones, which also have ecologically favorable attributes, may help us in analyzing contemporary obstacles to the development of a strong ecological commitment.

Notes

1. For a discussion of some of these positions, see Michael E. Zimmerman, *Contesting Earth's Future: Radical Ecology and Postmodernity* (Berkeley and Los Angeles: University of California Press, 1994); for selected readings, see *This Sacred Earth: Religion, Nature, Environment*, ed. Roger S. Gottlieb (New York: Routledge, 1996); *Reweaving the World: The Emergence of Ecofeminism*, ed. Irene Diamond and Gloria Feman Orenstein (San Francisco: Sierra Club Books, 1990); *Ecofeminism: Women, Culture, Nature*, ed. Karen J. Warren (Bloomington: Indiana University Press, 1997); and J. Baird Callicott, *Earth's Insights: A Survey of Ecological Ethics from the Mediterranean Basin to the Australian Outback* (Berkeley and Los Angeles: University of California Press, 1994 and 1997).

2. Karen J. Warren, "Feminism and the Environment: An Overview of the Issues," in *Philosophy of Women: An Anthology of Classic to Current Concepts*, ed. Mary Briody Mahowald, 3d ed. (Indianapolis: Hackett Publishing Co., Inc., 1994), 495–510. The eight (interrelated) types that she identifies are: historical and causal, conceptual, empirical and experimental, epistemological, symbolic, ethical, theoretical, and political.

3. Aldo Leopold, for instance, proposed that the concept of the ethical community be broadened beyond the sphere of human beings to include the land and all of the natural processes associated with it. See Aldo Leopold, *Sand County Almanac* (Oxford: Oxford University Press, 1949).

4. For an overall view on the various texts, see *Sources of Chinese Tradition*, comp. Wm. Theodore de Bary and Irene Bloom, 2d ed., vol. 1 (New York: Columbia University Press, 1999).

5. See Lynn White, Jr., "The Historical Roots of Our Ecologic Crisis," in *This Sacred Earth*, ed. Gottlieb, 184–93; Zimmerman, *Contesting Earth's Future*, 1–17; and Callicott, *Earth's Insights*, 1–43.

6. See *Sediments of Time: Environment and Society in Chinese History*, ed. Mark Elvin and Liu Ts'ui-jung (Cambridge: Cambridge University Press, 1998).

7. See *The Complete Works of Chuang Tzu*, trans. Burton Watson (New York: Columbia University Press, 1968), 105–6 and 165. Actually, it can be argued that even the countercultural narratives are not entirely countercultural, for they use many of the same "markers" of history, such as the ancient sage rulers and exemplary figures, even while they challenge conventional values. For instance, see the stories in *The Book of Lieh-tzu: A Classic of Tao*, trans. A. C. Graham (New York: Columbia University Press, 1960).

8. *Lao Tzu: Tao Te Ching*, trans. D. C. Lau (Baltimore: Penguin, 1963), chaps. 18 and 19.

9. Mark Elvin, introduction to Elvin and Liu, *Sediments of Time*, 2; also see Graham, *Lieh-tzu*, chap. 5, "The Questions of T'ang, 101; and John S. Major, *Heaven and Earth in Early Han Thought: Chapters Three, Four, and Five of the* Huainanzi (Albany: State University of New York Press, 1993), chap. 4, "The Treatise on Togography," 198.

10. Eduard B. Vermeer, "Population and Ecology along the Frontier in Qing China," in Elvin and Liu, *Sediments of Time*, 247–48.

11. Lau, *Lao Tzu*, chap. 32.

12. See Val Plumwood, "Androcentrism and Anthropocentrism: Parallels and Politics," in Warren, ed., *Ecofeminism*, 327–55; also Zimmerman, *Contesting Earth's Future*, chaps. 6 and 7, especially pp. 261 and 280–81.

13. "Oppression" is an important concept in social analysis, with a set of specific, but multiple, meanings. Iris Young has analyzed "oppression" in terms of "five faces": exploitation, marginalization, powerlessness, cultural imperialism, and violence. See her *Justice and the Politics of Difference* (Princeton: Princeton University Press, 1990), especially chap. 2.

14. Zhuangzi and Huizi argued this point as they watched fish swimming in the river. See Watson, 188–89.

15. See Watson, *The Complete Works of Chuang Tzu*, 45–46, 93, and 186–87.

16. Christian Lamouroux, "From the Yellow River to the Huai," and Helen Dunstan, "Official Thinking on Environmental Issues and the State's Environmental Roles in Eighteenth-Century China," in Elvin and Liu, *Sediments of Time,* 545–84 and 585–614, respectively.

17. See Major, *Heaven and Earth in Early Han Thought*, 164 and 167–68, for the passages referred to below.

18. For instance, see Nel Noddings, *Caring: A Feminine Approach to Ethics and Moral Education* (Berkeley and Los Angeles: University of California Press, 1984).

19. Zimmerman, *Contesting Earth's Future*, 310.

20. Ibid., 172–83, and passim.

21. Ibid., 285, quoting Jim Cheney.

22. Ibid., 277.

23. I am not suggesting that the choice is an either/or dichotomy. I am sympathetic, however, with Ellen Kaschak's view that gender involves a range of actions and "is something that one *does* repeatedly." See her *Engendered Lives: A New Psychology of Women's Experience* (New York: Basic Books, 1992), 43.

24. See Charlene Spretnak, "Radical Nonduality in Ecofeminist Philosophy," in Warren, ed., *Ecofeminism*, 425–36, especially 428–29.

25. Watson, *The Complete Works of Chuang Tzu*, 152.

26. Lau, *Lao Tzu*, chap. 16.

27. Watson, *The Complete Works of Chuang Tzu*, 372, and see Lau, *Lao Tzu*, chap. 28.

28. Watson, *The Complete Works of Chuang Tzu*, 372, and see Lau, *Lao Tzu*, chap. 28.

29. Graham, *The Book of Lieh-tzu*, 24.

30. See Watson, *The Complete Works of Chuang Tzu*, 84.

31. See, for example, Watson, *The Complete Works of Chuang Tzu*, 182–83, and Lau, *Lao Tzu*, chap. 28.

32. Zimmerman, *Contesting Earth's Future*, 286.

33. Spretnak, "Radical Nonduality," 433.

34. Ibid.

35. See Watson, *The Complete Works of Chuang Tzu*, 84.

36. See Lau, *Lao Tzu*, chap. 80.

37. Ibid., 94 and 64.

"Nature" as Part of Human Culture in Daoism

MICHAEL LAFARGUE

I would like to begin with some remarks on hermeneutics, a subject which I have thought about a great deal.[1] I want to advocate what might be called "confrontational hermeneutics," a hermeneutics oriented toward having a confrontation with a text or a tradition. The first step in this hermeneutics is to pay careful attention to the *otherness* of the text or tradition, its otherness from our own assumptions, concerns, and values. I mean not only its otherness from those aspects of our own cultural tradition which we already regard as problematic, but also from those assumptions, concerns, and values which are personally most dear to us, the interpreters. The historical people whose ideas we are studying are not around to confront us with their otherness, so our first task is to temporarily set aside our own views and to try to reconstruct their views, giving special attention to those aspects of their views which might present the strongest challenge to our own views. We should not, of course, immediately capitulate to their views, acceding to them because they belong to some authoritative tradition. This reconstruction is preparatory to actually wrestling with the ideas we have reconstructed, rationally weighing the strengths and weaknesses of their views in comparison with our own, weighing also their applicability to our present situation.

Confrontational hermeneutics contrasts strongly with traditional and still prevalent attitudes to the interpretation of classical and scriptural texts, in which the overwhelming emphasis has been on the task of drawing an edifying message from the text. "Edification" is a process of building—contributing to an edifice—and in this context one mines classical texts for potential contributions to an edifice already well in progress, whether that building be the Christian Church, the community headed by Daoist Celestial Masters, modern philosophy, New Age Zen, or modern environmentalism. In this approach, one is tempted to treat the Daoist tradition as a "resource," a grab-bag collection of ideas which one can pick from and interpret in such a way as to contribute to whatever cause one feels to be urgent and important. I do not think this approach is "wrong" in itself—many worthy edifices have been built up by this very process. It's rather that I think academic historical scholarship has something different to offer, pointing out *provocative* aspects of classical traditions, aspects that because of their otherness provoke critical reflection on those ideas that come more naturally to modern people.

In this spirit, I've chosen in this essay not to show how Daoism can be "used as a resource" giving added support to ideas about environmental protection we are already familiar with. Rather, I want to focus on one theme from the Daoist tradition that I think modern environmentalists might learn something new from, because it represents a view of the natural world that is different from, and in some respects opposed to, views common among environmentalists today. I do not present this as something that environmentalists should automatically accept as authoritative because it is associated with an authoritative tradition, but as a strong challenge worth wrestling with.

I have three texts I want to comment on. The first is not from a Daoist writing, but from a writing by John Blofeld, who traveled in the 1930s, visiting Buddhist and Daoist communities in China. Blofeld gives the following description of his visit to a remote Daoist hermitage in China:

> Not far from where the hermitage clung to the steep rock-face . . . where the path took a sharp turn towards a stone stairway leading to the main gate, it could be seen that the recluses' love of unspoiled beauty had not deterred them from lending nature a helping hand. The immediate environs of the Valley Spirit Hermitage gave the impression of a series of rocks and caverns, overhung by ferns and luxuriant plants, which just happened to emerge from the undergrowth in this vicinity, adding enormously to its picturesqueness. What aroused my suspicion was that no other section of the mountain, apart from the chasms and waterfall, looked so exactly like the original of a Daoist painting. There

was, of course, no obvious symmetry, but yet a sense of underlying harmony that was just a shade too pronounced to be altogether natural. Whoever had been responsible for making the "guided wildness" of the approach to the hermitage even lovelier than nature's untouched handiwork had surely been a master of subtlety, for there was not an object within sight of the stairway of which one could confidently affirm it had been tampered with.

Taoists, the ancient progenitors of several horticultural arts now widely associated with Zen, such as flower-arrangement, certain kinds of landscape gardening and the growing of dwarf trees, were wont to employ loving artistry in subtly modifying nature. . . . In landscaping, the underlying principle was to avoid artificiality not by refraining from improving on natural forms, but by bringing out or highlighting shapes—beautiful, amusing or grotesque—already inherent in the objects worked upon. A square should not be rounded, but a rough sphere could be made rounder; a shrub should be made to resemble a stork only if the stork already existed potentially in the plant's natural shape; water might be diverted from one pile of rocks to another to heighten the beauty of a cataract, but only if there were nothing inherently unnatural in the resulting flow and fall. Nature could be assisted to achieve masterly effects, but the concept in the improver's mind must in itself be based on intimate knowledge of nature's manifestations. In short, the aim in most cases was to assist nature to do what it might under more favourable circumstances have done for itself.[2]

I once injected the ideas from this passage into a conversation among college professors who were discussing Lynn White's now famous essay on Christianity's contribution to modern Western "dominating" attitudes toward nature. I meant this as a serious contribution to a discussion of what our attitudes toward nature ought to be, but it was greeted by dismissive laughter and amusement at these naïvely self-contradictory Daoist hermits. When I read this passage to students, invariably several are offended at the "hypocritical" attitude toward nature that they see reflected in Blofeld's description. The hermits seem to be pretending to be "nature lovers" while at the same time betraying that attitude most offensive to modern nature lovers—the arrogant, "anthropocentric" assumption that human values and human ideas about how the world ought to be are superior to nature in its pure, untouched state. A true nature lover would simply leave nature alone. Many people today have a strictly disjunctive view of the relation between "nature" and "human culture." Blofeld's hermits clearly do not share this disjunctive view, but favor a kind of "cultivated nature," a nature that is part of human culture.

I'm not sure we can take everything in Blofeld's rather journalistic book as representative of the Daoist tradition, but I would argue that what he says in this passage does reflect a view of things that closely matches the views expressed in the *Zhuangzi* and the *Laozi*, from which my next two passages are taken.

My *Zhuangzi* passage tells how Woodcarver Ching went about carving an awesomely beautiful bell stand:[3]

> When the bell stand was complete, those who saw it were amazed, [because it] seemed like [something belonging to the realm of] spirits. The Marquis of Lu went to see it, and then asked, "By what secret art did you make this?"
>
> [Ching] answered, "Your servant is [only] an artisan, how could I have a "secret art"? However, there is one thing. Suppose your servant is going to make a bell stand. I dare not let my qi dissipate. I make sure to fast, to still my mind. After fasting three days, I no longer presume to think of recognition and reward, of rank and salary. After fasting five days, I no longer presume to think of praise or blame, of skill or clumsiness. After seven days, I am so concentrated that I forget I have four limbs or a body. By this time [for me] there is no duke or court. The skill [for the work] concentrates and outside distractions disappear.
>
> Only after all this do I go into the mountain forest.
>
> [There I set about] observing the [inner] nature of Heaven ['s work].
>
> [When I discover wood whose] form and substance have reached perfection
>
> Only after this is [everything] complete, [enabling me to] see the bell stand.
>
> Only after this do I set my hand to work.
>
> [If things do] not [happen] like this, I give up.
>
> Thus I join Heaven to Heaven.
>
> This is probably the reason why the bell stand seems like something from the spirit-world.

The first thing I want to point out is the way that this passage seems, from a modern point of view, to illustrate a confusion between nature and human culture similar to that in the Blofeld passage. Clearly, Woodcarver Ching is a kind of "nature lover," in that he got the inspiration for his bell stand by a kind of careful and reverential observation of nature. On the other hand, the result of his nature observation was not a grasp of how nature works, but an ability to "see" in a tree a very useful human product, a bell stand. And his "reverence for nature" did not express itself in admiration of trees as they exist in their natural habitat themselves, but rather in chopping down and carving up a "natural" tree to produce a luxury item for the court of a local duke.

This is a good opportunity to illustrate what I mean by focusing on the "otherness" of some classical texts, the way they "do not make sense" when we see them within a framework set by categories dear to us. The confrontational hermeneutics I advocate recommends not glossing over or dismissing such "glaring confusions," but focusing on them as the kind of thing we ought to try to make sense of and then wrestle with, having potentially the most to learn by such wrestling.

Beyond this, several aspects of this passage are worth commenting on. In translating this passage I have tried for a rather literal rendering, and I have printed the last sentences on separate lines because the wording especially in these last lines is very interesting and significant in several places. First, the text says that after entering the forest, and before he was able to "see" the bell stand, Ching spent some time *guan tian xing* (*guan*, "observing"; *tian*, "Heaven's"; *xing*, "nature"). *Xing* is the word often translated as "nature." It usually refers to some core tendency of a being, not itself immediately visible, but which is the root of visible conduct. (As when, for example, Mencius says that people's desire to rescue a child from a well is a manifestation of an inner core tendency toward empathy and compassion, a manifestation of an internal "commiserating *xing*.") When the text says that Ching is observing *Heaven's* nature, "Heaven" in the context seems to refer to what we call "Nature" in a different sense, Nature as a personification of the forces operative in the natural world of the forest, outside human control. A literal translation of *guan tian xing* might then be "observing Nature's nature." The context seems to indicate that Woodcarver Ching spends time carefully observing the shapes of trees in the mountain forest, contemplating that kind of beauty observable in naturally occurring forms. He thinks of this as learning about "Nature's nature," the inner spirit of Nature. This contemplation is what eventually enables him to "see a bell stand" in a piece of wood "[whose] form and substance have reached perfection." That is, a piece of wood is said to have "reached perfection" (*chi*) when it exemplifies this "natural" kind of beauty, this "Nature's nature" to a very high degree—exhibiting this beauty in the form of a plausible bell stand.

Ching says that when everything comes together in this way, he is able to "join Heaven to Heaven." One "Heaven" here is clearly the *xing* of Heaven/Nature that Ching spent time observing, which he now sees in the wood. Following suggestions evident in other *Zhuangzi* passages, the other "Heaven" is most likely Ching's own mind, the mental state he was in as a result of conserving his qi, fasting, stilling his mind, and ridding himself of all the concerns for fame and social advantages that Zhuangzi rails against and to which Ching, as a court artisan, would usually have been susceptible. This kind of mental self-cultivation is what gave Ching a "Heavenly" mind. The finished bell stand was "like a spirit" (*gui-shen*) to people who saw it—it had that mysteriously awesome aura that Chinese associated with spirits—precisely because it was the result of combining Ching's "Heavenly" state of mind with the Heavenly beauty of natural forms he was able to perceive by his careful study of naturally occurring forms in the forest.

The final text I want to discuss is chapter 64 of the *Daode jing*. This chapter is crucial, I think, in understanding the themes of *wuwei* (not doing) and *ziran* (naturalness) in this work. These themes are notoriously paradoxical. Taken literally, *wuwei* seems to advocate not interfering in any way at all with the course of things in the world; from this point of view, the world would clearly be better off without any ruler. And yet it is difficult to ignore the fact that the *Daode jing* is (among other things) a manual for ruling well. It gives advice on how a ruler should go about gaining the allegiance and cooperation of the people (chapter 66). It advises the ruler to imitate water, whose "soft" strategy allows it to "overcome" opposing forces that are hard and strong, and chapter 36 gives examples of how this soft strategy can be used to "weaken" things and "bring them down." Other chapters advise the ruler to actively oppose the spread of knowledge and ambition among the people: "Empty their minds and fill their bellies, weaken their ambitions and strengthen their bones" (chapter 3); "ancients who excelled in doing Dao used it to keep people ignorant" (chapter 65). Chapter 37 advocates "restraining" people when they become "desirous and active." The last line of the last chapter (in the received Wangbi version) says that the wise person "works (*wei*) but does not contend."

Chapter 64 presents us with this paradox in a quite blatant form, and I think its last lines give us the key needed to resolve it. The chapter begins with some very interventionist lines:

> [When] sitting still, [things are] easy to hold down
>
> [when] there are no omens yet, [it is] easy to plan
>
> [when things are] fragile, [they are] easy to break
>
> [when things are] small, [they are] easy to scatter
>
> Work on it (*wei chi*) when it isn't yet
>
> Put it in order when it is not yet disordered.

This passage assumes that there are undesirable things happening that one wants to hold down, break, and scatter, and the advice is to "nip it in the bud," stop undesirable developments before they get a chance to start. This is explicitly described as "working on it," *wei chi*, a phrase sometimes used of governing (for example *wei kuo* means "govern the state"); the word I translate "put it in order" is also a common word describing the act of governing. The paradox comes only a few lines later, which says "Working (*wei*) ruins . . . and so the wise person does not work (*wuwei*)." The paradox is resolved in the final lines, which say that the wise person "assists (*fu*) the naturalness (*ziran*) of the ten thousand things, without daring to work (*wei*)."

Again, from a modern perspective, "assisting naturalness" doesn't make sense. "Natural" designates precisely what happens by itself, without any deliberate human "assistance." But this passage reflects a concept of "natural" similar to the two passages discussed above, and very different from the modern concept. For the *Daode jing* authors, *ziran* does not designate the actual state of affairs, whatever that might be, but an *ideal* state of affairs. It is assumed that things the way we find them are often not in their "natural" state and need to be assisted to become so. And just as Woodcarver Ching had to cultivate a particular mental state in order to see the "natural" bell stand in the wood and make it manifest by his carving, so chapter 64 of the *Daode jing* precedes the description of "assisting the naturalness" by a saying describing a kind of self-cultivation: the wise person "desires to be desireless . . . learns to be unlearned, turns back to the place all others have passed by, *so that* (*yi*) he can assist the naturalness of the ten thousand things without presuming to work."

The ideal ruler does not impose on a society ideas hatched completely in his own head to make his mark on the world for his own glory—this, I think, is the meaning of *wuwei* in the *Laozi*. But neither does the ideal ruler stand aside and *literally* "do nothing," no matter what is happening. He is carefully attentive to the subtleties of the unique structure and dynamics of the society in his charge and works hard to bring out the best in this particular society, "the best" being inevitably informed by his own feeling for what this society would be like at its best. Thus, the "naturalness" (*ziran*) of the society which he "helps along" does not represent society as it would function if it had no ruler at all. It is what we would perhaps think of as a rather "romantic" notion of naturalness, a state both in accord with the spontaneous impulses of the community, but also in accord with some human being's notion of an ideal society.

I've argued elsewhere[4] that the key value in the *Laozi* is that of organic harmony. By *organic* harmony I mean that kind of harmony that arises out of spontaneous mutual adjustment among many elements and forces in a given system, in contrast to that kind of order that is imposed by some dominant force or goal outside the system or that kind of order resulting from subordination of all elements and forces to one dominant center. Organic *harmony* refers to a stable, homeostatic *order* that arises out of the mutual adjustment of parts, in contrast to a random, disorderly, and unstable situation that might also sometimes be produced when different parts develop according to their own spontaneous (competitive and individualistic) impulses. An organic harmony tends to be more homeostatic and stable than an order brought about by a dominating external influence, because it does not require such a great degree of continued work to maintain. (A "low maintenance garden," for example, needs to form an organic, homeostatic system with its environment. Creating such a garden requires that the gardener understand and work with some given set of soil, water, and climate conditions. Gardens are "high maintenance" when the plants are not naturally suited to the given environment and so need constant intervention by the gardener.)

Deliberative human consciousness is seen in the *Laozi* as the primary threat to organic harmony. But this does not apply to deliberative human consciousness of any kind, in and of itself. The danger lies in the ability of human consciousness to separate itself off from a system on which it depends or is trying to manage and to vigorously pursue ends thus so separately conceived, working against given forces in the system.

The Daoist ideas discussed above obviously bear some close relation to modern ideas of nature and reverence for nature. One key difference seems to be that modern Western thought tends to conceive of nature in contrast to all human culture, but the corresponding Daoist ideas are clearly and straightforwardly conceived of as parts of a human cultural ideal. Woodcarver Ching's bell stand is not something that naturally occurred in nature, but reflects a particular kind of aesthetic sensibility that is part of "Daoist culture" (and eventually became an important strain in East Asian culture.) Conceiving of nature as something opposed to human culture makes people assume that the mistake of our ancestors was anthropocentric hubris, thinking that only what is important to humans is really important. We ought to take humans and human culture out of their place in the center and humbly recognize and honor the rights that other beings in the universe have which have nothing to do with human beings or human culture.

Many modern Western "interpretations" of Daoism have read this concept into Daoist writings, using a style of hermeneutics in which writings simply mirror back to the interpreter the concepts and assumptions she brings to them. Such interpretations generally gloss over the details of passages like those cited above. If one accords normative status to modern Western concepts of nature versus human culture, such passages can only appear confused and contradictory.

But the advantage of confrontational hermeneutics is that, first giving careful attention to the otherness of what our texts are saying, it makes us aware that categories like "nature" that seem to us simple, obvious, and universal are actually quite particular when set beside the categories other cultural communities have used. Rather than asking why Daoists failed to grasp what to us are obvious concepts, we should ask why it is that our thought is shaped by these particular categories rather than by the categories that other cultural communities have chosen to think in. Such reflections can reveal weaknesses in our own concepts that immersion in our own cultural assumptions have blinded us to. Specifically, in the present case, let us critically examine the idea that the "nature" we ought to reverence is something completely different from human culture, as contrasted with the Daoist idea that "nature" is itself part of a human cultural ideal.

Let us start with a typical experience that has inspired a great deal of modern ecological awareness. A person, used to city life, manages to hike into some expansive and beautiful area as yet completely untouched by human civilization. It is a wonderful feeling to be in such an area. The person articulates this feeling as a certain reverence for nature in its pristine state, something that ought to be protected against "human civilization," felt, by comparison, to be something of far less value, if not something of negative value. This experience calls to mind the way human beings have made a mess of things. The cities they have built are not pleasant places to live; human societies are beset with problems; and now this cancer is spreading to spoil the few places still not ruined by civilization. Whereas not so long ago our ancestors were proud of "the achievements of civilization," we should perhaps rather be embarrassed and ashamed and commit ourselves to defending the rights of nonhuman nature against what seems to us the anthropocentric hubris of previous generations.

However, the question now arises about the "nature" that this person perceives and reverences. Has this person gotten beyond the Kantian problem and experienced "the world as it is," uninfluenced by any structuring of this world by the categories of human thought? This seems unlikely. We can reflect further and see that the category "nature" that shapes this person's experience has in turn been shaped by the fact that it is part of a pair of contrasting categories that cultural anthropologists like Claude Lévi-Strauss have called our attention to. Nature in the wild has a meaning to our solitary hiker, and the meaning that it has is determined by the felt contrast between hiking in nature on the one hand, and relatively harried city life on the other. Isn't her perception of "nature's beauty" a perception produced partly by the particular state of mind produced by hiking in the wilderness— similar to the way that Zhuangzi's woodcarver's still mind is what allowed him to perceive a beautiful "Heaven-produced" bell stand in a tree?

Most people have probably always had some appreciation for the beauties of nature in the wild, but there have been historical periods in which the comforts of civilization have far outweighed the pull of nature's beauties. "Nature" did not have the meaning for such people that it has to nature lovers today. Should we conclude that such people were simply "wrong"? It seems more honest and plausible to say that they lived under different cultural conditions, and that it is different specific cultural conditions today that cause nature in the wild to have the meaning that it has for us.

This brings to mind Hegel's concept of an "antithesis." Certain cultural forces that become strong and dominant in a particular society tend to produce a contrary, "antithetical" reaction. A very orderly society tends to produce rebellious feelings; a period of chaotic rebellion tends to produce strong desires for order. Isn't it more plausible to regard the love of nature that drives the modern ecological movement as just such an antithesis? But, although an antithesis is usually a reaction against certain aspects of a dominant culture, it is itself a thoroughly "cultural" phenomenon, taking its particular shape from aspects of the culture in which it occurs, being, after all, partly defined by contrast to the cultural forces it opposes.

All this suggests that the "nature" that we love today is in some sense a thoroughly cultural construct. We do not really value nature as it exists in itself apart from any human culture. We value nature in the wild because of the *meaning* that it has for us, and like all meanings, this meaning is constituted by its relation to a large web of interrelated meanings that is our culture. We only seem to be contrasting nature with culture because our talk about nature takes place in the context of a multitude of cultural assumptions implicitly taken for granted, not explicitly thought about when we talk.

Furthermore, people motivated by experiences, such as hiking in the wilderness, conduct campaigns to protect nature—implicitly, at least, telling other citizens that nature is something *deserving* of our reverence, protecting it is a *good* for which we should be willing to sacrifice other goods. There is, after all, a moral judgement implied in the idea that we *should* reverence nature, that nature has "rights," and that it is a moral failing on our part to continue to "disrespect" nature by pollution, overbuilding, and so on. Moral judgments, shoulds, and moral failings are part of human culture, not part of the material world considered apart from culture.

If these observations are sound, the conclusion is not that our reverence for nature has no solid basis, only that we tend to analyze and interpret the experiences that lead to this reverence in a way that cannot be consistently sustained. That is, when we interpret these experi-

ences in terms of nature versus human culture *in toto*, we situate the
issues at too fundamental a philosophical level. We need instead to
think more carefully about what exactly nature means to us that we
feel reverence for it, and about which aspects of modern culture serve
as a point of contrast with this love of nature, which hopefully is becoming another aspect of our culture.

For example, we might reach a sounder analysis by distinguishing
between modern industrial and urban *civilization* on the one hand, and
human culture as a fundamental condition for all human being and
experience on the other. There can be valid reasons for preferring naturally occurring material environments to those produced by industrialization and urbanization. But intellectual honesty requires acknowledgment that such a preference does indeed reflect a value judgment
related to certain culturally specific ways of experiencing the world.

Or, we could recognize that it is a certain *kind* of anthropocentricity
we are opposed to—a strictly "using" attitude toward nature, using it
for human ends completely divorced from nature itself. But again, we
need to recognize that we are inevitably "anthropocentric" in a deeper
sense—in the sense that the surrounding material world only has meaning for us because it is part of that web of meanings that we call "human culture."

This of course brings us around to something like the conception of
"nature" and "natural" found in the Daoist writings commented on
above. There, too, we find an opposition to some aspects of "civilization" and to some kinds of human attempts to dominate the natural
world, and we find a respect for given conditions and spontaneously
occurring phenomena. But there is also a clear, if implicit, recognition
that the "naturalness" to be valued is a human ideal, something to be
brought about through careful human effort, not something that by
definition occurs only when no human intervention at all is involved.
It could be that reflection on what nature means to us will lead us in
directions different from that represented by the Daoist idea of organic harmony. For brevity's sake, let me suggest what it might mean
to take organic harmony, rather than untouched nature, as a guide to
the ecology movement.

Organic harmony has two aspects, one functional, the other aesthetic. On the *functional* side, it has become more and more obvious
that the character of the physical world surrounding our cities greatly
affects the quality of life of people living in those cities, the air they

breath, the water they drink, the food they eat, the diseases they are exposed to, and so on. We can reach a stable, homeostatic physical order of things only by taking a broader view of the health of the entire system over time, rather than retaining a narrow focus on individual objectives that have immediate and obvious impact on our lives this month or this year. We need to make this more of a priority and exercise a much more active concern about it, much more than was necessary for our ancestors, because the combination of more powerful technology and denser population makes the degradation of the sustaining physical environment a much greater possibility and danger. But, what we need to value here is not just any kind of nature, just because it exists. We need to work to bring about an enduringly stable harmony in a system that includes both the wild natural world and the civilized settlements of human beings.

This functional aspect of our concern for nature need not represent anything more than enlightened self-interest. In itself, it has no necessary connection with reverence for nature. The economist Lester Thurow suggests, for example, that the North American concern for South American rain forests is better cast in these terms. If we need these forests because we have already cut down the trees on our continent, we should rent them from the people in South America, rather than try to reap their benefits on the cheap by appealing to "reverence for nature."

Coming to the *aesthetic* aspect of organic harmony, I'm speaking here about that kind of aesthetic sensitivity to natural forms exhibited by Woodcarver Ching and by Blofeld's hermits, which is also expressed in some strains of the East Asian artistic tradition. This kind of sensitivity generally prefers irregularity to symmetry, since symmetry represents a subordination of individual elements to some master plan. It prefers what is unique to what is standardized. It prefers a kind of mysterious unity that comes from a balance of all elements to a unity that comes about through organization around one dominant element. The beauty in this kind of art seems similar to the kind of beauty that attracts many people to beautiful nature areas.

The appreciation of this beauty seems immensely enhanced because one does not observe it in artworks in a museum, but is able to be a participant in it by hiking, camping, kayaking, and so on. Such participation puts one in a particular state of mind and gives one particular sensitivities necessary to fully appreciate this kind of natural beauty—recall here the way Woodcarver Ching prepared himself by fasting to still his mind, conserving his qi, dismissing all the concerns of courtly civilization from his mind, and spending much time observing "Heaven's nature" manifest in the shape of trees. This participation also gives one a feeling of being a *part* of a beautiful organic harmony rather than just an observer of it. This kind of experience is what seems often to give rise to a feeling of "reverence for nature."

But if we take this kind of experience as a guide to our ecological concern, we probably need to recognize that not every place in wild nature has this wonderful organic harmony to an equal degree. Some places are more beautiful than others. And some places also allow, for the human participant, more of a feeling of harmony with nature than other places. It's difficult to feel harmony with nature while being bitten by black flies, making one's way through thorny underbrush, or wading through a smelly swamp.

This suggests more of a "gardening" approach to ecology than a purist, "let alone" policy favored by some. That is, we could select those areas of nature which are particularly beautiful and which are also particularly inviting to low-impact appreciative participation in this beauty by humans. We could work to enhance this beauty and also these capabilities for participation. Giving more people the opportunity to experience natural beauty, in circumstances that would induce a reverence for this beauty, helps greatly to solve the problem that the ecology movement needs widespread voter support to be successful. It seems that this approach is one already being adopted in many places, for practical reasons. For example, at this writing, the Massachusetts Audobon Society is planning to turn a portion of the mudflats in the Merrimac estuary into a marsh area carefully planned to provide habitats for various kinds of wildlife, and to provide the public with a means of participating in the natural beauty it hopes to create.

Population growth and technological advances have produced a situation where there are almost no areas of the globe left that are not affected by human civilization in one way or another. We don't really have a choice whether or not to have a significant impact on natural environments. Our choice is rather *what kind* of impact we are going to have. In this situation, it does not seem a bad idea to have some positive goals, such as enhancing nature's beauty and organic harmony—like Blofeld's Daoists "assist[ing] nature to do what it might under more favourable circumstances have done for itself."

Again, I don't offer these thoughts as doctrines backed up by the authoritative Daoist tradition, but as reflections stimulated by some aspects of the Daoist tradition worth wrestling with.

Notes

1. See my *Tao and Method* (Albany: State University of New York Press, 1994), chaps. 1–2; and my "Are Texts Determinate? Derrida, Barth, and the Role of the Biblical Scholar," *Harvard Theological Review* 81, no. 3 (1988): 341–57.

2. From John Blofeld, *The Secret and the Sublime: Daoist Mysteries and Magic* (New York: Dutton, 1973), 116–18.

3. Harvard-Yenching concordance to Chuang-tzu, Harvard-Yenching Institute Sinological Index Series, 20 (Cambridge, Mass.: Harvard University Press, 1956), p. 50, lines 54–59; my translation. See Victor Mair, *Wandering on the Way: Early Daoist Tales and Parables of Chuang Tzu* (New York: Bantam Books, 1994), 182–83; A. C. Graham, *Chuang-tzu: The Inner Chapters, A Classic of Tao* (London: Harper Collins, 1991), 135.

4. LaFargue, *Tao and Method*, 160–72.

Daoism and the Quest for Order

TERRY F. KLEEMAN

It is difficult to know how to apply a modern, Western term like ecology to a tradition like Daoism, which derives from an almost wholly disparate culture and time. One aspect of Daoism relevant to the modern conception of ecology, however, is the Daoist worldview in its most literal sense: what is this place where we live, how did it come about, and how is it maintained?

This essay is a modest contribution introducing to the West this Daoist view of the cosmos. It will differ from most similar attempts in three ways. First, I take the position that Daoism is not some nebulous, ill-defined mixture of mystical concepts pronounced by shadowy philosophers, but rather that Daoism is a major, living world religion, indeed the least understood and most poorly represented of all the major religious traditions. Daoism is a clearly defined social entity with an unbroken lineage of sacerdotal priests and continuously performed ritual stretching back two millennia, and it deserves all of the respect and attention accorded other world religions. Second, Daoism is fundamentally, even quintessentially, a Chinese religion, and it cannot be understood without careful attention to the Chinese intellectual and social context in which it arose and lives. Third, I will argue that the basic value of Daoism, far from the individualist freedom often propounded in Western popularizations of "Taoism," is order, the programmed, controlled order of a well-managed and closely regulated world.

Traditional Chinese Conceptions

A remarkable feature of the Chinese tradition is the near absence of true creation myths.[1] Most accounts of the ultimate beginnings of the world are late and abstract. A good example is the forty-second chapter of the *Laozi*, where we read:

> The Dao gave birth to the One. The One gave birth to the Two. The Two gave birth to the Three. The Three gave birth to the Myriad Things.

This is usually understood to denote a primordial unity, which is first divided into two opposing, dyadic forces, yin and yang, which interact, giving rise to a mediating third force; from this trinity flows all existence.[2] In early texts, however, this creator Dao is an abstract, indescribable force, never anthropomorphized. Moreover, there seems to be no teleology, no plan for humanity inherent in the design, but there is an orderly, mathematical progression that seems to follow an implied pattern.

The earliest Chinese culture heroes are not creators but teachers and discoverers who organize human society and teach the Chinese how to live. Even when we see in artistic representation remnants of a more explicitly mythical creation tale, such as the demiurge pair Fuxi and Nüwa, with their serpentine lower bodies intertwined, they hold in their hands as attributes a compass and and a square, with which they impart to their creation a mathematical order.[3]

From quite an early period, the conception of the world is explicitly political in nature. As is implied in the current name for China—the Middle Kingdom, or *Zhongguo*—China is at the center of this world, and the Chinese capital is at the center of China. There, the Chinese ruler, the Son of Heaven, occupied the throne, transforming all through the charismatic influence of his Virtue, or *de*. The area under his direct rule, the Royal Realm, was surrounded by five concentric squares (the earth, of course, being square in shape) of expanding size. Known as the Five Submissions, these squares marked the gradual decrease in influence of the Son of Heaven and the consequent decreasing level of civilization of their inhabitants. The outermost ring was the Wasteland.[4]

In another early conception, the Son of Heaven is flanked on all four sides by allies, chosen by the Chinese for their ability to keep the teeming barbarian peoples in check and act as a buffer against foreign invasion. During the Shang dynasty (ca. 1700–1050 B.C.E.), the Zhou fulfilled this function. The Zhou king held the title Elder of the West, and it was in this capacity of regional leader that King Wen and King Wu rallied the western peoples in revolt against the "evil" last king of the Shang. In the *Book of Documents* these regional leaders appear as marchmounts (*yue*), identified with the sacred, warding mountains that surrounded and protected China, and serve as counselors to the Son of Heaven. The magical mountains that were conflated with these border chieftains have roots deep in Chinese tradition; during the Shang, the marchmounts are among the most popular of nature deities, a class including the (Yellow) River, the Rainbow, and the Cloud, which could control meteorological phenomena, as well as bless or curse human beings.

As the limits of China's cultural domain expanded, the marchmounts shifted from liminal bulwarks on the borders to centers of numinous power within the Chinese realm. The mature system of marchmounts was a group of five, including the Central Marchmount, in a network of divine potentialities that defined China. The Five Marchmounts are among the pervasive and persistent elements of the Chinese sacred world. They were incorporated into both the imperial sacrifices and the Daoist pantheon and were primary objects of pilgrimage and worship on the popular level. There was a tendency to establish a hierarchy among these mountains. The Central Marchmount, associated with the color yellow and the legendary progenitor of the Chinese, the Yellow Monarch, seems originally to have had pride of place. When theories of the cyclical nature of the Five Phases led to a leveling of this group, the Eastern Marchmount, Mount Tai, rose to prominence. There were also regionally based systems of cosmological orientation that shifted through time.[5]

Traditional sources provide models of proper individual and state conduct during the year that also reflect these cosmological ideas. This tradition goes back to the poem "Seventh Month," in the *Book of Poetry*, which lays out a series of activities appropriate to each month of the year. The theme is much more explicit in the "Monthly Ordinances" (*Yueling*), a work now found both in the *Record of Rites* and distributed through twelve different chapters of the *Spring and Autumn Annals of Master Lü* (*Lüshi chunqiu*).[6] A number of provisions of these ordinances are relevant to ecology. In the first month of spring, the ruler is instructed to turn his own hand to the earth, ceremonially plowing three furrows to assure a good harvest. At this time, his inspectors repair the fields and assess what crop will grow best in each. Moreover, at this time:

> Prohibitions are issued against cutting down trees. Nests should not be thrown down; unformed insects should not be killed, nor creatures in the womb, nor very young creatures, nor birds just taking to the wing, nor fawns, nor should eggs be destroyed.[7]

Other months mention injunctions to "keep both the young buds and the more advanced from being disturbed," but to not "drain off all the water from dams and ponds" or "fire the hills and forests" (second month), to not "allow the cutting down of mulberry trees and silk-worm oaks" (third month), and so on.

The practical significance of many such injunctions is obvious: the intent was not simply to aid the specific living creatures involved, but, in a larger sense, to accord with the cosmic rhythms. Thus, offensive warfare was strictly prohibited during this period of new life, as was the execution of criminals; violation was "sure to be followed by calamities from Heaven." There was a sacrifice appropriate to each month as well as ritual vestments to be worn and ritual actions to be performed by the ruler. Each observance had to be properly timed:

> If in the first month of spring the governmental proceedings proper to summer were carried out, the rain would fall unseasonably, plants and trees would decay prematurely, and the states would be kept in constant fear. If the proceedings proper to autumn were carried out, there would be great pestilence among the people; boisterous winds would work their violence; rain would descend in torrents; orach, fescue, darnel and southernwood would grow up together. If the proceedings proper to winter were carried out, pools of water would produce their destructive effects, snow and frost would prove very injurious, and the first sown seeds would not enter the ground.[8]

There is, in a sense, an environmental ethic subsumed within this idea of a cosmic pattern that must be maintained through human action. The proper functioning of this world we live in depends upon an understanding of the cycles and rhythms of the cosmos and the appropriate actions based upon this understanding: interactions with the natural world of plants and animals; the administration of the human realm through governmental activity; and ritualized ministrations to the divine world of gods and spirits through sacrifice.

This view of the world as possessing a pattern or path, yet requiring the cooperation of human and divine forces to achieve it, is characteristic of China. It has sometimes been referred to as a Confucian worldview, other times as Daoist. It is simply Chinese. We will see that Daoism introduces some significant innovations into the Chinese religious world, but it never forsakes this basic viewpoint.

Daoism

Daoism is a vexing term that has been used to refer to a variety of intellectual, religious, social, and cultural phenomena. Here, I will use it in its most simple and straightforward meaning, to refer to China's indigenous higher religion (*Daojiao*). Daoism in this sense is an organized religion with an ordained clergy, a voluminous sacred canon, an extensive body of liturgy, an established eschatology, and an illustrious history of two millennia serving the Chinese populace. Daoism reveres Laozi in his divinized form as an exalted deity, makes liturgical use of the *Daode jing*, and draws occasionally upon the more mystical portions of the literary masterpiece attributed to Zhuangzi, but does not bear a special, direct relationship to the variety of Warring States and Han writings that are commonly lumped under the rubric of Daoist philosophy (which represents, in large part, the bibliographic category of "Dao lineages" [*Daojia*] established by Sima Tan in the second century B.C.E.). One great advantage of this usage is that we can clearly state who was a Daoist, a person ordained into a lineage of priests who donned sacerdotal robes and performed rituals, and what are Daoist texts, the religious literature created by these people.

The Daoist vision grows out of a wide variety of earlier traditions, including the classics, the Han dynasty apocryphal works, the practices of self-cultivation and alchemy, Han Confucianism, and Chinese occultism. It draws most directly on the literature of the "Dao lineages" in its theory of a primordial *hundun,* or "chaos," so ably studied by one of the editors of this volume, Norman Girardot.[9] *Hundun* is the term applied to the undifferentiated mass of potentiality before it was separated into the discrete elements of our mundane world. There is a cautionary tale in the *Zhuangzi* about an attempt to "improve" upon this unshaped power by opening up the seven orifices associated with all living beings that instead resulted in its destruction. It is perhaps ironic, therefore, that this *hundun* is associated with a world of perfect order.[10] In Daoist ritual, it is a wellspring of power that the priest calls upon to impose order on the universe.

In Daoism, the powers of this primordial world are embodied in the Three Pures (*sanqing*).[11] This trinity of high gods, the Primordial Heavenly Worthy, the Heavenly Worthy of the Numinous Jewel, and the Primordial Worthy of the Way and its Virtue, is different in kind from the popular gods of the Chinese pantheon. Popular gods were once human beings, and were elevated after death to positions of power and authority within the divine realm. Daoist deities are pure emanations of the Dao, unsullied by corporality. Their origin lies in the Anterior Heavens (*xiantian*), which existed before this material world took form. They rule over a bureaucratic hierarchy of transcendent officials and perfected, most of whom have their origin in the Latter Heavens (*houtian*) of the created world and have experienced life on earth.

The Daoist priest is not a carefree hermit questing after inner truth. He (or, more rarely, she) is a part of this bureaucratic enterprise. Ordination brings with it a transcendent office (*xianguan*). It is by virtue of this office that the priest enacts his rituals, issuing commands to the profane deities and spirits or the lower realms or journeying himself to the celestial precincts to present petitions before the high deities. On death, the Daoist priest assumes, full time, the office in the Daoist bureaucracy merited him by his official and private actions in this life. He lives a life of strict formality, appearing before fierce and unforgiving judges, submitting official documents to the highest lords, careful always to make no error in address or in orthography, lest he offend against the august powers of the universe and incur their punishment.

As pencil pushers extraordinaire, Daoist priests play a vital role in the functioning of the universe. Proper conduct is enforced through a system of divine justice. The lower level finks and informants are part of the profane pantheon, but it is Daoists, having succeeded to otherworldly office, who keep the records and pass judgment on their basis. Even after death, one might be accused by a fellow departed soul of some wrong while alive, and this has to be adjudicated most gravely, often to the detriment of the accused's living descendents. One of the primary functions of the Daoist priest, attested in rituals scattered throughout the Daoist canon, is to intervene with his colleagues in the divine world in order to alter the records, wiping out sins and prolonging the life span of the living and obtaining dismissal of pesky lawsuits affecting relatives in the other world.

Since the otherworldly strictures, in a continuity from pre-Daoist codes, contain rules that might be termed ecological in effect, if not necessarily in intent, such as the prohibitions on killing nesting animals in the spring, it is possible to see this entire Daoist moral bureaucracy as having an ecological function. Thus, the entire weight of the many-layered Daoist heavens, and the teeming corps of bailiffs, policemen, and jailers under their command, might be applied to one who violated the environment, dispensing corporal punishment, imprisonment, psychological torture (such as nightmares), and general misfortune for actions that in a temporal court would result in at most a civil action. The fly in the ointment here is the role of Daoist priests in aiding their followers in avoiding their just deserts. To be sure, these services are supposedly available only to bona fide members of the church in good standing, who deserve divine intervention because of their meritorious actions in supporting church activities, worshiping faithfully, and otherwise observing a strict code of conduct. Moreover, many rituals stress the honest repentance of their misdeeds and include a pledge that they will not repeat the proscribed act. Nonetheless, it is somewhat disconcerting that the Daoist priests themselves should provide the only loophole for escaping punishment from ineluctable justice.

Conclusion

The Chinese view of the cosmos was based upon the establishment and maintenance of order. Though the world had no clear creator and no teleological goal to attain, it did have a pattern of alternating forces and phases that shaped the rhythms of life. Actions in consonance with these patterns were deemed proper and morally good; those in opposition to the pattern were refractory, dangerous, and evil. This conception is found in a variety of writings of the Warring States period, writings usually associated with various of the supposed philosophical "schools" of that age. It functions as the dominant conception of the world and its operation throughout traditional China. This view has significant consequences for matters that we would today term ecological; many acts that we would see as harmful to the ecology were understood to be contrary to this pattern of life.

The Daoists, arising in the second century C.E., partake of this worldview. They see themselves as administering immutable codes of law deriving from transcendent, divine sources that embody these principles. They work tirelessly to keep in check the forces of chaos and disorder that would otherwise overwhelm this world. Occasionally, they come into direct ritual combat with these malefactors, but more often they act as magistrates and generals, dispatching troops and other minions to investigate and sieze them. Moreover, they perform grand rituals of cosmic renewal that revitalize the world and restore proper relations between the human and divine spheres. They do sometimes inhabit paradisiacal microcosms deep within the mountains, and some solitary searchers did live as recluses far from human society. But the dominant tradition in Daoism is one of service to the community of which the Daoist priest is an integral part, whether this be a village in the mundane world or among the celestial cohorts in one of the many Daoist heavens. Daoists were and are very much part of the world and played a key role in its continued operation. In this sense, they are an important element of the ecology.

Notes

1. On Chinese creation myths, see N. J. Girardot, "The Problem of Creation Mythology in the Study of Chinese Religion," *History of Religions* 15 (1976): 289–318.

2. The Han dynasty commentary attributed to Heshang Gong, for example, specifies that the yin and yang produce three energies, harmonious, pure, and turbid, which go on to form the triad Heaven, Earth, and Man. These three forces join in producing the myriad things. See Shima Kunio, *Rōshi kōsei* (Tokyo: Kyūko shoin, 1973), 145.

3. See Anne Birrell, *Chinese Mythology: An Introduction* (Baltimore: The Johns Hopkins University Press, 1993), 45, and the illustration from a Latter Han tomb tile on p. 70.

4. The themes in this paragraph are taken up in greater depth in my "Mountain Deities in China: The Domestication of the Mountain God and the Subjugation of the Margins," *Journal of the American Oriental Society* 114, no. 2 (April-June 1994): 226–38.

5. On the development of the marchmount system see, again, my "Mountain Deities in China," cited in the previous note.

6. There is also a version in *Huainanzi* 5. See the detailed discussion, as well as a translation of the *Huainanzi* version, in John S. Major, *Heaven and Earth in Early Han Thought: Chapters Three, Four, and Five of the* Huainanzi (Albany: State University of New York Press, 1993), 217ff.

7. *Li Chi: Book of Rites*, trans. James Legge, 2 vols. (1885; reprint, New Hyde Park, N.Y.: University Books, 1967), 1:256.

8. Legge, *Li Chi*, 1:257.

9. N. J. Girardot, *Myth and Meaning in Early Taoism* (Berkeley and Los Angeles: University of California Press, 1983).

10. In *Myth and Meaning in Early Taoism* (p. 15), Girardot notes that in these philosophical works, salvation "is a matter of the resynchronization of human periodicity with the cycles of cosmic time."

11. On the structure of the Daoist pantheon, see Kubo Noritada, *Dōkyō no kamigami* (Tokyo: Hirakawa, 1986); Kristofer M. Schipper, *The Taoist Body* (Berkeley and Los Angeles: University of California Press, 1993), 118–123. The present description is a simplification of a complex system that changed over time and varied from lineage to lineage. Nonetheless, the general nature of Daoist deities, and their contrast with profane gods, remains constant. For the moral basis of this distinction, see my "Licentious Cults and Bloody Victuals: Sacrifice, Reciprocity and Violence in Traditional China," *Asia Major*, 3d ser., 7, no. 1 (1994): 185–211.

Sectional Discussion:
What Can Daoism Contribute to Ecology?

JAMES MILLER

Introduction

The essays in this section are groundbreaking in that they represent the first serious and sustained attempt by scholars of Daoism to raise the significant theoretical question of how it is possible to relate Daoism and ecology. These essays are also groundbreaking in that they proceed by means of a sort of intellectual spadework or excavation to reveal a basic framework of questions that sets the agenda for the rest of this book. The framework is something like an alchemist's tripod of questions: 1) What is Daoism? 2) What is the relationship between Daoism and ecology? 3) How do we fit into all of this? These questions are academic, in a genuinely positive sense, in that they are intellectual, dealing with matters of culture, history, and hermeneutics. But they also indicate the practical and constructive orientation that this volume takes to heart: how can we construct a Daoist ecology? In this regard, the writers of these essays are fulfilling the obligations of responsible public intellectuals, a task that, as Jordan Paper points out, is characteristically "Confucian" rather than "Daoist." The goal of this brief response is to highlight the constructive nature of these three questions, and to point toward areas for further study.

What Is Daoism?

It is now evident that the relentless quest by Western intellectuals to project a taxonomic scheme upon the diverse phenomena of human religious cultures constitutes a cultural ecological catastrophe in its own right. The scientific drive to know and transform the world finds its cultural parallel in the drive to nail down the essential features of "Daoism." As Joanne Birdwhistell pertinently reminds us, this is a violent act in which cultural complexity is made to strip itself bare before the inquisitive eyes of the gatekeepers of the grand narratives. Zhuangzi resisted the imposition of order, moral and intellectual, upon a world of constant transformation. And yet, as Terry Kleeman clearly demonstrates, the Daoist religious tradition constitutes precisely one of these orderings or "economies."

The problem is that "Daoism" is in a certain sense an intellectual fiction that is convenient for exploring such topics as "Daoism and ecology" but not especially convenient for the historical evidence. Historically speaking, Daoism is best understood not as a body of ethical or religious doctrine, but as a collection of ways of relating to our cosmic ecosystem. Some of these traditions spawned clearly identifiable institutional structures, others diffused themselves more generally throughout Chinese culture. The question "How do we understand Daoism?" leads us to a deeper question: "How do we gain a Daoist understanding of the world?" At this point the activity of projecting intellectual order onto the world transforms into a scholarly "nonassertive-action" (*wuwei*), and our consciousness has become more properly receptive and attentive. This is the first step toward constructing a Daoist ecology.

What Is the Relationship between Daoism and Ecology?

If this book does one thing, it should be to drive a stake through the heart of the notion of Daoism as the soul mate of deep ecology. As Jordan Paper notes, deep ecologists in the West pointed an accusing figure at the concept of transcendence and found in the *Daode jing* the beautiful contrast of a philosophy of immanence and the valuing of nature. Michael LaFargue clearly explains, however, that "nature" is always at least partly a cultural construct, and that in Chinese cultural history value has always lain in the careful interaction of humans and their environment and never in the Western romantic notion of the wilderness or in the modern ethical concept of biocentric egalitarianism. To accept this fact, as we must, is to recognize that Daoism does not exist as a resource that we should mine however we see fit. Instead, it confronts us through the distinct otherness of its classical texts and as a living tradition found in the lives of millions of people throughout the world. Daoism is not a "useful" idea or a reified "entity" that can be made to address an "issue" in the same way that popes issue pastoral encyclicals.

Jordan Paper makes the argument that we should look to Daoism precisely as this living, religious culture if we are to discover anything to help the modern ecological movement. In the Daoist religious experience, the deep connections between human beings and their environment are forged and renewed time and again. The conclusion to be drawn from this is that we should pay serious attention to the function of religious rituals in shaping an ecological consciousness. This would be consonant with Daoism's focus on the local and the particular. While it is always helpful to formulate "Earth Charters," it would be characteristically Chinese to note that the existence of laws does not reform moral characters, but merely provides the means to criminalize and penalize those whose psychosocial flaws are manifest. To make the point another way, the mere existence of the Ten Commandments does not make people morally good; but the frequent recitation of the Ten Commandments in an appropriate ritual context is a different matter altogether. Jordan Paper is surely right: if Daoism is going to be of any practical help in constructing an ecological consciousness, we must direct our attention toward its biospritual technologies, whether in the *Zhuangzi*, in the Highest Clarity scriptures, or in Quanzhen monasticism.

How Do We Fit into All of This?

The answer to this question depends upon who "we" are. "We" who took part in the original conference and who wrote this book are scholars and students, Chinese and Western, practitioners and academics. But our motivation was simply the desire to pursue the application of our understanding to the complexity of the environmental situation, and, by publishing this book, to invite you to join this community. But, as Joanne Birdwhistell notes, this situation is problematized by the awareness of ourselves as actors shaped by narratives of culture, gender, and ideology. How is it possible to sort through the complexity of the Daoist experience, the complexity of the ecological situation, and the complexity of our roles in all of this?

By examining the original question, "What is Daoism?" we learned that clarity of perception does not correlate with clarity of taxonomy. We cannot simply "list" ourselves into right awareness. The reality is that we must become aware of how we include ourselves in the alchemical cauldron of Daoism "and" ecology. Our role, then, as writers and readers, is that of the alchemist who is perpetually forging a new reality out of the elements of cultural traditions and pressing political contexts. The nature of this reality depends upon what we ourselves bring to the crucible in forging this connection of Daoism "and" ecology.

Conclusions

These essays have presented the raw elements from which a practical intellectual consciousness of a Daoist ecology can be forged: 1) an awareness of the complexity of the Daoist tradition; 2) an awareness of how Daoism has been exploited as an intellectual resource; and 3) an awareness of our own roles as actors in, and shapers of, cultural narratives.

These three are the necessary ingredients for a practical, ecological, cultural consciousness. But to construct this "Daoist" ecology requires that the elements be brought together, compounded, and refined. However eloquently the authors have spelled out the recipe, this concoction can only take place in actuality and not in any purely theoretical sense. Future work in constructing a Daoist ecology would therefore do well to describe the practical nature of the Daoist ecological consciousness that evolves from these elements. The shape of this Daoist ecology is already becoming clear.

A Manifesto of Daoist Ecology

1) Daoist ecology is not an idea but a practical consciousness shaped by a living religious tradition.
2) Daoist ecology recognizes that nature is a construct of human culture, as well as something other than ourselves.[1]
3) Daoist ecology is not deep ecology, but "shallow." It insists upon the complexity of the particular and resists the simplicity of the universal.[2]

Notes

1. On the question of nature's otherness in classical Daoist philosophy, see Russell B. Goodman, "Scepticism and Realism in the *Chuang Tzu*," *Philosophy East and West* 35, no. 3 (July 1985): 231–37.

2. On the question of "the local and the focal," see the essay by Roger T. Ames in this volume.

II. Ecological Readings of Daoist Texts

Daoist Ecology: The Inner Transformation. A Study of the Precepts of the Early Daoist Ecclesia

KRISTOFER SCHIPPER

In this paper I would like to introduce evidence that there is more to Daoism and ecology than just a philosophy of nature. Indeed, Daoism did not only *think* about the natural environment and the place of human beings within it, but took *consequential action* toward the realization of its ideas. As early as during the first centuries of the common era, Daoism developed institutions and regulations (*The One Hundred and Eighty Precepts*; see below) with the purpose of protecting the environment and to ensure that its natural balance would not be destroyed. It purposely advocated respect for women and children, for all forms of animal life, for all plants, for the earth, for mountains, rivers, and forests and sought to preserve and protect them. These rules and institutions may be the earliest significant and conscious efforts of human civilization to protect the natural environment and to ensure the adaptation of culture to nature instead of the opposite.

However, before entering into all this, I feel some preliminary remarks are perhaps not out of place. As stated, we can see clearly that the protection of the natural environment and the adaptation of humankind to this environment was a matter of paramount importance to early Daoism. Yet it would be a mistake to consider that these early Daoists were "ecologists" as we define ecologists today. Ecology is a modern Western science, first advanced by the German zoologist Ernst Haeckel in 1866, that aims at studying the relationship between organisms and the environment in which they live. For almost a century, ecology was a scientific term known and used only by specialists and a purely technical term (I learned about it in 1950 when I joined a youth organization for the study of biology and tried to describe eco-systems). Since the 1970s, however, we have witnessed the advent of an "ecological movement" based on the idea that the industrialized world had provoked an ecological crisis. This "angst" of the Western world has been spread wide and far and with it a new set of ideas, not to say an ideology. This new Western doctrine is now also preached everywhere, to those who want to hear it and to those who do not, just as formerly other so-called great Western inventions like religion, science, democracy, development were disseminated. The history of this immodest imposition of Western concepts and ideologies on non-Western cultures that did not necessarily need them, as they had religions, democracies, sciences, and developments of their own but within different historical contexts, is not very edifying. We do not want to add to it. We may recognize from the start that there exists, indeed, something we may call Daoist ecology, but at the same time we should recognize that its basic tenets as well as its methods and aims may be very different from those of the boisterous political discourse that is current today in the Western ecological movements. One does not have to be a specialist of Daoist thought to see that someone like Zhuangzi (fourth century B.C.E.) would have had very little patience with the kind of intrinsically well-meaning but highly doctrinal and polemical "Earth Charter" composed by the Earth Charter Commission. To name an example, instead of telling us that we should impose that "All people have a right to potable water," the above-mentioned precepts say: "You should not contaminate water." Instead of trying to impose political changes by governments and powerful organizations, the Daoists recommend that you stay away from these instances as much as possible. Instead of speaking out loud in order to change others, Daoism speaks the soft language of the transformation of the inner self.

The One Hundred and Eighty Precepts

One of the most important documents concerning Daoist ecology is a short text called the *One Hundred and Eighty Precepts (Yibaibashijie).*[1] These are the guidelines laid down for those who, in the early Daoist movements, held the position of leaders of the lay communities. They were called libationers (*jijiu*). *Jijiu is* a very ancient term denoting the oldest and most respected member of a given community who, in that capacity, presides over the periodical community sacrifices. In Han times, *jijiu* also came to mean local leader. The term was adopted by the ecclesia of the Heavenly Master as well, and also, as far as we can tell, by other Daoist organizations. In the Way of the Heavenly Master, women as well as men could be *jijiu*. The *One Hundred and Eighty Precepts,* however, appear to concern mostly men.[2] This and other elements to which I shall return show that the *Precepts* were originally not produced by the Way of the Heavenly Master. The latter did, however, adopt them and transmitted them as part of its institutions, and it is for this reason that we have them today. There is ample evidence, as several recent studies have shown,[3] that the *One Hundred and Eighty Precepts* antedate the great scriptural renewal of the end of the fourth century. How far we can go back is at present anyone's guess. From a passage in the *Baopuzi neipian* we may conclude that these or similar precepts were known to Ge Hong (283–343).[4] Inasmuch as Ge Hong's claim that most of the texts he used in his *neipian* originally belonged to Ge Xuan (164–244) can be taken seriously, nothing would stand in the way, in principle, of a Three Kingdom or even late Han dating.

Among the *One Hundred and Eighty Precepts,* not less that twenty are directly concerned with the preservation of the natural environment, and many others indirectly. Here are some examples:[5]

14. You should not burn [the vegetation of] uncultivated or cultivated fields, nor of mountains and forests.
18. You should not wantonly fell trees.
19. You should not wantonly pick herbs or flowers.
36. You should not throw poisonous substances into lakes, rivers, and seas.
47. You should not wantonly dig holes in the ground and thereby destroy the earth.
53. You should not dry up wet marshes.
79. You should not fish or hunt and thereby harm and kill living beings.
95. You should not in winter dig up hibernating animals and insects.
97. You should not wantonly climb in trees to look for nests and destroy eggs.
98. You should not use cages to trap birds and [other] animals.
100. You should not throw dirty things in wells.

101. You should not seal off pools and wells.
109. You should not light fires in the plains.
116. You should not defecate or urinate on living plants or in water that people will drink.
121. You should not wantonly or lightly take baths in rivers or seas.
125. You should not fabricate poisons and keep them in vessels.
132. You should not disturb birds and [other] animals.
134. You should not wantonly make lakes.

Before making more detailed comments on these and other related precepts from our text, I would first like to address the fundamental question: what prompted Daoism to edict these rules? As those familiar with the Confucian classics will have noticed, some of these rules are already present in the "Yueling" chapter of the *Liji,* but there these rules apply for specific seasons only. Thus, in the spring season there should be no cutting of trees, no destruction of bird's nests, no upsetting of hibernating insects, and so on. In the Daoist *Precepts,* however, these rules apply in an absolute way and for all times and seasons.

Specific religious phenomena are related to a particular group of people in a specific context. Early Daoist cults and organizations centered around mountains, streams, seas, and uncultivated natural reserves, not only as places to retire from the world as hermits, but as places for their communities to congregate and perform rituals. For them, the natural environment was a "sanctuary" in both meanings of 1) a holy place dedicated and consecrated to the cult of gods and saints and where sacred things were kept, and 2) a place of refuge and protection, a place for those who wished to escape the calamities of the world and where they could dwell in peace.

It is often assumed that Daoist communities, and in the first place those of the Way of the Heavenly Master, developed on the outskirts of ancient China in regions where civilization had yet to be fully developed. Daoism would therefore be the product of some kind of "frontier society," through contact with aborigine tribes and their wild and primitive shamanistic religion. This does not correspond to the historical truth. The land of Shu in the great plain of present-day Northern Sichuan, where the movement originated, was the very first region to be developed by advanced hydraulic technology for the purpose of intensive agricultural exploitation. From the fourth century B.C.E. on, Shu was not only the key economical area that financed the military expansion of Qin and its ultimate conquest of all the Central States, it was also the place where the technology of wet rice farming saw its major advances, with all the consequences for the human and natural environment: an intensive economical exploitation with a rigidly controlled population of farmers, tied to their land by taxes and corvée labor, practicing a sheer monoculture of rice and agricultural products at the expense of pasture cattle farming. All the endemic ills connected with sedentary life and high economical pressure, the scourges of rural China, were experienced there: abusive taxation, raids by external nomadic tribes, high population density, epidemics, and famine. This is the social and economical background of the Way of the Heavenly Master. As to Zhang Ling, the reputed founder of the movement, he came from Pei, which corresponds to the northwestern tip of today's Jiangsu province. This is the very heartland of ancient Daoism, at a short distance from places like Bo Tai (nowadays northwest Anhui) and Meng At (nowadays Henan, north of Shangqiu), where Laozi and Zhuangzi are supposed to have lived. This region was the most developed economical area, with large merchant towns, like Tao (present-day Dingtao), which were the cultural and political centers of the early Warring States period. In short, neither classical nor medieval Daoism developed in primitive surroundings, but in places of highly developed culture.

It is against this background that we should consider the institutions of the Way of the Heavenly Master and first of all its network of "places of order" (*zhi*). As long surmised, and again demonstrated by recent research, the *zhi* were almost without exception situated on mountains and other natural reservations.[6] Here the communities assembled three times a year. Here they kept their scriptures, their archives, and their grain reserves. There were lodgings and, in later times, fortified congregation halls. It is impossible not to see a link between the political and economical situation in the plains of Shu and the establishment of these mountain sanctuaries. The preservation of these natural spaces must have been of vital importance for the population in times of crisis. Here, too, many similar instances can be quoted from later times and other places in China where sacred mountains and their temples played a vital role as sanctuaries and places for survival in times of contagious disease, famine, and war.

The ecological rules of the *One Hundred and Eighty Precepts,* with their emphasis on the preservation of the vegetation, the natural resources in water and plants, as well as the animals which inhabit the mountain sanctuaries, do well agree with the above-mentioned preoccupation. It is interesting to note that trees, plants, and animals should be spared only from "wanton" (*wang*) behavior. This leaves the possibility open for using these resources whenever truly necessary. In general, the *One Hundred and Eighty Precepts* do not enunciate hard and fast rules. Even the interdiction of eating meat is very much qualified, leaving much amplitude to individual situations. The key phrases are, often, "not too much," "not unnecessarily," or "if at all avoidable." The main theme is that of *respect,* not only for nature, but for all men, women, and children, their ways of life, customs, and culture.

The *One Hundred and Eighty Precepts* cover a very wide range of topics. The text does not, however, attempt any form of organizing them according to subject, or degree of importance, or any other principle. They are quoted without any order, for reasons we can only guess (see below). For the sake of their study, I have tried to rearrange them into categories, and these are the main groups I have isolated:

1. *Eating and dietary precepts*
Examples: no strong vegetables; no discussion about the quality or the tastiness of the food; not too many big banquets with food all over the place; etc.

2. *Sexual behavior; respect for women*
Examples: no adultery; no abuse of pupils; no debauchery; no drilling of holes in wall in order to peep at women; no talking with women in dark places; etc.

3. *Respect for seniors and juniors, for family and worthy people*
Examples: Do not slight elderly people; do not treat your pupils as if they were your own children; do not discuss and criticize your teachers.

4. *Respect for servants and slaves*
Examples: Do not buy or sell slaves, do not tattoo the faces of slaves; do not stand guarantee or mediate in transactions concerning houses and slaves; etc.

5. *Respect for animals*
Examples: do not kick domestic animals; do not watch animals copulate; do not without good cause make horses run and gallop while pulling a chariot.

6. *One's own possessions: avarice, grabbing, hoarding, sumptuary regulations*
Not too many male and female servants; not too many sets of clothing; no hoarding of treasures; do not seek out a too nice house and too comfortable a bed.

7. *Other people's possessions: stealing, and cheating*
Apart from stealing cheating, swindling, coveting other people's property, there are special precepts for not making use of religious services or religious status in order to have people make gifts.

8. *Killing living beings*
Not for consumption; not for others to eat; no abortion; no suicide; no murder.

9. *Protection of nature*
See above.

10. *Dealings with the state, civil, and military authorities*
Do not try to know about politics and military affairs; do not befriend officials; do not go in audience with the emperor or his officials; do not marry your children to them; do not collect taxes; do not carry weapons.

11. *Ritual conduct (as a priest) and personal religious observances*
This is a very long list: interdiction of "vulgar cults" and of veneration of "other" ancestors; only use appropriate healing and exorcist practices; offering incense; prayer; remuneration of services; attitude proper for a libationer; etc.

12. *Divination, prognostication, geomancy*
Do not practice divination; do not practice geomancy.

13. *Proper behavior and conduct, etiquette*
Another long list: Do not swear at people; stick out your tongue; make big eyes; play nasty jokes; be polite; never get angry; take abuse with calm; always smile; do not run around naked; do not startle people; etc.

14. *Respect for writing*
Do not use cursive script in letters; do not write too often for nothing; do not throw away or pollute paper with characters on it.

Of course, these categories are arbitrary and very often overlapping. Also, many of the precepts, in addition to those quoted above, have a bearing on ecology. They concern, for instance, frugal living: to not eat from gold or silver recipients, not waste food, not lust after tasty dishes, not possess too many clothes, not long to live in luxury. The precepts concerning the respect one owes others are especially numerous: Women should not be approached for dalliance, should not be spied upon, nor forced into unwanted intimacy. Children should not be abused sexually, nor spoiled by favoritism. The creeds and customs of others should be respected, as well as their private lives and marriage secrets. In fact, all precepts deal with the relationship of the individual with his or her natural or cultural environment and, as such, can be deemed "ecological," in the sense of a *social ecology*. Though intended for the religious elite, they nevertheless concern all members of the Daoist community, as the libationer was not a priest but an elder, a laymen supposed to possess outstanding virtue and to be an example to all others.

Origins and Antecedents of the Precepts

The present version of the *One Hundred and Eighty Precepts* comes with a long preface of doubtful authenticity.[7] It tells us that the *One Hundred and Eighty Precepts* were given by Laojun to Gan Ji. As is known, the immortal Gan Ji is linked to the origin of the *Taiping jing*. According to the preface, once Gan Ji had received the *Taiping jing,* his followers adopted it, but later their zeal slackened and corruption set in. Thereupon, Laojun gave him the *One Hundred and Eighty Precepts,* so that the libationers should know right from wrong, and the Way could again prosper.

As is also known, there are several different traditions surrounding the *Taiping jing*. There are two that imply the immortal Gan Ji. The first is that of the *Taiping jing* version mentioned in the *Houhan shu,* and named *Taiping qinglingshu,* presented to the throne by Xiang Kai in 168. It is reputed to have contained all kinds of recommendations for the emperor, his house, and the country and therefore must have been a Confucian apocrypha (*chenwei*). It is said to have been 170 *juan* (scrolls) long and divided into ten sections numbered according to the Heavenly Stems.

The second is a Daoist work which Laojun is supposed to have revealed to Gan Ji at a much earlier time. Daoist works of the early

Middle Ages do give a prominent place to this revelation, but differ as to the place it occupies in the sequence of the revelations of the Dao. For the fifth-century *Santian neijie jing,* Gan Ji's *Taiping jing* was the very first of revelations, to be followed by the *Daode jing* and later by the revelation to Zhang Daoling. Other works have a different chronology and put the *Daode jing* first, then the *Taiping jing,* and finally Zhang Daoling. The second tradition places the transmission to Gan Ji at the very end of the Zhou dynasty, under King Nan.

This is also the chronology found in our preface. It states that once Laojun had given the *Taiping jing* to Gan Ji, he went to the west and converted the inhabitants of that region. He came back "under the reign of King You." This is, of course, impossible, as Nan Wang reigned from 314 to 255 B.C.E. and You Wang from 781 to 770 B.C.E. It is difficult to say whether the You Wang episode is a later addition to an original preface dealing only with the revelation to Gan Ji under Nan Wang. In fact, the inconsistencies of the preface are by no means limited to this sole instance, and therefore, in my opinion, the whole text must be spurious. The fact remains, however, that someone must have thought, at some time during the Six Dynasties period, that the *One Hundred and Eighty Precepts* were related to the Daoist *Taiping jing* and that this opinion gained widespread acceptance. The preface is not only given with the existing versions and in the Dunhuang manuscript, but also quoted in relation to the origins of the Daoist *Taiping jing* in the preface of the Taiping part of the Daoist canon of the Sui and early Tang.[8] Also, from other sources we see that the link between the *Taiping jing* and the *One Hundred and Eighty Precepts* was universally accepted during the Tang, even by eminent scholars, such as Chen Xuanying, the commentator of the *Zhuangzi.*

The problem is made more difficult by the fact that the presently extant version of the *Taiping jing,* which has become widely accessible and studied today, is in fact a late remake by a certain Zhou Zhixiang from around 569 and that this remake probably was based not on the Daoist *Taiping jing* but on the Confucian *Taiping qinglingshu.*[9] It is therefore not surprising that the two texts have little in common, except perhaps for a certain congruence in the themes and topics being discussed, albeit with very different conclusions. The possibility that our *Precepts* were originally based on the Daoist *Taiping jing* therefore remains only a possibility, and nothing more.[10] That the *Precepts* were based on an existing book is, however, very probable. As said above, the different rules are given without any attempt to order them according to subject or importance. This may well be due to the fact that the author has abstracted a holy scripture, noting down its recommendations and commandments in the order they appeared therein, so as to get at the gist of the sacred revelation. If this hypothesis is correct, then the book in question must have been a very important canon, inasmuch as the *Precepts* ranked as the highest set of rules to be followed by the Daoist clergy of the Middle Ages. What book this was, if not the *Taiping jing,* we will probably never know.

That it must have been a Daoist work is, however, beyond doubt. Indeed, one of the most interesting features of the *One Hundred and Eighty Precepts* is its relation to ancient great texts, in particular the *Zhuangzi.* The very first rule, "You should not keep too many male and female servants," is reminiscent of the story of Gengsang Chu in the *Zhuangzi* (chapter 23). The sage discarded the more capable of his male and female servants,[11] and thereafter his household and the country prospered. Rule 98, recommending that birds should not be trapped in cages, reminds us of other passages where such an instance is discussed. *Zhuangzi* 3, "The Secret of Caring for Life" (*Yangsheng-zhu*), says: "The pheasant in the wilds walks ten steps in order to eat something and one hundred in order to drink. Yet it does not ask to be put in a cage and to be fed. Although its spirit is hailed as a king, it does not like it." As so often, this passage has a more elaborated, parallel story in the later chapters. This is found in chapter 18, "Supreme Happiness (*Zhile*), and tells the anecdote of a sea bird that alighted in the suburb (the dynastic altar) of Lu. The bird was captured and feasted with sacrificial fare and music. However, it looked very sad, it did not partake of the sacrificial delicacies, and after three days it died. The author then goes on to explain that men should live as men but should let animals live as they like, birds in deep forests, fish in water, and not transform the world according to what we as humans deem beautiful or important. To let all creatures live according to their own nature is, says the author in conclusion, "to sustain happiness by applying the principle of order."

Simple life, each being according to its natural constitution and needs is, indeed, a main theme in the *Zhuangzi,* and also in the older chapters. Each organism—animal, man, tree—can only thrive in its natural environment: fishes in the water, trees free from the hatchet, leeches in the mud, pheasants in the wilderness. Each time, however, the philosophical reasoning behind the criticism of disturbing nature is more than an idea of simple "non-interference" (*wuwei*). Fishes not only thrive in water, they *enjoy* it, and the joy of fishes is also the joy of men. Animals like oxen, pigs, and turtles do not like to be sacrificed, even after being bred and fattened and bedecked with beautiful decorations. Trees like to grow freely and even will assume holiness if that can save them from the ax, inasmuch as they hate to be exploited and made useful for mankind. Underlying all this is a critique of the human search for transcendence and the transgression of boundaries. When humans approach nature from their anthropocentric point of view and begin to impose their ideas on nature, there is already a lack of respect. That this lack of respect is intimately linked to human religion, and especially to its sacrificial practices,[12] is a very profound and important insight. The same ideas of rejecting sacrificial religion and showing respect for the biospheres of other beings and for nature in general is very much present in the *One Hundred and Eighty Precepts*.

How can we explain that so much of the ecology of the *Precepts* was already present in the *Zhuangzi*? Did the Daoists of the Han or later periods take their inspiration from ancient philosophical texts, or were the antisacrificial opinions already current in the Chinese mystery religion of Zhuangzi's time and is his treatment of them a philosophical elaboration? My own answer would be the second alternative. As such, the ecological *Precepts* are rooted in the rejection of the feudal society and the ritual practices of the public cults of the city-states.

Implementation and Influence of the Precepts

The *One Hundred and Eighty Precepts* were the rules by which the leaders of the Daoist communities, the libationers, had to abide. The late-fourth-century Lingbao scripture *Tayi zhenrenfu Lingbao zhaijie weiyi zhujing yaojue* (*Daozang* 532:17a; hereafter DZ) reads:

> The Taiji zhenren says: Libationers should respectfully abide by the Laojun's *One Hundred and Eighty Precepts*. Only then can they be said to be Libationers. Therefore it is said: if one has not received the *One Hundred and Eighty Precepts*, then one should not be honored by the

people and the pupils. Those who have received them should think of them in their hearts and respectfully put them into practice, and only then can they be the Libationers of today. Libationers should rid themselves of (spurious) considerations and cast away all that is wrong, but practice the rules with determination, like this. These are the marvelous rules of the Way of Huanglao, the methods of the Seed People (*zhongren*), which no one who is not true can ever reach. Even if there is one in ten thousand that will be much!

Such was the prestige of the *Precepts* that the early Lingbao scriptures did incorporate them and adapted them to their needs. This adapted version is the *Taishang dongxuan lingbao sanyuan pinjie gongde qingzhongjing* (DZ 456). Later, but also during the Six Dynasties period, this adapted version and the original *Precepts* serve to create the largest set of rules of Daoism ever, the "Great Rules of Wisdom in Self-Examination" (*Shangqing dongzhen zhihui guanshen dajie wen;* DZ 1364). Yet, the original *Precepts* remained the standard code of conduct for the libationers, as is shown by the fact that the great Lu Xiujing (406–477), in his *Lu xiansheng daomen kelüe* (DZ 1127:17b), states that:

> Therefore the scriptures say: When a Daoist master has not received Laojun's *One Hundred and Eighty Precepts*, then his body will have no virtue, and he cannot be deemed to be a Daoist master and receive the homage of the people, nor can he rally and administrate the gods and ancestors.

We do not know at present when the *Precepts* lost their position as the foremost code of conduct for the Daoist masters. Probably they went into disuse at the end of the Tang and the Five Dynasties period, when Daoism underwent great changes and the Way of the Heavenly Master lost its position as the very foundation of the Daoist institutions. Yet, the spirit of the precepts were not lost, as can be seen from the fact that in many stories, such as the *Daojiao lingyanji* of Du Guangting (850–933), pollution of springs and rivers and cutting of trees are noted as major crimes against the Dao.

Much further study is needed to link the rules of Daoist ecology with the ecological practice of Chinese traditional rural communities in premodern China. One could no doubt point to several indications that much of the spirit of the code of conduct we have examined continued to be adhered to until modern times.

The Inner Transformation

Having looked at the relationship between Daoism and the natural environment as codified by the early Daoist communities, there is still one question we have to discuss: What was the community framework of the libationers who followed these rules, and what was the ecclesiastical authority that could enforce them? The remarkable thing is that any clear reference to such an organized religious structure is absent. To be sure, the text makes distinctions between adepts and "vulgar people," between the Dao and the world, but these indications are very vague. There is no reference to what might be seen as an order of priesthood, and, indeed, the text does suggest that such a thing did not exist. The libationers who observed the rules were not clerics but laymen, and could be farmers, traders, and even soldiers. They should not occupy official positions, collect taxes, or have political functions within the official hierarchy of the state. We do not see in the *Precepts* any reference to some higher religious authority, nor to some set of beliefs and concepts, some ecclesiastical authority. The only allusion to the Daoist ecclesia is in rule 144, which says: "You should take refuge in the One and Orthodox (*zhengyi*) and not practice vulgar cults." But there is never a trace of the kind of stipulation which would say: "you should not cut trees except when authorized by the competent hierarchy." Each person has a responsibility of his or her own; there is no church authority to enforce the rules.

In ways of religious practice, the *Precepts* do state that propitiatory rites should not be performed for one's own benefit, but always for others. Healing practices are mentioned, in the same spirit. When indications are given for personal religious practice, these are breathing exercises, dietary practices ("cutting the cereals"), prayer, burning incense, and the like. All these are personal observances. Here, as in Daoism in general, what matters above all is every person's personal relationship to the Dao. As said over and over again, and in many texts: "the Dao is in ourselves, not in others." The important scripture *Zhengyi fawen Tianshi jiaojie kejing,* in the part called "Dadao jialingjie," makes this remark with respect to the practice of the *Book of the Yellow Court (Huangting jing)*, that fundamental guidebook of the inner landscape.

The emphasis on the self, on the personal relationship to the Dao, implies, also with respect to the preservation of the natural environment, that each person is responsible for the Dao, each person embodies the Dao. The preservation of the natural order therefore depends absolutely on the preservation of this natural order and harmony within ourselves and not on some outside authority. The environment is within us.

This priority of the inner world is one of the great tenets of Daoism. The outside crises and dangers can only be overcome by transforming them within us, by purifying and reshaping them through the harmony of our body. All beings are transformed through it. When it has reached perfection, the body radiates harmony that is beneficial to its environment. Thus, the *One Hundred and Eighty Precepts* never speak of protests to the higher authorities, of political actions, revindications, demands for justice and peace, but only of respiration exercises, of inner harmony and individual peace. This is the only way to save the environment. True perfect nature can only be found within oneself. To regulate the world, we have to cultivate ourselves, to tend our inner landscape. Beyond, beneath, behind, and inside the *Precepts* of the Daoist libationer, we find a whole new world of spiritual ecology.

Notes

1. This text is not preserved as an independent work but is incorporated into several collections: 1) *Taishang Laojun jinglü* (DZ 786: 2a–20b), a late Six Dynasties collection of precepts; 2) *Yunji qiqian* 39: 1a–14b, the Daoist encyclopedia of the Song (compiled ca. 1025), based on Tang materials; 3) *Shuhuang juanzi* P.4562, P.4731: "Laojun yibaibashi jie," an early Tang manuscript; 4) *Yaoxiu keyi jielü chao* (DZ 463: 5:14a–19b), a Tang encyclopedia.

2. A note in *Taishang Laojun jing* indicates that the 180 precepts were intended for men, whereas a now lost *Taiqingyinjie* were intended for women.

3. The studies are: H.-H. Schmidt, "Die Hundertachtzig Vorschriften von Laochün," in *Religion und Philosophie in Ostasien: Festschrift für Hans Steinger*, ed. Gert Naundorf et al. (Würzburg: Königshausen & Neumann, 1985), 149–59; Benjamin Penny, "Buddhism and Daoism in *The 180 Precepts Spoken by Lord Lao*," *Taoist Resources* 6, no. 2 (1996): 1–16; Barbara Hendrischke and Benjamin Penny, "The *180 Precepts Spoken by Lord Lao—A Translation and Textual Study*," *Taoist Resources* 6, no. 2 (1996): 17–28. This last article contains a complete translation of the text. I am presently engaged in a study concerning the dating and origins of the 180 precepts.

4. *Baopuzi neipian* 6, "Weizhi," page 126 (Wang Ming ed.). I owe this reference to Wang Zongyu of Peking University.

5. The *One Hundred and Eighty Precepts* have been translated by Hendrischke and Penny in the article mentioned above in n. 3. Theirs is an excellent translation. However, for reasons of consistency, the translations in this article are my own.

6. See Wang Chunwu, *Tianshidao ershisi zhi kao* (An investigation of the twenty-four *zhi* of the Way of the Celestial Masters) (Cheng-tu shih: Sichuan University Press, 1996). The Yuqu zhi situated in Chengdu seems to have been the only exception.

7. The preface and its dating have been the object of an excellent article by the young Japanese scholar Maeda Shigeki, "Rōkun setsu ippyaku hachiju kaijo no seiritsu ni tsuite" (The formation of the preface of the 180 precepts spoken by Lord La o), *Tōyō shisō to shūkyū* 2 (1985): 417–424.

8. See the Dunhuang ms. S.4226 (*Taiping jingbu*, j. 2).

9. See *Xuanmen dayi*, app. *Yunji qiqian*, j. 6.

10. I leave the question of yet another *Taiping jing* called the *Dongji taiping jing* out of this discussion as it is not linked to Gan Ji but supposedly revealed by Laojun directly to Zhang Ling (*Xuanmen dayi*, app. *Yunji qiqian*, j. 6).

11. The female servants are called *qie* in both instances.

12. "Foodstuffs should not be thrown into the fire." Later, this rule was taken over by the three hundred *guanshenjie* as an interdiction of sacrificing animals to the gods and ancestors of the Six Heavens.

The Daoist Concept of Central Harmony in the *Scripture of Great Peace*: Human Responsibility for the Maladies of Nature

CHI-TIM LAI

Introduction

The *Scripture of Great Peace* (*Taiping jing*) is one of the most important Daoist texts of the Eastern Han dynasty (25–220 C.E.), the period in Chinese history when Daoism began to establish itself as a formal religious system. It advertises the dawning of an era of social and cosmic harmony, "Great Peace," and describes the conditions under which this era might be ushered in. The text thus contains a wealth of information about the ideological milieu from which Daoism emerged and has been the focus of considerable attention by leading scholars of Chinese religion. One of the most important textual questions that is still under debate is to what extent the present text (available in the modern reprint of the Ming dynasty Daoist canon) completely derives from the Han dynasty original.[1] This essay cannot do justice to the complexity of these scholarly debates, though I have treated this question at length elsewhere.[2] In brief, most Chinese scholars accept the view that the text that we have in our hands is indeed the very text of the Eastern Han.[3] At the very least, it is safe to say that the present text agrees in many respects with the original Han dynasty version.[4] While admitting the existence of different textual layers, as well as their different sources,[5] in this essay I shall focus on the concept of "Central

Harmony" (*zhonghe*), one of the basic themes in the text, which is of
particular relevance to understanding the "ecological" cosmology that
has pervaded Daoism since ancient times.

In the *Taiping jing*, Central Harmony is viewed as a way, a condition, and a factor that may potentially contribute to the formation of
harmonious relationships between Heaven, Earth and humanity. The
text is very much concerned with how to form the condition of Central
Harmony out of which will come the era of Great Peace (*taiping*) or
Great Harmony (*taihe*).

There is no doubt that the text's general worldview is imbued with
cosmological theories and utopian ideas that are characteristic of
many Han dynasty philosophical and religious texts. Han theories on
the origin of Heaven and Earth, Primordial Energies (*yuanqi*), yin and
yang, and the correspondence of Heaven (and Earth) and humans all
appear in the text. The concept of Central Harmony calls special attention to an important aspect of "reciprocal communication" (*xiangtong*)
which, according to some scholars, is intended to supplement the bipolar theory of yin-yang.[6] The *Taiping jing* thus formulates a theory of
three qi, or vital energies: the great yang (*taiyang*), the great yin
(*taiyin*), and Central Harmony. Equally important is the related idea
that humans belong to the domain of Central Harmony and are assigned the unique duty of protecting, circulating, and cultivating the
qi of Central Harmony (*zhonghe qi*) between Heaven and Earth. Thus,
the concept of Central Harmony indeed argues for the view that it is
the great responsibility of humankind to maintain harmonious communication with Heaven and Earth in order to bring about cosmic harmony and social peace. Accordingly, human evil deeds and the resulting "inherited guilt" (*chengfu*),[7] which might threaten to eradicate the
harmony of cosmos, society, and family, are considered the major cause
of the interruption of the qi of Central Harmony between Heaven and
Earth. When human evils deplete the qi of Central Harmony, Heaven
and Earth are said to become sick, resentful, and furious. Cataclysms,
wars, diseases, bad harvests, and other "strange and calamitous phenomena" (*zabian guaiyi*) occur. They are the means by which Heaven
and Earth express their discontent.

In what follows, I will try to present the major elements of the
text's interpretation of the concept of Central Harmony in relation to
the vision of the ideal communicative and harmonious relationship
between humans and Heaven and Earth. Although the discussion of

the expression of Central Harmony has frequently been the key issue in both Daoist and Confucian traditions, the subject under discussion in this essay demonstrates the unique conception of the relationship between humans and nature that Daoism held right from its beginnings as an institutional religion in the second century C.E. This essay will suggest that the core aspects of this concept of Central Harmony accord with the contemporary ecological concerns for promoting a positive relationship between humans and their natural environment. In contrast to the modern industrial drive for unlimited economic growth and resource development, the *Taiping jing* lays down an explicit and important ecological rule that calls for a harmonious relationship between cosmos, society, and human beings, as well as the human responsibility for giving birth to and nourishing life.[8]

The Natural Order and the Will of Heaven and Earth

As is the case with so many Han dynasty texts, such as the *Huainanzi* and the *Chunqiu fanlu*, the balance of Heaven and Earth (what we may call the order of nature) is of primary concern. The irregularity of natural phenomena is understood as a means by which to know and judge the human deeds that cause Heaven and Earth to issue forth calamities and disorders in the human realm.[9] In the *Taiping jing* the origin of Heaven and Earth is accounted for in terms of the two forces of yin and yang. Yin and yang are the two essential aspects of the Primordial Energies (*yuanqi*), out of which all beings in the universe are formed.[10] So the text states: "All things in Heaven and Earth are characterized as yin and yang, [the movement of yin and yang] enables them to give birth to and nourish one another";[11] and "Primordial qi embraces Heaven and Earth and the eight directions, which are born from primordial qi."[12] There is no doubt that the Chinese model of the order of nature is founded upon the image of organismic process in which parts of the entire universe interact and transform under the self-generating principles of qi and yin and yang.

Although the Chinese ideal model of nature "is an ordered harmony of wills without an ordainer," to use Joseph Needham's phrase,[13] the *Taiping jing*'s understanding of nature does not quite affirm the claim made by some scholars that the Daoist conception of nature simply implies impersonal, noninterfering and spontaneous attitudes.

For these scholars, Daoism is principally characterized by the values of "non-action" (*wuwei*) and "self-so" (*ziran*).[14] From this perspective, "For the Taoists, in order to be in consonance with the Tao in nature one must withdraw from active involvement in social and political affairs and learn how to preserve and nourish nature and human life."[15] While this generalization may be largely true of the earlier (and better-known) Daoist texts, such as the *Laozi* and the *Zhuangzi*, the *Taiping jing's* view of nature is embedded in Han dynasty theories of correspondence between humans and Heaven and is thoroughly anthropocosmic. As symbols of significations, Heaven and Earth give messages to the human world.[16] Although they are described as self-generating (*ziran*) and primordial,[17] they are also intensely personal and intervene in history out of pity for the predicament of humans. Natural phenomena provide human beings with "voices of Heaven and words of Earth" (*tiantan diyu*)—divine messages that convey heavenly agreement with or criticism of human conduct.

Heaven, then, is not simply a "spontaneously self-generating life process," but, more or less, a superhuman entity whose prime function is to direct the cyclical rhythms of nature. Heaven is intensely concerned, happy or unhappy, with human conduct and offers corresponding rewards or punishments.

Although Heaven and Earth are imagined as personal, the text follows Han dynasty thought by acknowledging that they cannot, like a person, communicate directly with humans through words.[18] Heaven and Earth have two indirect ways of conveying their will: implicit and destructive omens and catastrophes, and explicit and constructive sages whom they send to the world to instruct people.[19] In order to establish great peace in the world, the authors of the text identify themselves as Heavenly teachers and announce that humans should follow the will of Heaven and Earth and be of one heart with Heaven. Yet, to stay in accord with the will of Heaven is not simply a question of valuing one's relation with nature, engaging in noninterfering action, or achieving harmony with nature. The will of Heaven calls for the concrete mission of giving life to and nourishing all the ten thousand beings.[20] By equating this call with the fundamental law governing the universe, the text states, "Heaven grows out of yang and holds the principle of giving birth to life; Earth grows out of yin and holds the principle of nourishing life."[21] To justify this law of nature as Daoist, the text goes on to parallel the triad of Dao, *de*, and *ren* (the Confucian virtue of humaneness) with *sheng* (giving birth), *yang* (nurturing), and *cheng* (completing):

> That which gives life is Dao. That which nourishes is *de*. That which completes is *ren*. When not a single thing is born the one Dao is blocked and cannot penetrate. When not a single thing is nourished, not a *de* has been cultivated. When not a single *de* is completed, not a single *ren* is carried out. If you want to know if there is any Dao, *de* or *ren*, you can know just by observing things.[22]

> Dao is heaven. It is yang, and is in charge of giving birth. *De* is the earth. It is yin, and is in charge of nurturing. . . . When a Dao that flourishes is in charge of giving birth, all things come to life. When a *de* that flourishes is in charge of nourishing, all creatures and citizens are nurtured and there are no grudges.[23]

Hence, an individual's performance of circulating the Dao of Heaven, the *de* of Earth, and the *ren* directly refers to one's potential contribution to helping living creatures to give birth to and nourish each other.

The equation of the will of Heaven and Earth with the principle of giving birth to and nourishing life is also a key in understanding the text's definition of nature (*ziran*) in view of its relation to Great Peace (*taiping*). The *Taiping jing* uses the term nature in three ways. First, when combined with the term denoting the Primordial Energies (*yuanqi*), nature points to the will of Heaven and Earth.[24] As we have mentioned, the will of Heaven and Earth is to give birth to one another. Hence, when nature refers to the will of Heaven and Earth, undoubtedly it means the natural law of giving birth to and nourishing life.

Second, the text claims that the ten thousands things are "completed" (*cheng*) by nature.[25] In such a state they are in the unique place that suits them.[26] More importantly, we are told in another regard that when each of the ten thousand beings is in the place that suits them, this denotes the era of Great Peace in which the ten thousand beings suffer no harm. Put in a positive sense, in the era of Great Peace the myriad things are able to follow their inner capacity to grow,[27] that is, they attain their nature. Nature then conforms to the order of Great Peace in which all beings can find a place that suits them.

The third way in which the text uses the term nature is to denote a means of communication between Heaven and humans. Despite the fact that Heaven and Earth do not directly speak to humans, their law must harmonize with the requirements of nature;[28] therefore, the will of Heaven is equally to demand that humans be in harmony with nature. Hence, the principal concern of the authors is to learn how to be in consonance with Heaven and Earth by complying with the law of nature.

Nature actually suggests itself to be not only the will of Heaven, but also an attitude of heavenly judgment upon humans. Only the presence of nature in all modalities of being determines whether humans remain in accord with the will of Heaven and Earth. In this sense, the call for harmony with nature certainly contrasts with other accounts of the Daoist conception of nature. Relying solely upon the idea of non-assertive action (*wuwei*), they suggest that the Daoist view is that "one must withdraw from active involvement in social and political affairs" and "learn how to be non-egocentric."[29] In contrast, the *Taiping jing* insists that humans have to take responsibility for giving birth to and nourishing life and positively helping all creatures not to be harmed.

The Meaning of Central Harmony

In order to establish the mutuality of Heaven, Earth, and humanity, the authors of the text do not simply confine themselves to the popular Han dynasty theories of yin and yang, the correspondence between cosmic and human realms, and the cosmogony of primordial qi. More than that, it is possible to suggest that the importance of the text lies in the uniqueness of its interpretation of Central Harmony (*zhonghe*).[30] Relevant to this concept of Central Harmony are the ideas of the triadic structure of Primordial Energies—the three qi of the great yang, the great yin, and Central Harmony; the union of the three qi (*sanhe*); and, most importantly, the harmonious relationship of "reciprocal communication" (*xiangtong*). Indeed, one of the unifying theses presented by the text is the assertion that to establish the era of Great Peace or Great Harmony, it is necessary to have the essential component (the qi) of Central Harmony as the proper means to preserve and circulate the harmonious communication between Heaven, Earth, and human beings.

As mentioned earlier, the idea of qi greatly shapes the Chinese model of the universe, which is frequently characterized by the metaphor of organismic process.[31] Chapter forty-two of the *Daode jing* serves as a good example of the cosmogonic function of qi with which the theory of yin qi and yang qi is linked:

> Dao gave birth to the One. The One gave birth to the Two. The Two gave birth to the Three. And the Three gave birth to the ten thousand things. The ten thousand things carry yin on their backs and wrap their arms around yang. Through the blending of qi they arrive at a state of harmony.[32]

Here, the harmony of the myriad things is defined by the fundamental alternation of yin qi and yang qi when the two complementary opposites are in dynamic balance. A similar account of the fruition of harmony in the order of nature brought by the balance yin qi and yang qi can be seen in the *Huainanzi*:

> Thus it is said, "*Dao* begins with one." One (alone), however, does not give birth. Therefore it divided into yin and yang. From the harmonious union of yin and yang, the myriad things were produced.[33]

The viewpoint that the union of yin qi and yang qi will give rise to the birth (*sheng*) and harmony (*he*) of myriad things is certainly seen in the *Taiping jing*. For example, it states: "All things in Heaven and Earth belong to yin and yang. [Because of their mutual interaction] all things are able to give birth and nourish one another."[34]

However, in addition to the presupposition of the language of yin qi and yang qi, Harada Jiro is right to point out that the authors of the text actually employ another theory of "the union of three qi" as the dominant interpretive framework for discourse on Heaven and Earth.[35] The three qi refer to the great yin, the great yang, and the *zhonghe,* or Central Harmony. By definition, they originate from the primordial energies, or simply are the different names of the One of primordial energies.[36] Not only that, but they are also seen as the archetype for the triadic structure of primordial energies manifested in many dimensions of nature. For example, Heaven originates from the yang qi, Earth from the yin qi, and human beings from the qi of Central Harmony. Other beings modeled on the triad of great yin, great yang, and Central Harmony qi are grouped in similar ways: Sun, Moon, and stars; mountains, rivers, and plains; father, mother, and child; sovereign, minister, and people.[37] In sum, the phenomenal world is represented and grouped according to the tripartite model of the three qi, which originate from the undifferentiated One of primordial energies.

Established upon the tripartite model of primordial energies, which cannot be reduced to the bipolar model of yin-yang, the *Taiping jing* squarely emphasizes the qi of Central Harmony and its function as the unique and vital medium for the goal of achieving the harmonious order of nature. The essential role that the qi of Central Harmony plays is to preserve and circulate harmonious "communication" (*tong*) between the yin qi and yang qi. So, the authors tell us that "As for yin and yang, the essential is in the Central Harmony. When the qi of Central Harmony gains, the ten thousand beings will flourish; people will live harmoniously and the rule of the sovereign will bring peace."[38] In these words, the qi of Central Harmony is given an importance equal to that of the yin qi and yang qi in forming one great harmony.[39] By leading the yin qi and yang qi into united harmony (*hehe*) and by establishing reciprocal communication between them, the qi of Central Harmony uniquely brings about the dynamic balance and spontaneous growth of nature, referred to in the text as Harmony (*he*) or Great Harmony (*taihe*).

Many Han dynasty texts show that it was indeed one of the basic ideas of the time to assert the harmonious union of Heaven and Earth (*tiandi hehe*) as a necessary condition for the spontaneous growth and harmony of nature. So we read in the *Lu shi chunqiu*: "For Heaven and Earth to unite harmoniously is the great principle of life."[40] The *Huainanzi* says: "As for the qi of Heaven and Earth, nothing is greater than its harmony (*he*). As for harmony, by it yin and yang are regulated, day and night distinguished, [and so] it produces beings."[41] Thus, *he* basically refers to two things: a state of the harmonious union of Heaven and Earth; and the outcome of the spontaneous growth of myriad things in the world.

Understandably, since the *Taiping jing* inherited this Han dynasty worldview, harmony (*he*) is also seen as the state defined by the spontaneous growth of myriad things. The *Taiping jing* states: "[If] there is only connection between Heaven and Earth but without harmony, beings cannot accommodate each other and nourish themselves."[42] While *he* is seen as the natural state of all beings in which each can spontaneously grow according to its own nature,[43] the concept of *he* in this text is linked to the idea of Great Peace (*taiping*). As mentioned earlier, one of the basic characteristics of Great Peace is that all beings are not harmed in their natural processes of growth. In this sense,

peace (*ping*) and harmony (*he*) more or less denote similar cosmological and social utopias, the concepts of which the authors of the
text shared with many other Han dynasty thinkers.

Despite a certain similarity between this and many other Han dynasty writings, the *Taiping jing*'s concept of Central Harmony is unique.
First, the concept of the Central Harmony qi is described as an essential link that leads to the supreme vision of mutual communication
between Heaven, Earth, and human beings: it is the lack of mutual
communication that causes cosmic disorder and the interruption of
proper functioning of the qi of heaven and the qi of earth. Conversely,
if the qi of Central Harmony is present, if there is mutual love and
communication between the three elements in the various triadic
structures of the cosmos, then the era of Great Peace will come. As
Max Kaltenmark has already pointed out, it is a basic idea of the text
that only when the mixture of yin qi and yang qi is united with the qi
of Central Harmony, which forms the conduit for mutual communication with what is above and below, will harmony and spontaneous
growth prevail, free of evil.[44]

> As for the laws of qi, it circulates under Heaven and above Earth; if yin
> and yang obtain each other, mingling and attaining harmony, they unite
> as three with the qi of Central Harmony, together nurturing all things.
> If the three qi cherish each other and mutually communicate, there will
> no longer be any harm.[45]

Hence, the concept of the Central Harmony qi points to the interconnected relationship between the vision of harmony (*he*) and the idea
of communication (*tong*). Central Harmony is a *necessary* condition
for preserving harmonious communication between Heaven and Earth
in order that the myriad things will spontaneously grow according to
their own nature.

Second, the tripartite model of Primordial Energies and the linked
vision of *he* and *tong* is known as an ideal model for the other various
tripartite structures. According to the authors of the text, Heaven, father, sovereign are attributed to yang; Earth, mother, minister are yin;
human beings, child, people are Central Harmony. This way of grouping various triads, therefore, suggests that each group must emulate
the Primordial Energies so that all three components harmoniously
communicate with each other in order to bring about Harmony and
Great Peace.

Let us consider the paradigm of Heaven, Earth, and human beings (*tian-di-ren*). Humans are said to stand in the central position of the cosmic order and therefore are uniquely designated to conduct the qi of Central Harmony: "What is pure adheres to Heaven, what is turbid adheres to Earth, and the Central Harmony adheres to humanity."[46] The central position of human beings has to be in relation to Heaven and Earth, the whole universe. Humans never stand isolated from nature.[47]

Furthermore, by designating humans as conductors of the qi of Central Harmony, the text asserts that it is our human mission to preserve, protect, and circulate harmonious communication between the realms of cosmos and humanity. Here again, we see humans as the medium for the cultivation of harmony, through which Heaven and Earth lavish their blessings of Great Peace or Great Harmony upon the world. It is humanity's great responsibility to put the qi of Heaven and Earth into harmony and thereby permit the spontaneous growth of all beings.

When circulation of the qi of Heaven and the qi of Earth is interrupted and they cannot unite with the qi of Central Harmony, humans are blamed for their inability to bring about harmonious communication between Heaven and Earth. In a sense, they do evil to Heaven and Earth. Consequently, the "triunity" (*sanhe*), or "triunion" (*sanhe*), of Heaven, Earth, and humanity is disrupted; the heavenly Dao of life and the nourishing *de* of Earth are interrupted, and the wrath of Heaven and Earth are the result. Subsequently, major disasters are bound to occur. All sorts of "evil forces" (*xieqi*) arise, and the qi of Great Peace, or the qi of Great Harmony, disappears.

In the chapter "Instruction for Construction Work" (*qitu chushu jue*), it is said that humans offend the will of Heaven and Earth by wantonly gouging the earth and carrying out massive construction work.[48] The chapter accuses people of digging the earth in such arbitrary ways that "those who dig deep reach down to the yellow springs; those who dig shallow reach down several *zhang*."[49] Here, the guilt of humans lies in their arbitrarily harming Mother Earth. When her body is exploited and gouged, the qi of Earth leaks and, as a result, she ails and grieves. In response to her misfortune, suffering, and resentment, Heaven becomes furious and issues forth strange and calamitous phenomena.

Viewed from this perspective, humans resist the call that Heaven,

Earth, and Central Harmony should be a family, love one another, and mutually communicate in order to give birth to and nourish the myriad things. Humans' excessive gouging of the earth makes them responsible for interrupting the circulation of the qi of Central Harmony, and thus renders them enemies of the qi of Heaven and Earth.

The Maladies of Heaven and Earth and the Doctrine of Inherited Guilt

The *Taiping jing* transposes cosmological language into ethical reasoning.[50] If Heaven and Earth are guided only by the law of nature, with the prime concern for giving birth to and nourishing one another, human beings must accordingly bear the responsibility for expanding this vision of life in the world by facilitating united harmony (*hehe*) and reciprocal communication (*xiangtong*). In order to be in accord with the will of Heaven and Earth, there is no way that humans can avoid or be free of this responsibility. On the contrary, when there is a lack of harmony and communication between the qi of Heaven, qi of Earth, and qi of Central Harmony, Great Harmony or Great Peace disappears, the maladies and distress of Heaven and Earth appear, and the "inherited guilt" (*chengfu*) of human generations piles up and brings about calamity and hardship.

The doctrine of inherited guilt is at its explanatory best in dealing with the impact of the interruption of harmony and communication between nature and humans and of impending cataclysms for which only human inherited guilt is to be blamed.[51] First, addressing the idea that wars, diseases, disasters, bad harvests, and cataclysms occur because of the conjunction of cosmic cycles,[52] the text explains that these events are, indeed, messages from Heaven and Earth sent to human beings as warnings and as expressions of the maladies, anger, and pain experienced by Heaven and Earth.

> People are given life by Heaven and are reared by Earth. Heaven and Earth grieve over their ailments, and thus produced strange phenomena to punish their children.

> The calamities and strange phenomena under Heaven are of myriad varieties. They are all the reformative words of Heaven and Earth and the yin and yang.[53]

Second, on the causes that bring about the maladies and distress of Heaven and Earth, the text asserts that these are due to a human factor. People constantly offend the will of Heaven with their misdeeds. The gouging of Mother Earth, female infanticide, excessive punishments, and the neglect of the needs of the poor by the rich are considered enemies of the three qi of Heaven, Earth, and Central Harmony and as interrupting the union of the three qi.

Finally, communal guilt accumulates due to the chain of continuous transmission of the misery, maladies, and wrath of Heaven and Earth caused by human offenses against the will of Heaven and Earth. Starting from the original actions of the ancestors (*xianren*), inherited guilt is not only passed on from one generation to the next but grows epidemically, thus threatening to eradicate the human world. Once let loose, this epidemic is contracted by an ever larger group of people (*gengxiang chengfu*).[54] The doctrine of inherited guilt is therefore used to illustrate the spread of evil deeds that harm the harmonious order of Heaven and Earth. As a consequence of receiving and transmitting heavenly punishment in the form of calamities, both current and later generations suffer.[55]

In the *Taiping jing*, the doctrine of inherited guilt is presented as follows:

> *Cheng* refers to "before" and *fu* to "after." *Cheng* means that the ancestors originally acted in accordance with the will of heaven, and then slowly lost it; after a long time had elapsed, [their mistakes] had amassed and those of today living afterwards, then through no guilt of their own succeed to [the formers'] mistakes and culpability and so continuously suffer from the catastrophes engendered by them. Therefore that which is before is *cheng* and that which is afterwards is *fu*. *Fu* means that the various catastrophes do not go back to the government of the one man, but to a successive lack of balance. Those who live before put a load on the back of those who come later. This is why it is called *fu*. *Fu* means that the ancestor puts a load on the descendent.[56]

The *Taiping jing*'s doctrine of inherited guilt clearly shows that moral principles are not absent from the Chinese understanding of nature. It is understandable how this Daoist text intends to entertain the thought that once people learn the truth of inherited guilt, they will take it to heart, and so the anger and maladies of nature will disappear on their own. Otherwise, the guilt interrupting the heavenly Dao of life and the nourishing *de* of Earth can be inherited over the generations. Human evil deeds only increase the load of misery transmitted to later generations. Hence, not only is one bearing the responsibility of protecting and circulating the qi of Central Harmony in the universe, but the reverse is also true. Acts of good will harmonize the myriad beings and, in some way, cancel the burden of compounded evil and the misery resulting from inherited guilt.

Conclusion

The concept of Central Harmony has frequently been a key issue in both Daoist and Confucian traditions.[57] For instance, the idea of *zhonghe* in the the *Doctrine of the Mean* (*Zhongyong*) is used in Neo-Confucianism as a paradigm for interpreting "an experiential confirmation of the inner self."[58] The first chapter of the *Zhongyong* states:

> Before the feelings of pleasure, anger, sorrow, and joy are aroused it is called equilibrium (*zhong*, centrality, mean). When these feelings are aroused and each and all attain due measure and degree, it is called harmony. Equilibrium is the great foundation of the world, and harmony its universal path. When equilibrium and harmony are realized to the highest degree, heaven and earth will attain their proper order and all things will flourish.[59]

According to Tu Weiming's interpretation of *zhonghe* in his work on the *Zhongyong*, "Centrality" is a "state of mind wherein one is absolutely unperturbed by outside force." Explaining *zhong* as such a state of mind, Tu argues that when the centrality inherent in each human being can become united with Heaven and Earth, harmony (*he*) is its outcome, "an actual human achievement."[60]

In contrast to the Neo-Confucian "psychological" way of understanding, the *Taiping jing*'s concept of Central Harmony deals with how to harmonize the qi of Heaven and Earth. Instead of describing *zhong* as an ideal state of mind, it emphasizes the outcome of the union of Heaven and Earth and yin and yang that leads to the continuity of spontaneous growth of all myriad things in the world. In this understanding of Central Harmony, the center of gravity is not a discourse on human selfhood. Rather, it is conceived as a way, a factor, a method, and a vision for the harmonious union and reciprocal communication between Heaven, Earth, and humanity. In order to achieve harmony of nature, the text asserts that "it is necessary that the three qi love one another and mutually communicate in order that there may be no more evil."[61]

"Central" here refers to a central, harmonizing position between Heaven and Earth and the qi of yin and yang. Equally important, humans are in the domain of the qi of Central Harmony. Each human being has the responsibility for protecting, preserving, and conducting the qi of Central Harmony in the universe. A definite criterion in the text for human fulfillment of this communicative task is whether or not all myriad things are able to give birth to and nourish one another. If each thing is not hurt by outside forces, it is said to be in the state of naturalness (*ziran*), that is, in the place that is suitable for it.

Nevertheless, the authors of the *Taiping jing* understand that harmony is certainly not an intrinsic characteristic of Heaven and Earth, but perhaps occurred in the long-lost Golden Age. The increase of human errors only increases the misery, resentment, and discontent on Earth, disrupting the harmonious communication between Heaven, Earth, and humanity. Thus, the doctrine of inherited guilt explains why Heaven and Earth are sick, angry and furious, which then causes Heaven and Earth to issue calamities. Moreover, this doctrine expresses a genuine religious belief that people receive and transmit the mistakes of their ancestors, and that calamities of former and later times will combine to hurt them.

Finally, in view of the Daoist concept of Central Harmony in this text, we have argued that the interdependence of human being and nature is self-evident. An important component of the close relation between humans and nature is spiritual and ethical. Human beings are responsible for causing either deterioration or harmonization of the organismic processes of nature, Heaven, and Earth. The spread of human evil deeds that are hurtful to harmony and communication between humans and nature leads to the maladies, anger, and pain of Heaven and Earth. Conversely, we may also suppose that the merit achieved by actively caring for nature is also handed down from one generation to the next.[62]

Notes

1. The edition of the *Taiping jing* (hereafter TPJ) in the modern Daoist canon, the Ming *Daozang jing* of 1445, contains only 57 *juan* (scrolls). In the Han dynasty there is said to have existed a *Taiping* scripture, the *Taiping qingling shu*, in 170 *juan*. It is controversial and difficult to prove that the extant version in the Ming Daoist Canon represents the authentic remnants of the Han dynasty *Taiping* scripture. The *History of the Latter Han* says: "Prior to this [the memorial of Xiang Kai of 166 C.E.], during the reign of Emperor Shun (r. 126–144) Gong Chong of Langye went to the palace gates to submit the 'divine book' (*shenshu*) his teacher Yu Ji had obtained at the Spring of the Crooked Yang. The Book was in 170 *juan*, all of white silk with red lines, blue headings, and red titles, called *Taiping qingling shu*" (*Hou Hanshu* 30b [Beijing: Zhonghua shuju, 1962], 1084).

In 1961, Yoshioka Yoshitoyo's discovery of a manuscript from Dunhuang (S. 4226), which contains the complete table of contents of the *Taiping jing* in 170 *juan*, proves that at least the present text must be dated no later than the sixth century. See Yoshioka Yoshitoyo, "Tonkōben Taiheikyō ni tsuite" *Tōyōbunka Kenkyūjo Kiyō* 22 (1961): 1–103; Max Kaltenmark, "The Ideology of the *T'ai-ping ching*," in *Facets of Taoism*, eds. Holmes Welch and Anna Seidel (New Haven: Yale University Press, 1978), 19; B. J. Mansvelt, "The Date of the *Taiping Jing*," *T'oung Pao* 66 (1980): 151.

2. See Chi-tim Lai, "Shiping zhongguo xuezhe guanyu *Taiping jing* de yanjiu," *Zhongguo wenhua yanjiusuo xuebao*, n.s., 5 (1996): 297–318.

3. On the other hand, some Japanese and Western scholars do not support the view that there existed a 170-*juan*-long *TPJ* in the Eastern Han, although they do not come to the conclusion that the *TPJ* was a text originating from the sixth century. See Mansvelt, "The Date of the *Taiping Jing*," 149–82; Jens Ostergard Petersen, "The Early Traditions Relating to the Han Dynasty Transmission of the *Taiping Jing*," part 1, *Acta Orientalia* 50 (1989): 133–71.

4. Kaltenmark, "The Ideology of the *T'ai-ping ching*," 20, 45.

5. The obvious differences and, sometimes, contradictions among the ideas of the *TPJ* suggest that the received text would not be the work of a single author or a unified composition. As first established by Xiong Deji and recently refined by Takahashi Tadahiko and Barbara Hendrischke, three different textual layers are identified in the fifty-seven *juan* of the received text. The identification of three different textual layers infers that there were different sources of composition of the text. Cf. Xiong Deji, "*Taiping jing* de zuozhe he sixiang ji qi yu huangjin yu tianshidao de guanxi," *Lishi yanjiu* 2 (1962): 8–25; Takahashi Tadahiko, "Taiheikyō no shisō kōzō," *Tōyō bunka kenkyūjo kiyō* 95 (1984): 295–336; Barbara Hendrischke, "The Concept of Inherited Evil in the *Taiping jing*," *East Asian History* 2 (1991): 1–29.

6. Harada Jiro, "Taiheikō no seimeikan. nagaikishin ni tsuite," *Nihon Chōgokugaku kaihō* 36 (1984): 72.

7. See further explanation of the *TPJ*'s doctrine of *chengfu* in the last part of this essay.

8. For recent discussion of the ethical idea of giving birth to and nourishing life in other Daoist texts, see *Daofa ziran yu huanjing baohu*, ed. Zhang Jiyu (Beijing: Huaxia chubanshe, 1998), 26–34.

9. Tung Chung-shu's *Chun qiu fanlu* is a paradigm for such Han thought. See Jin Chunfeng, *Handai Sixiang shi* (Beijing: Zhongguo shehui kexue chubanshe, xiu ding ban, 1997), 142–211; Mikisaburo Mori, *Jōko yori kandai ni itaru seimeikan no tenkai* (Tokyo, 1971), 203–19.

10. For the discussion of the discourse of *yuanqi* in the *Taiping jing*, see Li Jiayan, "*Taiping jing* de yuanqi lun," *Zongjiao xuebao*, 1983, 11–16.

11. Wang Ming, *Taiping jing heijiao* (hereafter *TPJHJ*) (Beijing: Zhonghua shuchu, 1960), 221.

12. *TPJHJ*, 78.

13. Joseph Needham, *Science and Civilisation in China*, vol. 2 (Cambridge: Cambridge University Press, 1969), 287.

14. For example, see Mary Evelyn Tucker, "Ecological Themes in Taoism and Confucianism," in *Worldviews and Ecology*, ed. Mary Evelyn Tucker and John A. Grim (Maryknoll, N.Y.: Orbis Books, 1994), 150–60; Graham Parks, "Human/Nature in Nietzsche and Taoism," in *Nature in Asian Traditions of Thought*, ed. J. Baird Callicott and Roger T. Ames (Albany: State University of New York, 1989), 79–97.

15. Tucker, "Ecological Themes in Taoism and Confucianism," 152.

16. The term "anthropocosmic" is first used by Mircea Eliade to denote manifestations of cosmic realities as symbols of significations giving meanings to the human world. Tu Weiming uses this word to denote the Chinese model of nature in his study of the *Zhongyong* in his *Centrality and Commonality: An Essay on Confucian Religiousness* (Albany: State University of New York Press, 1989), 9, 78, 102, 107–9.

17. *TPJHJ*, 17.

18. Ibid., 651.

19. This classification is borrowed from Jens Ostergard Petersen, "The Anti-Messianism of the *Taiping jing*," *Studies in Central and East Asian Religions* 3 (1990): 9.

20. *TPJHJ*, 515.

21. Ibid., 221.

22. Ibid., 704.

23. Ibid., 218.

24. Ibid., 17.

25. Ibid., 701.

26. Ibid., 398.

27. Ibid., 399.

28. Ibid., 154.

29. Tucker, "Ecological Themes in Taoism and Confucianism," 154.

30. The doctrine of *chengfu* (inherited guilt) is another aspect showing the originality and uniqueness of the *TPC*.

31. Tu Wei-ming, "The Continuity of Being: Chinese Visions of Nature," in *Nature in Asian Traditions of Thought*, ed. Callicott and Ames, 67–78.

32. *Daode jing*, chapter 42.

33. *Huainanzi*, chapter 3, *Tianwenxu* (Beijing: Zhonghua, 1989), 112. Translation is of John S. Major, *Heaven and Earth in Early Han Thought* (Albany: State University of New York, 1993), 108–109.

34. *TPJHJ*, 221.

35. Harada Jiro, "Taiheikō no seimeikan. nagaikishin ni tsuite," 72.

36. *TPJHJ*, 19, 236.

37. Ibid., 236.

38. Ibid., 20.

39. Li Jiayan, "*Taiping jing* de sanhe xiangtong shuo," 28.

40. *Lu shi chunqiu*, chapter 13 (*youshi pian*).

41. *Huainanzi*, chapter 13, 432.

42. *TPJHJ*, 149.

43. Ibid., 203.

44. Kaltenmark, "The Ideology of the T'ai-ping ching," 26.

45. *TPJHJ*, 148.

46. Ibid., 205.

47. For instance, the idea of "Three Lineages" (*santong*) is representative of the text's view that the world of nature and the world of humans are intimately connected. The Three Lineages (*santong*)—Heaven, Earth, and human beings—depend upon each other for their existence and develop giving form to each other, just as a human being has a head, feet, and an abdomen. If one lineage is annihilated, all three lineages will be destroyed. *TPJHJ*, 373.

48. *TPJHJ*, 114–122.

49. Ibid., 114.

50. Barbara Hendrischke, "The Concept of Inherited Evil in the *Taiping jing*," 6.

51. For recent studies of the *TPJ*'s doctrine of *chengfu* inherited guilt, see Hendrischke, "The Concept of Inherited Evil in the *Taiping jing*"; Kamitsuka Yoshiko, "Taiheikyō no shōfu to taihei no riron ni tsuite," *Nagoya daikyōyōbu kiyō* A-32 (1988): 41–75; Chen Jing, "*Taiping jing* zhong de chengfu baoying sixing," *Zongjiaxue yanjiu* 2 (1986): 35–39; Chi-tim Lai, "The Early Celestial Masters' Rite of Repentance (*shou-kuo*) and the *Taiping jing*" (paper presented at the annual meeting of the Association of Asian Studies, Chicago, 1997).

52. *TPJHJ*, 370–1.

53. Ibid., 321.

54. Ibid., 96.

55. Petersen, "The Anti-Messianism of the *Taiping jing*,"14–15.

56. *TPJHJ*, 70.

57. For the discussion of theories of *zhonghe*, Central Harmony in the traditional history of Chinese philosophy, see Zhang Liwen, "Zhonghe lun," in *Zhongguo zhexue fanchou fazhan shi* (Beijing: Zhongguo renmin chuban she, 1995), 146–78.

58. Tu Wei-ming, *Centrality and Commonality*, 7.

59. Translation is from *A Source Book in Chinese Philosophy*, trans. and comp. Wing-tsit Chan (Princeton: Princeton University Press, 1963), 98.

60. Tu Wei-ming, *Centrality and Commonality*, 20.

61. *TPJHJ*, 148.

62. Although Barabara Hendrischke (The Concept of Inherited Evil in the *Taiping jing*," 15) and Jens Ostergard Petersen (The Anti-Messianism of the *Taiping jing*," 19) have pointed out that the *TPJ*'s term *chengfu* is only applicable to the explanation of transmission of evils and the result of misery, and not to the transmission of good deeds, it is reasonable to assert that the inheritance of merit is often put forward by Daoism. See, for example, Ge Hong's *Baopu zi neipian*, chapter 6 (*weizhi pian*).

"Mutual Stealing among the Three Powers" in the *Scripture of Unconscious Unification*

ZHANG JIYU and LI YUANGUO
Translated by Yam Kah Kean and Chee Boon Heng

Introduction

"Mutual stealing among the three powers" is the translation of the four-character Chinese phrase *san cai xiang dao* and sums up the interdependence of humans, Earth, and Heaven. It is thus a significant theoretical model in Daoism for understanding the status and function of human beings in relation to the natural world. It reveals a genuine depth to Daoism's triadic harmonization of the empirical world, the spiritual world, and human beings. The evolution of the concept of "mutual stealing among the three powers" from earlier Daoist concepts, such as "Dao follows natural spontaneity" (*Daode jing* 25) and "the equalizing of things" (*Zhuangzi* 2), indicates a corresponding evolution in the understanding of the connection between Heaven and human beings.[1]

The *Huangdi yinfu jing* (The Yellow Emperor's Scripture on "Unconscious Unification")[2] reflects this later stage in the evolution of Daoist thought and attempts to "expose heaven's mysteries and reveal divinity's workings." It became one of the most important classics of Daoism, second only in significance to the *Daode jing*. Zhang Boduan (987–1082), in his *Wuzhen pian* (An Essay on Realizing Perfection),[3] said: "The treasured *Yinfu jing* consists of more than three hundred words whereas the inspired *Daode jing* has five thousand characters. All those who attained immortality in the past and attain it in the present have comprehended the true meaning of these scriptures."[4]

There are different opinions about the origin of the *Yinfu jing*. Although traditionally ascribed to the Yellow Emperor (circa 2697 B.C.E.), we know that the text was only made known to the public during the early Tang dynasty. The text has a total of about three hundred Chinese characters,[5] and the style is simple and assertive. Discarding the convention of erring on the side of mildness and moderation, the *Yinfu jing* abounds with a direct and startling language of life and killing, crisis and opportunity, production and conquest.

Content

The theory of "mutual stealing among the three powers" refers to the symbiotic relationship that exists among Heaven and Earth, the myriad things, and human beings. To ensure a peaceful world, these three powers must remain in a state of coordinated harmony. The text first describes how "the myriad things steal from Heaven and Earth,"[6] meaning that they derive their vital energy (qi) from Heaven and Earth in order to survive. Li Quan (fl. mid-eighth century) explains in his *Huangdi yinfu jing shu* (A Commentary on the Yellow Emperor's *Yinfu jing*):[7]

> Heaven is the overall name for yin and yang. The light and clear yang qi floats upward to form heaven, and the heavy and turbid yin qi sinks downward to form earth. Heaven and earth are joined together and cannot be parted. The functioning of heaven and earth occurs through the Five Phases of yin and yang qi. Being generated by the Five Phases, all the myriad things are considered as their children. If human beings understand the way of heaven and earth, the operation of yin and yang qi, as well as the sequence of the Five Phases, they will be able to perceive the rise and decline of society, the life and death of themselves and all the myriad things.[8]

In the ancient Chinese worldview, human beings and the myriad things are formed from vital energy, or qi. Yang qi pertains to Heaven and yin qi pertains to Earth. All "things," or the myriad things, are formed from vital energy into a concrete form, the former understood as "heaven" and the latter as "earth." Thus, the *Zhuangzi* says, "I realize myself as a production of nature: Heaven and Earth have given me my body and appearance through my reception of the vital energies of yin and yang" (*Zhuangzi* 17). Since all the myriad things ultimately derive their existence from the interaction of Heaven and Earth, the *Yinfu jing* declares, "The myriad things steal from Heaven and Earth."[9] This is the first meaning of mutual stealing.

The word "stealing" (*dao*) implies "silence" and "mystery." Heaven and Earth work silently and the myriad things flourish mysteriously. Human beings are born and approach death without warning. Suddenly, time has run out, and everything seems to come to an end. This process resembles the work of a skilled thief who steals but does not make a show of it.

The word "mysterious" relates to moments of crisis and decision. The fact that it is difficult to observe the transformations of Heaven and read the signs of human affairs points toward a fundamental quality of "mystery" inherent in things. The *Yinfu jing* records: "Eat at the right time and the body is properly nourished. Act at the opportune moment and all transformations are appropriately settled."[10] Jian Changchen explains, "'Mystery' refers to the critical moment of gain and loss, change and transformation. Heaven and earth, the myriad things, and human beings are located between the poles of hindrance and success, rising and falling, increase and decrease, prosperity and decline." The philosopher Zhu Xi (1130–1200) also observes that "If human beings manage to take advantage of the opportunities given by heaven and earth, then their bodies will be in good shape. If human beings are able to master the mystery of the transformation [of things], then the world will be at peace. This is the principle of stealing."[11]

Human Morality

In this structure of Heaven-human relations, only human beings possess moral subjectivity. Thus, the moral intentions and value orientations of human beings have a decisive effect upon the consequences of this "mysterious stealing." Liu Haichan (fl. 1031) writes:

> The profound person knows the great Dao, incorporates all good, and thus pursues harmony and perfection. All his actions have good motives so that they may conform with the great Dao. He thinks swiftly, observes the Dao, and adopts the subtle [transformations] of Heaven and Earth, in order to nourish his nature. He steals the vital essence of

the myriad things to become immortal. Thus it is said "When the profound person possesses it, his body is safe and everything is accomplished."

On the other hand, when the base person possesses it, he persists in doing evil and cunningly seizes every opportunity to profit himself. He is dishonest in his duty, and aims to enrich himself with gold and silk without working. Sometimes he delights in sensuality and schemes to have a life of ostentatious luxury.

The result is that the higher he climbs, the heavier he falls; through his greed for money and sex he [only] harms himself. Although he may be rich and powerful, it will not last long, for he is not free from calamity. He will come to this end because he does not understand the true purpose and profound mystery of the great Dao. This is what is meant by "When the base person possesses it, he makes light of his life."[12]

Human morality is crucial in the attempt to transform nature and make use of the myriad things.

Wisdom and Folly

In a world of "mutual stealing among the three powers," it is up to human beings to take the initiative to engage nature and participate wholeheartedly in the transformation of the myriad things, rather than do nothing. The *Yinfu jing* says, "If one observes the Dao of Heaven and maintains its functioning, his duty is accomplished."[13]
    Li Quan understands these lines as follows:

No different from birds, animals, and plants, human beings come into existence through the working and transformations of yin and yang qi. Nonetheless, since human beings possess intelligence and wisdom, they may be considered as the master of the myriad things. They are able to examine their own nature, and also to plumb the depths of the origins of things and events. Thus they can steal and exploit the energies of the Five Phases and yin and yang.[14]

In other words, in addition to knowing the laws of nature, human beings are self-conscious subjects who manage to realize their own nature by making use of the myriad things to serve their purposes. However, the process of stealing from and exploiting Heaven, Earth, and the myriad things is both the means by which humans come to recognize nature and the history of how human beings have acted successfully in accordance with natural law.

It is certain, however, that the inner mysteries of the world are too profound to be conceptualized. As Laozi puts it, "There is nobody under heaven who is able to know [the Dao]" (*Daode jing* 70). Faced with this incomprehensibility, the foolish person is simply perplexed by the transformations of nature. He reveres Heaven, Earth, and the myriad things as divine entities, without inquiring into their hidden nature. Hence, the *Yinfu jing* states, "A fool only marvels at the superiority of astronomy and geography, but by depending on the opportune moment I become intelligent."[15] Astronomy refers to the patterns of celestial transformation evident in the seasons and the astrological calendar, while geography refers to the patterns of transformation in the earthly landscape. Human beings tend to believe that astronomical and geographic phenomena are abstruse and unpredictable; therefore, they worship them as prodigious elemental powers. Liu Haichan quotes Zhang Zifang's explanation in his *Collected Commentaries*:[16]

> When a foolish person sees a bright star, a clear cloud, lucid dew, a kylin, a phoenix, or feels a mild breeze and gazes at fine crops, he treats them as divine entities and feels happy. When he sees a solar or lunar eclipse, a comet, a violent storm, or an earthquake, he regards them as though the great sovereign has lost his temper. Then he becomes terrified.

> When a sage perceives some special astronomical features, he acts according to circumstances by abiding in the great Dao. Thus, he will not be harmed even though there are natural calamites.[17]

In other words, the fool treats changes of nature as manifestations of a hidden divinity, whereas the wise person considers the changes not as prodigious events but as part of the normal transformation of things. The difference between folly and wisdom lies in whether a person is able to understand the principle of nature and the function of Heaven and Earth. By grasping an opportunity and acting at the correct time, a wise person will achieve some success in bringing prosperity to the country, and the people will enjoy a peaceful life. However, the wise person always acts silently and stays away from the limelight. Thus, everyone reveres him as a sage.

The Sage

When a sage fully employs his initiative in the process of "stealing the myriad things," he must also respect the law of the phenomenal world by following the principle of the natural way of heaven. This is because the natural, moral, and social dimensions of human existence are being constantly determined in relation to the transformations of the natural environment. The *Yinfu jing* says:

> The natural Dao is still and tranquil, and so it is that Heaven, Earth, and the myriad things are generated. The way of Heaven and Earth proceeds without being noticed, and thus yin and yang overcome each other by turns; the transformations proceed in an orderly way. Therefore, the sage knows that the natural Dao cannot be opposed, and acts according to circumstances in order to utilize it.[18]

This means that at the origin of the universe, the Dao is silent and immutable. However, things that are produced by the Dao are placed in a state of gradual change, going through the process of production and loss in the form of yin and yang slowly superseding each other. Everything exists and functions according to this ecological system.

The evolution and supersession of yin and yang is the natural way of the Dao. The sage acts according to the Dao, obeys the transformations of yin and yang, and bases his laws and regulations upon these. This is what "act according to circumstances in order to utilize it" means. Jian Changchen says:

> The natural Dao has its own rule. The sage knows it cannot be violated and thus utilizes it accordingly. So it is said, "Human beings follow the way of Earth; Earth follows the way of Heaven; Heaven follows the way of the Dao; and the Dao follows its own spontaneity."[19]

Thus, the operation of the Dao places no restrictions on Heaven, Earth, and the myriad things as regards their mutual stealing or mutual utilization. Daoism, as represented in the *Yinfu jing*, sees nature and the myriad things as human beings' best teacher, and as an unlimited vehicle for human enlightenment and social development.

The *Daode jing* declares that "the highest good is like water. Water benefits the myriad things but never strives against them. It rests at the lowest place which everyone dislikes. Hence it is similar to the Dao" (*Daode jing* 8). The *Guanyinzi*[20] says:

There are hundreds of millions of things in this world, but Heaven and Earth never take forcible possession of them. As such, when a sage rules a country, he has to be like the sea: he is willing to stay downstream and endure all manner of filth; thus, all under Heaven support him.[21]

These words adduce the way of Heaven in order to elaborate the humility that ought to be the cherished possession of all human beings. In another passage the *Guanyinzi* says:

The myriad things under Heaven have adapted to their own ways without any blockage or obstruction. Therefore the sage follows the great Dao, observes [the principle] of non-action, responds to changes through his virtue, and everything becomes useful.

Human beings, moreover, may learn from the myriad things as well as from Heaven and Earth. As the *Guanyinzi* has it:

The sage learned from bees how to establish the relationships between ruler and subject. He learned from spiders how to weave nets. He invented ceremony and ritual after seeing an erect mouse cupping its forelimbs. He mastered the tactics of war by observing fighting ants.

Thus, the sage is the master of learning from the myriad things, educating virtuous persons, and helping ordinary people. Only the sage manages to abide with nature, become one with the myriad things, and make things convenient for people as well as benefit the whole world without prejudice.

The Four Obediences

The core meaning of the principle of "mysterious stealing" that the *Yinfu jing* advocates is that human beings must master the key to the transformations of things and events. Its purpose is to ensure the prosperity of the myriad things by ordering the harmonious, symbiotic coexistence of human beings and nature. The principle of "the four obediences" in the *Guanyinzi* helps to explain this doctrine.

Heaven could not make a lotus flower in winter, nor cause chrysanthemums to blossom in spring. Thus the sage does not disobey the cycle of the four seasons or go against the flow of time. This is called "obeying time."

Earth could not make citrus grow in Henan, nor raise foxes and raccoons south of the Changjiang River. Thus the sage does not disobey the customs of a place or go against the transformations of energies. This is called "obeying custom."

A sage could not make people walk on their arms or grasp things with their feet, and everything serves its own purpose. This is called "obeying one's strength."

A sage could not make fish fly or birds dive. This is called "obeying the strength of [other] things."

As a result, Heaven, Earth, and the myriad things are able to change, stay still, conceal, and manifest themselves. In terms of acting according to circumstances, they aim to utilize and nourish the myriad things, without being scrupulous. This is the way of the Dao.

If human beings managed to observe these four basic principles, without disobeying nature's transformations or adhering rigidly to worldly rules, then they would be able to deal with any unforeseen circumstance, and thus attain the profound Dao.

Conclusion

Clearly, the *Yinfu jing* adopts its theoretical structure from the traditional Daoist ideal of "the unification of Heaven and human beings." It differs from the *Daode jing* and the *Zhuangzi* in that it distinctly emphasizes human initiative in this relationship. This resulted in a tremendous change in the traditional Daoist idea of "the unification of Heaven and human beings." By eliminating the intrinsic passivity of the original idea, it raised the status of human beings in engaging and appropriating their natural environment. We can make the following conclusions as to the characteristics of this way of Daoist thinking:

1. The concept of the unification of Heaven and human beings differs fundamentally from concepts of subjectivity and objectivity in Western philosophy. From Protagoras's "man is the measure of all things" to Kant's transcendental idealism, Western philosophy has di-

vorced objective nature from subjective experience. It has thus stressed the ability of human beings to conquer and manipulate nature. By contrast, the *Yinfu jing*'s "mysterious stealing" insists upon the harmony of nature and human beings as its major premise and then makes possible the initiative of human beings to take advantage of the principles of transformation among the myriad things. This means that human beings have to make rational decisions when it comes to matters concerning Heaven, Earth, and the myriad things. Although Heaven and human beings are situated in contradictory positions with respect to the concept of "mutual stealing," this opposition is predicated on an inviolable essential unity. "Mutual stealing" connotes a system of compensation between Heaven and human beings. This system makes both members come together, the "unconscious unification." Therefore, there is no trace of absolute separation or contradiction between Heaven and human beings in the *Yinfu jing*. In this respect, it differs from the traditional Daoist idea of "the unification of Heaven and human beings" as well as from the Confucian idea of "the interaction of Heaven and human beings."

2. The practices of obeying the way of Heaven and transforming nature become organically integrated in the *Yinfu jing*. Obeying the way of Heaven is the premise, but the arising and diminishing of the Five Phases are the basic content of the way of Heaven. Human beings may follow their own inclinations in utilizing nature, provided that they have understood how these Five Phases work. As such, obeying the way of Heaven itself implies the active modification of nature. One must, in turn, adhere to the way of Heaven while transforming nature. This theory is not restricted to a particular set of circumstances, but rather, generally applies to the relationship between nature and human beings. This thinking in the *Yinfu jing* suggests that human beings should adopt an active stance toward their natural environment, while placing human transformation and the utilization of nature within a more acceptable theory. Thus, the *Yinfu jing* has frequently been quoted as the intellectual foundation for the transformation of nature and the control of human destiny. For Daoists, the *Yinfu jing* and the *Daode jing* reveal the path to immortality. As Zhang Boduan mentioned,

> Mutual stealing among the three powers makes everything appropriate for its time. This is an opportunity to attain immortality and morality.

> Since the myriad things have come to rest and are freed from worries, their bodies would be properly nourished and would accomplish the state of non-action.[22]

Wang Chongyang, the founder of Quanzhen Daoism,[23] also said, "[To attain immortality] one must fully understand the three hundred characters of the *Yinfu jing* and read up on the five thousand words of the *Daode jing*."[24]

3. The idea of "mutual stealing among the three powers" emphasizes the importance of human beings' morality and self-discipline. The saying "When the profound person possesses it, his body is safe and everything is done. When the base person possesses it, he makes light of life"[25] clearly points out that different moral orientations can result in entirely different consequences while pursuing the same process. Liu Bowen of the Ming dynasty (1368–1644) says in his *Yulizi*:[26]

> Only a sage knows the real meaning of stealing. Stealing does not mean "robbing treasure," but "executing one's authority and using one's abilities." Plant in the spring and harvest in the autumn, build houses at high places and dig ponds at low grounds, make junks sail in the water and control the sail according to wind.

> An ordinary person does not understand the real meaning of stealing, and so he "expends the opportunities and misuses his energies." He even goes to the extent of exploiting natural resources without limit, until they are exhausted.

Does ancient Daoist wisdom provide any insight into the role and status of human beings in nature? We leave this question to the sages and philosophers of the future. Nevertheless, our feeling is that this crystallized wisdom can definitely play a fruitful part in the spiritual evolution of the human species.

Notes

Editors' note: We are extremely pleased to be able to publish a translation of this paper by leading Daoists from mainland China. The translation as originally presented to us did not come equipped with the scholarly apparatus that is conventional in Western academic publications. For this reason we asked Louis Komjathy at Boston University to locate and annotate as many quotations from Chinese texts as possible. He accomplished this difficult task with extraordinary success, and for all of his efforts we are deeply grateful. All the notes below are his work.

Texts in the Daoist canon (*Daozang*, hereafter abbreviated DZ) are given according to the number of the reduced sixty-volume edition published in Taipei and Kyoto. These numbers coincide with those found in Kristofer Schipper's *Concordance du Tao Tsang* (Paris: Publications de l'École Française d'Extrême-Orient, 1975). "Fasc." stands for "fascicle" and refers to the volume number of the 1925 Shanghai reprint of the original canon of 1445 (*Zhentong Daozang*).

1. The notion of the "mutual stealing among the three powers" does not occur in the *Yinfu jing* itself. Rather, Zhang Jiyu is drawing a connection between this phrase, which is quoted from the *Wuzhen pian* (see note 22 below) and parallel notions of "stealing" (*dao*) and the three powers (Heaven, Earth, and humanity) as they occur in the *Yinfu jing*. Note also that the literal rendering of the title is the *Scripture of the Hidden Talisman*.

2. The *Yinfu jing* appears in DZ 31, fasc. 27. For a translation, see volume two of James Legge's *The Sacred Books of China: The Texts of Taoism* (New York: Dover, 1962). For a translation with Liu Iming's commentary, see Thomas Cleary, *Vitality, Energy, Spirit* (Boston: Shambhala, 1991). For a translation and discussion of the *Yinfu jing* in terms of Chinese military thought, see Christopher C. Rand, "Li Ch'üan [Li Quan] and Chinese Military Thought," *Harvard Journal of Asiatic Studies* 39, no. 1 (1979): 107–37. For a discussion of authorship, possible dates of composition, and its place in the history of Daoism, see Florian C. Reiter, "The 'Scripture of the Hidden Contracts' (Yin-fu ching): A Short Survey on Facts and Findings," *Nachrichten der Gesellschaft für Natur- und Volkerkunde Ostasiens* 136 (1984): 75–83. For other references, see Fabrizio Pregadio, "Chinese Alchemy: An Annotated Bibliography of Works in Western Languages," *Monumenta Serica* 44 (1996): 439–76.

3. The *Wuzhen pian* appears in DZ 263, fasc. 122-31. For a translation with Liu Iming's commentary, see Thomas Cleary, *Understanding Reality* (Honolulu: University of Hawaii Press, 1987). For references related to Zhang Boduan and his other writings, see Pregadio, "Chinese Alchemy." For the place of Zhang Boduan and his writings in the context of the Daoist tradition of internal alchemy, see "Inner Alchemy (*neidan*)" by Fabrizio Pregadio and Lowell Skar in Livia Kohn's *Daoism Handbook* (Leiden: Brill Academic Publishers, forthcoming).

4. DZ 263: 128.17b (chapter 58). See Cleary, *Understanding Reality*, 117.

5. For an overview of some of the surviving editions of the *Yinfu jing* see Rand, "Li Ch'üan." The text actually varies in length: the shorter edition consists of 332 characters, while the longer edition has 445 characters. The editions are quite similar when the final/additional 113 characters are dropped. Rand argues that the shorter text is the older version.

6. DZ 31: 1.1b. The *Daozang* text reads: "Heaven and Earth steal from the ten thousand things. The ten thousand things steal from humanity. Humanity steals from the ten thousand things. When the three thieves are correctly ordered, Heaven, Earth, and humanity are at peace."

7. The *Huangdi yinfu jing shu* appears in DZ 110, fasc. 55.

8. DZ 110: 2.1a–3a.

9. Ibid. 31: 1.1b.

10. Ibid.

11. For information on Zhu Xi's commentary on the *Yinfu jing*, identified as the *Huangdi yinfu jing kaoyi*, see Reiter, "Scripture of the Hidden Contracts."

12. For a brief discussion of Liu Haichan, see Julian F. Pas, *Historical Dictionary of Taoism* (Lanham [Maryland]: Scarecrow Press, Inc., 1998). For his place in the internal alchemy tradition see Pregadio and Skar, "Inner Alchemy (*neidan*)." I am grateful to Livia Kohn for the latter reference.

13. DZ 31: 1.1a.

14. Ibid. 110: 1.1a–1b.

15. Ibid. 31: 1.2a. A more literal translation is the following: "Ignorant people consider the patterns and principles of Heaven and Earth sacred; I consider the patterns and principles of time and substance wisdom."

16. This is the *Huangdi yinfu jing jijie* (Collected Commentaries on the Yellow Emperor's Scripture of the Hidden Talisman). DZ 111, fasc. 55.

17. Ibid. 111: 3.8b–9a.

18. Ibid. 31: 1.2a.

19. I have been unable to locate the source for this quotation.

20. The *Guanyinzi* appears in the *Daozang* as the *Wenshi zhenjing* (The True Scripture on the Beginning of Writing). For a discussion of the *Guanyinzi*, see Livia Kohn, "Yin Xi: The Master at the Beginning of the Scripture," *Journal of Chinese Religions* 45 (1997): 83–139. I am indebted to Livia Kohn for providing this reference. I am uncertain if this is in fact the same text from which Zhang Jiyu and Li Yuangouo are quoting.

21. For this quotation and all others attributed to the *Guanyinzi*, I have been unable to locate the source.

22. DZ 263: 28.17a (chapter 57). See Cleary, *Understanding Reality*, 116. "Mutual stealing among the three powers" translates *sancai xiangdao*.

23. For Wang Chongyang (Wang Zhe) and Quanzhen (Complete Perfection) Daoism, as well as the place of the *Yinfu jing* in this lineage, see Tao-chung Yao, "Ch'üan-chen [Quanzhen]: A New Sect in North China during the Twelfth and Thirteenth Centuries" (Ph.D. diss., University of Arizona, 1980). Also see Yao's "Quanzhen—Complete Perfection" in Kohn, *Daoism Handbook*.

24. *Chongyang quanzhen ji* 3 13/7b. I am grateful to Tao-Chung Yao at the University of Hawai'i for this reference.

25. *Daode jing* 41.

26. I am unfamiliar with this text as well as the quotation which follows.

Ingesting the Marvelous:
The Practitioner's Relationship to Nature
According to Ge Hong

ROBERT FORD CAMPANY

Introduction

Ecology, in our appropriately angst-ridden and global sense of the word, was not a recognizable problem for early medieval Chinese authors. There are consequently no directly analogous responses to ecological crises that the historian of this period can point to as exemplary. More broadly speaking, I am somewhat wary of (though certainly open to considering) attempts to draw on texts and ideas from premodern, non-Western religious traditions as in some sense directly appropriable resources for the solution of the distinctly modern and unprecedented ecological crisis. One can imagine that the argument might even be mounted—though I myself do not now wish to do so— that such attempted appropriations themselves participate in a pattern of thought and action that is itself a contributor to the predicament we find ourselves in. It is my personal opinion that, short of a worldwide transformation in human nature resulting in people's wanting fewer consumable goods and conveniences, our best hope for a way out of the ecological crisis is to retrace the path that led us into it, relying on careful, thoughtful deployments of technology guided by political action. We cannot turn back the historical clock, nor will contemplating ancient worldviews in itself halt the degradation of our environment that is proceeding apace every hour of every day.

Something of the tension between nostalgia for a return to a pre-modern, pretechnological relationship with nature, on the one hand, and an urgent need to deploy technology in order to manage and master natural processes, on the other, is mirrored in alchemy. An examination of some implied understandings of humans' relationship with nature as found in the literary remains of a figure such as the fourth-century writer-practitioner Ge Hong (283–343 C.E.) will bear out this proposition. And, if nothing else, a review of some alchemical stances toward nature may serve to correct a pervasive twentieth-century tendency among scholars and the reading public to identify the Daoist tradition exclusively with the *Zhuangzi* and the work attributed to Laozi, the *Daode jing*, or else simply to ignore all currents of the Daoist tradition save these texts. Alchemy, no less a part of Daoist tradition, is founded on a stance toward nature and the natural that is sharply at odds with the ones favored by Zhuangzi and Laozi.

I begin with a story.

During the reign of Emperor Cheng of the Han, hunters in the Zhongnan Mountains[1] saw a person who wore no clothes, his body covered with black hair. Upon seeing this person, the hunters wanted to pursue and capture him, but the person leapt over gullies and valleys as if in flight, and so could not be overtaken. The hunters then stealthily observed where the person dwelled, surrounded and captured him, whereupon they determined that the person was a woman. Upon questioning, she said, "I was originally a woman of the Qin palace. When I heard that invaders from the east had arrived, that the King of Qin would go out and surrender, and that the palace buildings would be burned, I fled in fright into the mountains. Famished, I was on the verge of dying by starvation when an old man taught me to eat the needles and nuts of pines. At first they were bitter, but gradually I grew accustomed to them. They enabled me to feel neither hunger nor thirst; in winter I was not cold, in summer I was not hot." Calculation showed that the woman, having been a member of the Qin King Ziying's harem, must be more than two hundred years old in the present time of Emperor Cheng.

The hunters took the woman back in. They offered her grain to eat. When she first smelled the stink of the grain, she vomited, and only after several days could she tolerate it. After little more than two years of this [diet], her body hair fell out; she turned old and died. Had she not been caught by men, she would have become a transcendent.[2]

Ge Hong tells his poignant version of this tale[3] to underline the theme of the eleventh chapter of his book, *The Master Who Embraces Simplicity, Inner Chapters* (*Baopuzi neipian*): the efficacy of naturally found medicinal substances in enhancing and prolonging life. But the tale's outcome, here quite different from the earlier version that was probably Ge Hong's source, also draws a stark contrast between a natural or wild diet based on uncultivated flora, conducive to life, and the culturally dominant cultivation- and grain-based diet, here revealed to lead to death. The one diet is practiced in the mountains, without clothing, away from society; the other is practiced alike at the palace and in the hamlet, clothed, in one or another sort of captivity, that of the harem or that of a party of hunters who have taken a human quarry.

Unfortunately for those quick to draw a Lao-Zhuang-zian moral from this story, its neat nature/culture dichotomy—the Daoist siding with nature—shows only one of the many facets of Ge Hong's complex attitude toward nature and the natural.

The heart of the particular tradition or style of early practice represented by Ge Hong[4] could perhaps be summarized as "salvation by ingestion." Dietary guidelines were certainly integral to all of the Daoist traditions in the early medieval period, but in hindsight it appears that only in the lineage of teaching and practice represented in Ge Hong's writings was ingestion the *key* soteriological activity.[5]

It is when we ask *what* was ingested and *how* it was procured and prepared that the complexity of Ge Hong's stance toward nature begins to show itself.

Ingesting the Marvelous/Natural, Shunning the Ordinary/Cultural

Many of the vegetable and mineral substances which are to be gathered and prepared by the seeker of transcendence, according to Ge Hong, share two sets of traits: first, they are rare, hard to obtain, and located in barely accessible places; second, they harbor marvelous qualities—their effects of health and longevity, of course, but also their strange appearance. Ge Hong's protracted discussion of *zhi* in the eleventh book of *Inner Chapters* offers many cases in point. In his usage the very term *zhi* is redolent of the numinous; it clearly does not mean "mushroom" but is a generic word for protrusions or emanations from rocks, trees, herbs, fleshy animals, or fungi (including mushrooms).[6] These exudations all have striking shapes, and most are striking in the same way: they "resemble" or "are like" beings or objects of other classes; in other words, they are visually and morphologically anomalous, straddling taxonomic boundaries. "Stony cassia exudations" (*shigui zhi*), for example, whose habitat is the caves of noted mountains, "resemble cassia trees but are really stones"; one catty of their powder allows one to live one thousand years. "Fungi exudations" (*junzhi*) in their shape "may resemble palaces and chambers, carriages and horses, dragons and tigers, humans, or flying birds"; ingested, they enable one to ascend as a transcendent. And so on. These are not the sorts of growths found in domestic gardens. Perhaps the most telling example of inaccessibility is the "stony honey exudations" (*shimi zhi*)—from the description, a set of dripping stalactites—said to grow on the Lesser Chamber peak of Mount Song, the Central Marchmount. The outcropping on which these *zhi* are produced, we are told, juts out over a chasm so deep that a stone dropped from it can be heard tumbling down for half a day. Despite the inaccessibility of the site, there is an inscription in stone over the outcropping that reads, "Anyone able to ingest one *dou* of this stony honey exudation will live ten thousand years!" Ge Hong continues:

> Many practitioners of the Dao have given long thought to this place but feel that it is unapproachable. Perhaps only by affixing a bowl to the end of a bamboo or wooden pole, and stretching it across, might one catch and collect some [of the *zhi* as it drips down]; but no one has been able to do even this. But, judging by the inscription in the rock over the chamber, someone in an earlier age must have been able to procure some.[7]

And indeed, extant fragments of Ge Hong's hagiography, *Traditions of Divine Transcendents* (*Shenxian zhuan*), mention two otherwise obscure adepts, the masters Xian Men and Feihuang zi, who accomplished this feat.[8] The link between inaccessibility and efficacy is here nicely captured by the inscription, promising untold benefits, yet carved in a seemingly unattainable spot: the reach required to get the transcendence-inducing mineral nectar just exceeds the grasp of all but the most fortunate and best qualified aspirants.[9]

Another indication that Ge Hong's adept dines on marvels is the curious business of the "traveling canteen," "mobile kitchen," or "*cui-*

sines de voyage" (as *xingchu* has been variously rendered), a sumptu-
ous banquet of delicacies served up anywhere on command by spirits
to those who know how to summon it. The ability to summon the traveling canteen allowed for easy procurement of distant, rare foodstuffs.
In *Traditions of Divine Transcendents* we read that the adept Li Gen
"could sit down and cause the traveling canteen to arrive, and with it
could serve twenty guests. All the dishes were finely prepared, and all
of them contained strange and marvelous foods from the four directions, not things locally available."[10] Again, in Ge Hong's hagiography of Wang Yuan and Maid Ma we find this passage:

> When they were both seated, they called for the traveling canteen. The
> servings were piled up on gold platters and in jade cups without limit.
> There were rare delicacies, many of them made from flowers and
> fruits, and their fragrance permeated the air inside [their host Cai Jing's
> home] and out. When the meat was sliced and served, [in flavor] it
> resembled broiled *mo*,[11] and was announced as *kirin* meat.[12]

It is crucial to note that Ge Hong recommends such marvelous
foodstuffs not as mere supplements to, but as substitutes for, the staple
foods of mainstream culture. Foremost among the cultural foods Ge
Hong targets for replacement are the entire class of cultivated grains;
as in the story of the Qin palace woman above,[13] grains constitute a
synecdoche of the whole cultural diet. His writings are peppered with
methods for abstaining from grains and hagiographic stories of the
benefits (health, longevity, even transcendence) of abstention.[14] He
even claims that he personally was "able to abstain from cereals and
not to eat [ordinary food]."[15] His rationales for this avoidance (in
which, of course, he was hardly unique or original in the larger Daoist
tradition[16]) vary,[17] but what is clear is that many of the natural, uncultivated substances he recommends are explicitly credited with substi-
tuting for grains and with abetting appetite. Wild herbs to which he
attributes this function include asparagus root,[18] Solomon's seal,[19] and
atractylis.[20] But the more basic stuffs on which Ge Hong's adept may
feed as an alternative to grains are the pneumas (*qi*) and essences
(*jing*) of various spatio-temporal aspects of the cosmos: for example,
the fresh pneumas that circulate between midnight and noon,[21] or the
pneumas of planets,[22] or the essences of spirits of the sexagenary cycle
of time.[23] Given enough of this special sort of ingestion, the adept can
eventually practice "embryonic breathing" (*taixi*) and not require any

further outside source of nourishment at all, subsisting on his own purified breath and saliva.[24]

This pneumatic diet is not a human invention; it is something humans can learn to do in imitation of certain animals noted for their longevity and their ability to go long periods without eating. Stories circulated in the early medieval centuries of people lost in the wilds who observed tortoises or snakes stretching their necks eastward at dawn and dusk and swallowing their breath (thus suggesting *xingqi* or *daoyin*, or both), and Ge Hong quotes one of these from a now otherwise almost completely lost collection of anomaly accounts titled *Records of Marvels Heard* (*Yiwen ji*) by Chen Shi.[25] A man named Zhang Guangding, meeting with disorder, has to flee quickly; he must abandon his little daughter, whom he cannot carry and who is too small to keep up. He expects that she will die of starvation, but, not wishing her bones to lie exposed, he lowers her into an old grave mound, leaving her with a few months' rations and planning to return later to collect her bones for burial. Three years later, when he is able to return, he finds her alive in the tomb (though at first he takes her for a ghost). When asked how she managed, she explains that she noticed a creature in the corner of the tomb that stretched out its neck and swallowed (its?) breath. Imitating it daily, she gradually lost her hunger. The creature turns out to be a large tortoise. As we would by now expect, the story ends by noting the wilderness dweller's trauma upon cultural and dietary reentry: "When the girl left the tomb and ate grains, at first she had stomach pain and nausea; only after a long while did she acclimate herself."

But it was not solely to avoid eating ordinary foods and seek out marvelous ones that so many of the adepts Ge Hong held up for emulation withdrew from society and family to live in the mountains. It was also because the overall condition of society had deteriorated in latter times: from a primordially "pure and simple" (*chun pu*) state, the social world had fallen into one of increasing "cleverness and fakery" (*qiao wei*), in which the Dao and those who sought it were subject to doubt and slander.[26] (All of this is Ge Hong's version of a familiar Daoist narrative of devolutionary sociocosmogony.) Tellingly, however, it is by a culinary-based metaphor—the contrast between rotten, stinking meat and the absence of it—that Ge Hong characterizes the gulf between the social and natural worlds: "It is not that the Dao is located [exclusively] in mountains and forests, but one who would seek the Dao must enter the mountains and forests, drawing near to their clarity and purity, in his earnest desire to distance himself from the rottenness."[27]

To this point, Ge Hong's stance toward nature and the natural would appear to be summarizable as a sort of culinary "back to nature movement," where "nature" stands for a source of purer, superior nourishment. But what has been passed in review so far is only a prelude to what follows. Further analysis will show how inaccurate it would be to leave matters where they here stand.

Processing the Natural: The Practitioner's Labors at Methodical Modification

> If you have a heart faithful to the Dao but engage in no pursuits to augment yourself, then, if your allotted length of years falls under a vacuous sign[28] and your body comes to be at risk of injury, the Three Corpses will summon in perverse pneumas and invite demons to do you harm in accordance with the perilous months and days, the places and times when you are vulnerable to a life-destroying illness.[29]

In this and similar passages, Ge Hong warns those disinclined toward acting to lengthen their lifespan that they are fools. He roundly rejects the attitude, based on what he considers a facile reading of the *Laozi* and *Zhuangzi* texts but bandied about in the fashionable "obscure learning" (*xuanxue*) salons by those who "turn away from the declarations of the scriptures to attend to the books of the philosophers," that we should simply "rejoice in Heaven and acknowledge our allotted span" and that we should accept "the theory that death and life should be looked upon with equanimity." He notes wryly: "Nowadays I observe that people who talk this way rush off for acupuncture and moxa when taken ill, and when confronted with danger are quite afraid to die"; and he points out that Zhuang Zhou, after all, counseled non-involvement in political affairs (for the sake of self-preservation) and was himself unable to regard death and life as the same.[30] The bleakness of death—"in the endless night deep beneath the Nine Springs, first becoming food for crickets and ants and finally merging one's body with dust and dirt"—should spur us to "abandon our non-urgent affairs and cultivate the mystic, wondrous enterprise."[31]

Transcendence is possible. But it must be earned; much labor is required, labor which none can perform on the adept's behalf.[32] When Ge Hong makes his interlocutor ask, "Did the transcendents of antiquity achieve their state by dint of study or by a special natural endowment of distinct kinds of pneumas?" the rebuke is swift: "What sort of question is that?"[33] No mere natural lottery of traits accounts for transcendence; it must be actively pursued—and it can be by anyone (here lies a seeming contradiction) destined to do so.[34] Hence the often-cited, bold declaration—but as much a call to self-responsibility as to self-empowerment—"My allotted lifespan resides with me, not with Heaven."[35] And hence Ge Hong's insistence that transcendents are not different in kind from other humans: "I have evidence to confirm that long life is attainable and that transcendents are not a species unto themselves."[36]

As we have seen, Ge Hong sends transcendence-seekers out to the mountains and wilds, where marvelous, natural substances may be found and where the practitioner may escape harmful social entanglements. But this turn from culture to nature is hardly the whole story. Ge Hong also insists that natural substances lack the efficacy of methodically synthesized elixirs, that adepts require the protection of written talismans to enter mountains safely, and that they must find another kind of social context to succeed in their quest for transcendence.

On the superiority of elixirs over other ingestables and other methods Ge Hong is clear.

> Someone asks: "In the world there are some who consume medicinal substances, circulate pneumas, and 'guide and pull,'[37] yet do not avoid death. Why?" The Master who Embraces Simplicity replies: "If you do not obtain gold or cinnabar, but only ingest medicinals from herbs and trees and cultivate other minor arts, you can lengthen your years and postpone death, but you cannot attain transcendence."[38]

But in making this claim, Ge Hong opens himself to criticism from the partisans of "naturalness" so numerous in his time. He directly addresses this criticism in chapter sixteen of the *Inner Chapters*. In the course of this discussion, he records himself as having raised this objection to his teacher, Zheng Yin: "Why not ingest the gold and silver that exist in the world? Why fabricate them by transformation? After all, if one fabricates them then they are not genuine, and if they are not genuine then they are counterfeit." Zheng Yin at first responds by citing the expense of natural gold and silver, but he then makes an astonishing claim: "Only gold fabricated by transformation harbors the essences of various medicinal substances, and in this it is superior to the natural sort."[39] If gold is a natural marvel, fabricated gold is a greater marvel. Sharing this can-do attitude is the adept Huangshan zi, who is quoted as having claimed, "Since Heaven and Earth contain gold, I too can fabricate it."[40]

Ge Hong marvels at humankind's nature-transcending powers. He wonders, "What is it that the arts of transformation (*bianhua zhi shu*) cannot do?" He elevates humanity above other life-forms: "Among creatures, humanity possesses an honorable nature that is most numinous"—although he immediately goes on to note that even humans in some cases change sexes and transform into other species (not to mention the many other cases he cites of cross-species animal transformations).[41] From such cases he reasons as follows: "Since transformation occurs spontaneously in Heaven and Earth, why should it be doubted that gold and silver can be fabricated out of things different from them in kind?" And since even humans, noblest of creatures, transform into other kinds, why may not gold and silver, noblest of metals, be fashioned by transformation from other kinds?

His other main defense against the anti-alchemy argument from naturalness is to point out (several times in chapter sixteen) that transformed members of a class of animals, plants, or minerals are no different in their qualities and capacities than nontransformed members. Fire and water obtained by means of mirrors behave the same as ordinary fire and water; successfully fabricated gold reacts as does natural gold, and "from this [you will know that] you have obtained the Dao of naturalness; and since it possesses [the same] capacities, why should it be called 'counterfeit'?"[42] The term "counterfeit" should be reserved, Ge Hong says, for things merely disguised on their surfaces as other things.

In short, natural ingredients are necessary but not sufficient for attaining transcendence by the highest routes. They work best when their several powers are combined and augmented by the alchemist's labors. As Mircea Eliade observed more than forty years ago:

Alchemy prolongs and consummates a very old dream of *homo faber*: collaboration in the perfecting of matter while at the same time securing perfection for himself. . . . In taking upon himself the responsibility of changing Nature, man put himself in the place of Time; that which would have required millennia or aeons to 'ripen' in the depths of the earth, the metallurgist and alchemist claim to be able to achieve in a few weeks.[43]

By speeding up time, the alchemist, though drawing on nature, improves upon it.

The key role of human artifice and the manipulation of natural substances in Ge Hong's tradition is also clear when we examine the place in that tradition of written texts and of the lineages of master-disciple filiations along which they are passed down.

For Ge Hong and his tradition, mountain recesses harbored not just numinous minerals and plants but also sacred books; the world's outlying extremities were sources not just of rare creatures and stones but also of rare texts.[44] Hence, in antiquity the Yellow Thearch, during his tours of inspection to the periphery of his realm and the mountains within it, gathered both potent herbs and scriptures, at the head of which was the *Inner Writings of the Three Sovereigns (Sanhuang neiwen)*, by which he could "control and summon the myriad spirits."[45] The alchemical scriptures *Nine-Reverted Elixir (Jiuzhuan dan)* and *Scripture of Potable Gold (Jinye jing)*, along with esoteric instructions on how to use them, were recovered from a jade casket deep within Mount Kunlun.[46] Indeed, Ge Hong writes that

> all the noted mountains and the Five Marchmounts harbor books of this sort [i.e., like the *Sanhuang neiwen* and the *Charts of the Perfect Forms of the Five Marchmounts (Wuyue zhenxing tu)*], but they are hidden in stone chambers and inaccessible places. When one who is fit to receive the Dao enters the mountain and meditates on them with utmost sincerity, the mountain spirits will respond by opening the mountain, allowing him to see them. . . .[47]

He proceeds here to mention the case of Bo He, narrated more fully in *Traditions of Divine Transcendents*. In *Traditions*, Bo He is instructed by his teacher to

> ". . . remain here in this cave, carefully regarding the north wall. After a long time you will see that there are written characters on the wall, and if you study these, you will attain the Way." Only after Bo He had regarded the wall for three years did he see the words of the *Scripture of Grand Clarity (Taiqing jing)*,[48] which some person of ancient times had carved there. He studied[49] them and thereby attained transcendence.[50]

But if the inner chambers of mountains are sources of sacred texts, the adept needs other sacred texts to gain safe entry to them. He cannot simply bound in with only his body as resource: that may work for those fortunate few, such as the Qin palace woman and the little daughter of Zhang Guangding, who happened upon instruction on how to preserve themselves in the wilds, but the seeker of transcendence is advised to be much more cautious, to gird himself with apotropaic armaments, for "many meet with disaster and harm because they do not know the method (*fa*) for entering mountains"; in short, "mountains may not be entered lightly."[51] It cannot be emphasized too heavily that "nature" as Ge Hong pictured it in wild mountainous zones was no serene landscape painting; it was a terrifying, dangerous environment, a terrain of combat into which only idiots would venture without weapons. Foremost among these are the talismans (*fu*), among which the ones handed down by Laozi are most efficacious because they are in celestial script (*tianwen*) and were transmitted to him by gods.[52] The talismans' power is what causes hidden excrescences and herbs to reveal themselves; without it, these treasures would remain concealed. And it is talismans that ward off all manner of attacks from wild creatures—tigers, wolves, deadly insects, and fierce spirits, all jealously patrolling their territory—and allow the adept to subjugate the local mountain gods. Mirrors are also essential for decoding mountain phenomena: in them, transcendents and beneficent mountain gods are reflected in human form, while marauding demon-spirits of birds and beasts appear in their true forms.[53]

Finally, it is to written texts of formulas and procedures—*fang* and *fa*, special, hardly obvious, often rather weird "methods," or else "arts" (*shu*—all of these terms in Ge Hong's usage connoting "*esoteric* methods/arts")—that the adept must look for guidance in his practices. Without them, he is helpless; Ge Hong never imagines it possible for an adept simply to make up procedures from scratch or to happen upon any but the simplest preparations by accident or trial-and-error. And he may receive these texts only after having qualified himself by a long period at the feet of a teacher, and then only after swearing an oath of secrecy; only then does he receive the most carefully preserved arcana of all, the oral instructions needed to complete the instructions entrusted to writing.[54]

Nothing could be more "natural" than to merge completely with nature via death, one's body devoured by other bodies and reduced to dust mixed with other dust,[55] but Ge Hong's whole reason for writing is to steer his readers away from their headlong rush to this fate and toward the alternative he believes possible, the arts which allow transcendence of death. Fabricated gold is superior to natural gold; the alchemist improves upon nature. Mountains harbor not merely the herbs and minerals necessary for the practitioner's work but also the texts needed to guide it. Other texts cause all the dangerous spiritual forces and hidden potencies of mountains and forests to reveal themselves and submit before the texts' possessors. If at first Ge Hong seemed to urge a dietary return to nature, preferring nature's purity over culture's decadence and a foraging/gathering over an agrarian way of life, he now seems to have arrived at the opposite side of the nature/culture dichotomy, foregrounding the various humanly transmitted, humanly practiced "methods" for transforming what is natural.

In fact, neither side of the nature/culture dichotomy is an adequate basis for characterizing Ge Hong's stance toward nature because, of course, the dichotomy itself is false as a characterization of his way of thinking. I conclude by pointing out some reasons why.

Transcending the World by Means of the World

Ge Hong's adept methodically transforms naturally available materials into products that, ingested, confer the capacity to transcend the limits of *both* the natural *and* the human worlds. But he is able to do so only by virtue of naturally engendered materials and culturally transmitted techniques, all of which he must derive from the very realms he seeks to transcend. If we summarize the main ways in which the adept is beholden to the interlaced structures and interlocked rhythms of space and time, it becomes clear that only within a framework of correlative thinking is it possible to construct practices for transcending that framework.

The Daoist's main working medium, his very body, first of all, as Kristofer Schipper and Isabelle Robinet have made us well aware, is microcosmic in structure—in more than one way. Ge Hong shares this ubiquitous religious Daoist vision. In his writings, the body's microcosmic nature is manifest in at least two senses: one passage describes the body as parallel in its structure to a state; another imagines the body as composed of five jades, their colors corresponding to the seasons of the year.[56]

Second, all of the actual alchemical procedures, as well as the theory behind them—the mixing of particular substances of particular colors and names,[57] the careful adherence to phased cycles of time, the spatial orientation of the workspace—were built upon the systems of five-phase, yin-yang, space-time correspondences dating back to the pre-Han period. This is well known and needs no elaboration here.[58]

Third, every phase of the practitioner's transformative work—not merely the actual alchemical steps, but also the harvesting and gathering of ingredients,[59] the adept's ritual self-purification, and the schedule for ingesting the product—was carefully timed to coincide with the natural rhythms of sun and moon.[60]

Fourth, the adept's very prequalification to enter into these protracted labors depends on his having been born at an auspicious time and thus having received the appropriate allotments of qi and *ming*.[61] In effect, the scheduling of his self-cultivational work began at conception, when the correlative building blocks of his particular set of aptitudes were first assembled.

Despite, then, the bold claims made for the transcendence-inducing, supranatural results of alchemical transformations, those transformations as processes are possible only in and through the matrix of space and time and essentially depend at every step upon the environing natural and social worlds—the natural world as the source of the ingredients, the social world of textual filiation and master-disciple relationship as the source of the operations performed on them. Ge Hong's practitioner cannot transcend the world, both natural and social, simply by leaping clear of it in a single moment, but is instead radically beholden to it at every turn as he labors toward his goal. And even after he has attained the goal of transcendence, we as readers know that he has done so only through the "traces" he has left behind in this world (both natural and social), traces which Ge Hong as essayist and hagiographer collected and transmitted to future generations.[62] All of the tracks leading beyond this world are to be found in this world; the only way to get There is from Here.

Ingesting the Marvelous: End Matter

Abbreviations

HY   Wang Tu-chien, ed. *Combined Indices to the Authors and Titles of Books in Two Collections of Taoist Literature.* Harvard-Yenching Sinological Index Series, no. 25. Beijing: Yanjing University, 1925. Works in the Daoist canon are cited by the number assigned them in this index.

NP   Wang Ming, ed. *Baopuzi neipian jiaoshi.* 2d ed. Beijing: Zhonghua shuju, 1985. Cited by chapter number/page number.

T   *Taishō shinshū daizōkyō.* Tokyo, 1922–1933. Cited by text number.

Ware   James R. Ware. *Alchemy, Medicine and Religion in the China of A.D. 320: The Nei P'ien of Ko Hung.* Cambridge, Mass.: MIT Press, 1966. Reprint, New York: Dover, 1981. Cited by page number.

Conventions

All translations are my own. Traditionally formatted Chinese works are generally cited by fascicle number followed by folio page number, with *recto* and *verso* indicated by "a" and "b," respectively (thus "*Taiping yulan* 857/2a" indicates the second page, *recto*, of fascicle 857 of *Taiping yulan*). When citing Ge Hong's *Inner Chapters*, although I do not quote Ware's translation, I do provide the location of *Inner Chapters* passages in it so that the reader may easily consult it for a sense of the context of the cited passage.

Notes

My thanks to Edward Davis, Kristofer Schipper, and Mary Evelyn Tucker, in particular, for their encouraging responses to my paper.

1. Located west of Chang'an.

2. NP 11/207; compare Ware, 194.

3. Ge Hong's version resembles the story of "the Hairy Woman" found in the Han-period *Liexian zhuan* (HY 294, 2/7b–8a); see Max Kaltenmark, *Le Lie-sien Tchouan (Biographies légendaires des immortels taoïstes de l'antiquité)* (Pékin: Université de Paris, Publications du Centre d'Études Sinologiques de Pékin; reprint, Paris: Collège de France, 1987), 159–61. The main differences are that, in the *Liexian zhuan* account: 1) both the woman and her teacher are named: she is Yu Qiang, and he is Gu Chun (he receives a hagiography in *Liexian zhuan* 2/6b–7a; see Kaltenmark, *Lie-sien*, 157–58); and 2) the woman is not captured or fed grains, and the implication of her inclusion in the hagiography is that she did, in fact, achieve transcendence as envisioned in that text. *Sandong qunxian lu* (HY 1238) 20/9a6 ff., erroneously citing *Shenxian zhuan* as its source, gives a late and slightly altered version of the *Liexian zhuan* story.

4. On the extent to which Ge Hong and the tradition he advocated are properly "Daoist," see part one of my *To Live as Long as Heaven and Earth: A Translation and Study of Ge Hong's "Traditions of Divine Transcendents"* (Berkeley and Los Angeles: University of California Press, forthcoming). But later Daoists certainly regarded his tradition as ancestral, and in this I follow, not the tendency to utterly marginalize him that is exhibited in some writings of T. H. Barrett ("Ko Hung," *Encyclopedia of Religion*, ed. Mircea Eliade [New York: Macmillan, 1987], 283–343) and Nathan Sivin ("On the Word 'Taoist' as a Source of Perplexity," *History of Religions* 17, no. 3-4 (1978): 323–27), but the rich studies by Isabelle Robinet; see Isabelle Robinet, *La révélation du Shangqing dans l'histoire du taoïsme,* 2 vols. (Paris: Publications de l'Ecole Française d'Extrême-Orient, 1984), 1:7–104, and *Taoism: Growth of a Religion*, trans. Phyllis Brooks (Stanford: Stanford University Press, 1997), 78–113. Ge Hong's "tradition," then, represented in the library of texts his teacher owned, stretched back through several distinct lineages into the early Han period and perhaps earlier, and encompassed a loose congeries of practices and systems developed by *fangshi* in the eastern regions (while also on occasion asserting its own superiority to these). Its scriptural centerpieces were the *Numinous Treasure Five Talismans* (*Lingbao wufu,* now at least partially preserved in HY 388), containing various herbal and other procedures as well as talismans for entering mountains to gather ingredients; the *Grand Clarity Scripture of the Divine Elixir of Potable Gold* (*Taiqing jinye shendan jing*), an alchemical text; the *Writings of the Three Sovereigns* (*Sanhuang wen*); and the *Charts of the Perfect Forms of the Five Marchmounts* (*Wuyue zhenxing tu*).

5. By this I mean that what was ingested, and how, was taken as the necessary and centrally important practice leading toward achievement of the religious goal, however defined. For the Celestial Masters, for example and by contrast, the centrally soteriological activities would seem to have been the making and renewing of cov-

enant relations with deities, constant living according to the precepts, and the rite of petitioning (including petitions to undo the consequences of inevitable failures to keep the precepts). For the Shangqing tradition—closest in spirit to Ge Hong's tradition, yet innovative in crucial ways—the central soteriological activity would seem to have been the formation, primarily by techniques of visualization, of personal relationships with Perfected members of the divine hierarchy; ingestion of elixirs and medicinals was merely one means of conveyance from this world to the next-world niche these higher beings had prepared for the adept. (The argument could be made that what Gregory Schopen in the Buddhist case has called "the cult of the book"— the possession and cultic treatment of sacred texts, with the promise of salvation as a direct result—held an equally important place in Shangqing practice.) For the Lingbao tradition, the key soteriological acts were liturgical in character, where the liturgies also encompassed a cult of the book. And so on. None of this is to deny that, for example, dietary restrictions had a place in the Celestial Master precepts and prohibitions (e.g., Stephen R. Bokenkamp, *Early Daoist Scriptures* [Berkeley and Los Angeles: University of California Press, 1997], 42, 50), or that alchemy continued to play an important role in Shangqing practice (e.g., Michel Strickmann, "On the Alchemy of T'ao Hung-ching," in *Facets of Taoism*, ed. Holmes Welch and Anna Seidel, 123–92 [New Haven: Yale University Press, 1979]; Bokenkamp, *Scriptures*, 331–39).

6. Thus Ware's rendition as "excrescences" is not a bad choice.

7. NP 11/198; cf. Ware, 180–81.

8. The relevant Xian Men fragment appears in *Taiping yulan* 988/5b; the Feihuang zi fragment (one of two concerning him), in *Taiping yulan* 857/2a. The often-cited, recompiled editions of *Shenxian zhuan* lack hagiographies of either figure. Other *Traditions* fragments credit Xian Men with other practices, and he is mentioned at three points in NP, once in connection with an elixir recipe. He was also accepted into the Shangqing hierarchy of transcendent-officials, for the Lady of the Southern Marchmount explained to Yang Xi (at *Zhen'gao* [HY 1010] 14/16a ff.) that he was currently "a Transcendent Chamberlain of the [Lord of the] Central Prime" and also revealed the location of his tomb—no doubt (judging from the passage) a local cult site in the vicinity of Lushan.

9. This trope of "display of knowledge of the periphery and the mastery of esoteric skills" figures importantly in texts associated with *fangshi* traditions; see Robert Ford Campany, *Strange Writing: Anomaly Accounts in Early Medieval China* (Albany: State University of New York Press,1996), 280–94. The point, as in much of the Han Emperor Wu cycle of legends, is always to show to the reader (par excellence, the ruler) just enough of a marvel to convince him of its wondrousness and desirability while also reserving access for those with the proper esoterically transmitted qualifications.

An additional point that bears emphasizing in this section, and that will be returned to below, is the specialness of *mountains* as mysterious, dangerous, powerful sources of nourishing herbs and minerals. Aside from NP 11 and 17, a good example of mountains as living sources of potent substances is the *Traditions* hagiography of Wang Lie, in which he happens upon a rice-flavored "green mud" oozing "like bone marrow" from an avalanche-induced rock crevice—a vivid example of what Ge Hong means by "exudations" (*zhi*).

10. This passage, however, is not attested in any of the earlier strata of sources of *Traditions* fragments.

11. This *mo* can mean either a fanciful, black-and-white colored, bear-like beast that feeds on bronze and iron (as mentioned in the *Classic of Mountains and Waters*), or, as a cognate, a domesticated animal resembling the donkey; see *Dai kanwa jiten*, comp. Morohashi Tetsuji, rev. ed. (Tokyo, 1984), 7/698b.

12. This scene is described in the following early sources (not to mention later ones not listed here): *Beitang shuchao*, comp. Yu Shinan (558–638); published in a block-print ed. re-cut from a traced Song ed. in 1888; facsimile reprint in 2 vols. of 1888 ed. (Taibei: Wenhai chubanshe, 1962), 145/5b; *Yiwen leiju*, comp. Ouyang Xun (557–641) et al.; modern critical recension ed. Wang Shaoying, 2 vols. continuously paginated (Beijing: Zhonghua shuju, 1965), 72/1242; *Taiping yulan*, facsimile reprint of Shangwu yinshuguan 1935 printing from a Song copy (Beijing: Zhonghua shuju, 1992), 862/6a; *Chuxue ji*, comp. Xu Jian (659–729) et al., modern critical recension by Si Yizu, 3 vols. continuously paginated (Beijing: Zhonghua shuju, 1962), 26/642; *Yongcheng jixian lu* (HY 782) 4/10b–13a. See my forthcoming *To Live as Long as Heaven and Earth* for a complete listing of sources and variata.

13. Her story points up the link between grain-avoidance and dwelling in mountain wilderness areas: it is in those areas that the adept needs to rely on foods other than grains. Compare these similar statements by Ge Hong: of those who are foolish enough to enter mountains without the necessary protections, he says that they are liable (among other dangers) to "become famished but lack a method for abstaining from grains" (NP 6/124; Ware, 114); elsewhere he advises, "If you meet with unstable times and seclude yourself in the mountain forests, you will by knowing this method [i.e., the method of avoiding grains] be able to avoid starving to death" (NP 15/266; Ware, 244).

14. His major disquisition on the subject in NP comes at the head of chapter 15 (Ware, 243–49); other mentions in that text come in chapters 5 (Ware, 104), 11 (Ware, 196), 12 (Ware, 210), and 20 (Ware, 320). In *Traditions*, adepts who successfully abstain from grains and benefit thereby include Bo He, Ge Xuan, Jie Xiang, Kong Anguo, Kong Yuanfang, Lu Nüsheng, Shen Jian, Taishan laofu ("the Old Man of Mt. Tai"), Yan Qing, and Zuo Ci. A small number of exceptions proves the rule: Wu Mu is described as meeting a certain Master Sun who raised grains and sesame outside his mountain cave (but note that sesame is elsewhere in Ge Hong's corpus prescribed as a *substitute* for grains; also, the text is questionable here), and Baishi xiansheng ("Master Whitestone") is said to have broken this and other dietary restrictions, liberally consuming meat, liquor, and various cereal foods.

15. NP 3/50; Ware, 60.

16. For a survey of the topic, see Jean Lévi, "L'abstinence des céréales chez les Taoïstes," *Études Chinoises* 1 (1983): 13–47. I have also expanded on the symbolic significance of the three frequent (but hardly universal) Daoist food avoidances— grain, meat, and liquor—vis-à-vis the cult of the dead, in two unpublished papers, "Comparing Traces of Transcendence and Eminence," paper delivered at the 1995 meeting of the American Academy of Religion, and "The Special Dead, the Ordinary Dead, and the Undead in Early Medieval China," paper delivered at the 1996 meeting of the American Academy of Religion. Kristofer Schipper ("Purity and Strangers:

Shifting Boundaries in Medieval Taoism," *T'oung Pao* 80 [1994]: 61–81) and Terry F. Kleeman ("Licentious Cults and Bloody Victuals: Sacrifice, Reciprocity, and Violence in Traditional China," *Asia Major*, 3d ser., 7, no. 1 [1994]: 185–211) also address aspects of this theme.

17. At this writing I have yet to find in Ge Hong's writings the claim, common elsewhere, that grains should be avoided because they are what the parasitic "three corpses" or "three worms" of the body feed on. His only clear statement on the subject, in NP 15/266 (Ware, 243), is that clean intestines and absence of feces are prerequisites to attaining long life. See part one of my *To Live as Long as Heaven and Earth* on reasons for thinking that "grains" in such contexts really connoted "ordinary foods" in general.

18. Credited at NP 11/2a (Ware, 178) with rendering consumption of grains unnecessary.

19. At NP 11/197 (Ware, 179) he notes the usefulness of *huangjing* for this purpose, though he says it is less effective than atractylis.

20. *Shu* was claimed already in the early Han *Huainan wanbi shu* to be a substitute for grains, a claim Ge Hong repeats. Kaltenmark (*Lie-sien*, 65–67) points out that sesame (*huma*, also known as *jusheng*) seems, at least soon after its introduction to China, to have served a similar purpose.

21. Ge Hong's most frequent general term for breath practices is *xingqi*; he seems to use it generically for any practice in which breath is swallowed and then systematically circulated (sometimes guided by visualization) throughout the body. A good description of part, at least, of what *xingqi* meant to Ge Hong can be seen in NP 8/149–50 (Ware, 139–40); moreover, his discussion at the outset of NP 15 shows that he considered pneumas a dietary substitute for grains. (Compare also his more general statement on the importance of pneuma-cultivation in NP 5/114 [Ware, 105].) Some forms of regulated breathing for the enhancement of health and longevity had been practiced in China for at least five centuries prior to Ge Hong's era; references appear in early parts of the *Zhuangzi* text, in the *Laozi*, *Huainanzi*, and elsewhere (see Bokenkamp, *Scriptures*, 69 n. 38, 70 n. 39). In any Daoist context, such practices must usually be understood in conjunction with cosmological and cosmogonic notions of pure, primordial pneumas (*yuanqi*; see Edward H. Schafer, *Pacing the Void: T'ang Approaches to the Stars* [Berkeley and Los Angeles: University of California Press, 1977], 21–33; Isabelle Robinet, *Méditation taoïste* (Paris: Dervy Livres, 1979), 32ff., 129ff.; Bokenkamp, *Scriptures*, 125 n., 188–94, 207 n.): the adept replaces the impure, stale pneumas of these lower regions and latter days with the purer, fresher pneumas of earlier cosmic eras and higher cosmic zones, of which gods partake. For a sense of the range of practices, see Ute Engelhardt, *Die klassische Tradition der Qi-Übungen (Qigong)* (Wiesbaden: Franz Steiner Verlag, 1987); Ute Engelhardt, "*Qi* for Life: Longevity in the Tang," in *Taoist Meditation and Longevity Technigues*, ed. Livia Kohn, Michigan Monographs in Chinese Studies, no. 61, 263–96 (Ann Arbor: Center for Chinese Studies, 1989) (on early Tang texts preserved in *Yunji qiqian* [HY 1026] 33 and 57); and Henri Maspero, *Taoism and Chinese Religion*, trans. Frank A. Kierman, Jr. (Amherst: University of Massachusetts Press, 1981), 460–541.

22. As described in NP 15/267; Ware, 246.

23. Also described in NP 15/267; Ware, 246.

24. In *Traditions*, Ge Hong ascribes skill in *taixi* to the adept Huang Hua. The expression is alternatively translatable as "embryonic respiration" or "womb breathing." In the larger Daoist tradition, *taixi* is a technique of pneuma-circulation the nature and understanding of which underwent considerable change through history. Ge Hong (NP 8/139; Ware, 139) seems to have understood *taixi* as the culmination of long-term daily practice of "pneuma-circulation" (*xingqi*)—a state of achievement in which one finally "does his breathing as though in the womb, without using nose or mouth." He also claims that his great-uncle, Ge Xuan, was able to remain underwater for long periods by means of *taixi*. For more on the details and history of the practice in various Daoist traditions, see Maspero, *Taoism*, 459–505; Engelhardt, *Qigong*, 109–21; Engelhardt, "*Qi*," 287; Kristofer Schipper, *The Taoist Body*, trans. Karen C. Duval (Berkeley and Los Angeles: University of California Press, 1993), 157; Bokenkamp, *Scriptures*, 315.

25. The quotation appears in NP 3/48 (Ware, 57–58). NP 3 treats as well of other ways in which adepts may live long by imitating animals. On the *Yiwen ji*, of which only one other fragment survives, see Campany, *Strange Writing*, 43. A structurally similar tale is recounted in *Bowu zhi* 10/111 (item no. 322), and a virtually identical version is attributed to the *Youming lu* in *Taiping yulan* 69/1b (this is item no. 172). It runs as follows: a man lost in the mountains falls into a deep ravine, almost starves, but then begins imitating the way in which tortoises and snakes stretch their necks eastward at dawn and dusk, and so loses his hunger and experiences a lightening of his body. Eventually he finds his way out and returns home, ruddier in complexion and cleverer than when he left. But he then resumes his old diet of grains and spicy foods, and in a hundred days reverts to his previous condition.

26. NP 3/186 (Ware, 169–70). The interlocutor's question here and Ge Hong's answer in fact helpfully explain one of the largest differences between Ge Hong's hagiography and that found in the *Liexian zhuan*: quite a few figures in the Han-period text are portrayed as serving one or another ruler, sometimes with, but often without, an implied tension between their goals and their rulers'; in Ge Hong's text, by contrast, a marked divide has opened between the political and the transcendence-quest spheres. The difference was briefly remarked upon in Campany, *Strange Writing*, 305, and will be treated at more length in my study *To Live as Long as Heaven and Earth*.

27. NP 10/186 (contrast Ware, 171). Ge Hong at several other points in NP discusses the necessity of abandoning high social status to pursue transcendence: see, for example, chapter 12 passim (esp. passages translated in Ware at pp. 200–201, 205, 209), and chapter 14 (Ware, 227). Linked with this theme is the even more basic one of "confusion by externals," which eventually leads, according to Ge Hong, to the making of excessive and unauthorized blood sacrifices as a way to try to solve one's problems; see the seminal discussion at the outset of NP 9. In contrast to such "externalist" reliances on outside spiritual forces, Ge Hong insists that "people's allotted lifespans lie within themselves and are not entangled with things outside" (NP 9/170; Ware, 152). Finally, Ge Hong elsewhere elaborates on the reasons for the need to withdraw into mountains, away from ordinary society, to perfect elixirs; see, e.g, NP 4/84–85 (Ware, 93).

28. This only loosely renders the expression *nianming zai guxu zhi xia*, for an explanation of which see NP 16/297 n. 43.

29. NP 15/271; Ware, 252.

30. NP 14/253–54; Ware, 228–29. Compare his (and his teacher's) interpretation of the *Laozi*'s injunction against valuing things that are hard to get, in NP 16/286 (Ware, 267).

31. NP 14/254 (Ware, 229).

32. Ge Hong thus rejects all attempts to "buy" longer life by means of petitions, prayers, or offerings to gods, a fact which distinguishes his position not only from popular temple cults (against which he rails in NP 9) but also from Celestial Masters petitions for longevity (on which see my "Living Off the Books: Fifty Ways to Dodge *Ming* in the Fourth Century C.E.," paper delivered at the 1998 meeting of the Association for Asian Studies). Compare his insistence that "allotted lifespan is situated in humans and is not entangled with externals" (NP 9/170 [Ware, 152]). Anna Seidel astutely noted ("Traces of Han Religion in Funeral Texts Found in Tombs," in *Dōkyō to shūkyō bunka*, ed. Akizuki Kan'ei, 21–57 [Tokyo: Hirakawa, 1987], 232) that Daoism is, at base, a *jiriki* religion in which one must achieve salvation by one's own power—although in the same article she goes on to show the importance of the development of liturgies for salvation of the dead, who are no longer able to save themselves.

33. NP 13/239 (Ware, 212).

34. The contradiction would seem to dissolve in the common *post hoc ergo propter hoc* claim that mere possession of a sacred text or meeting with a teacher already self-referentially confirms a suitable fate. In *Traditions*, senior masters, upon encountering a neophyte in the mountains, often say something to the effect of "Since you were able to meet me, you must be fated to attain transcendence"; and many texts, including some of the ones in Ge Hong's possession, claim that only those who are qualified are able to come into possession of them or read them—a revelation that must have satisfied any who owned the books or were reading the passage in question! Statements to the effect that a fitting "preallotment" (*ming*) is a necessary (but never a sufficient) condition for transcendence occur at NP 11 (Ware, 197), 12 (Ware, 203–4), and 14 (Ware, 227), as well as in *Traditions*. The mechanics are explained, e.g., at NP 12 (Ware, 203–4); on this topic, see also my unpublished paper, "Living Off the Books."

35. Cited with approval from a *Guijia wen* in NP 16/287 (Ware, 269). Compare Ge Hong's statement (as in NP 14/252 [Ware, 227]) that no one ever finds a teacher without seeking one.

36. NP 5/110 (Ware, 97). Compare his reason for rejecting the extravagant claims of Laozi's divinity in his *Traditions* hagiography of that figure (which treats him as one particularly skilled transcendent among others, not as an originally divine, celestial being): "These sorts of speculations are the product of recent generations of practitioners, lovers of what is marvelous and strange, who have created them out of a desire to glorify and venerate Laozi. To discuss it from a basis in fact, I would say that Laozi was someone who was indeed particularly advanced in his attainment of the Dao, but that he was not of another kind of being than we. . . . The view that Laozi was originally a deity or numen must stem from practitioners of shallow views who

wished to make Laozi into a divine being of a kind different than we, so as to cause students in later generations to follow him; what they failed to realize was that this would cause people to disbelieve that long life is something attainable by practice. Why is this? If you maintain that Laozi was someone who attained the Dao, then people will exert themselves to imitate him. If you maintain that he was a deity or numen, of a kind different than we, then his example is not one that can be emulated by practice." As explained in the text-critical notes to my forthcoming translation and study of *Traditions*, this translation is based on a comparison of the quoted passages in *Fayuan zhulin* (T 2122) 31/520b and *Taiping guangji* (reprint, Shanghai: Saoye shanfang yinhang, 1930), 1.1.

37. *Daoyin*, a term attested in late Warring States and early Han documents excavated in the past two and a half decades (see He Zhiguo and Vivienne Lo, "The Channels: A Preliminary Examination of a Lacquered Figurine from the Western Han Period," *Early China* 21 [1996]: 81–124. 87 n. 12 and 116–22), encompasses a wide range of kinetic and gymnastic exercises (on which see Catherine Despeux, "Gymnastics: The Ancient Tradition," in *Taoist Meditation,* ed. Kohn, 225–62.; Maspero, *Taoism*, 542ff.). In *Traditions*, Ge Hong makes Peng Zu a primary practitioner of these arts. His description there makes clear that what he understood to be "guided" and "pulled" is one's mental attention and one's breath. In NP, Ge Hong comments several times on these arts. In chapter 3 (Ware, 53, 58) he avers that they are based on the imitation of long-lived animals such as tortoises and cranes. In chapter 4 (Ware, 81) he notes their inferiority to alchemy as a way to achieve long life; similarly, in chapter 5 (Ware, 103) he lists them among several other hygienic and magical practices conducive to avoiding harm but clearly inferior to the major arts of transcendence. A passage in NP 15/274 (Ware, 257) seems to be a list of several specific, named *daoyin* exercises.

38. NP 13/243 (Ware, 219). Similar sentiments are expressed elsewhere in NP and in the *Traditions* hagiographies of Lü Gong and Liu Gen.

39. NP 16/286 (Ware, 267–68).

40. NP 16/287 (Ware, 269). The figure Huangshan zi is obscure but is mentioned elsewhere in Ge Hong's extant writings, including fragments of the *Traditions* hagiography of Peng Zu, where he is said to have inherited and practiced that sage's methods and thus become an earthbound transcendent.

41. For more on the significance of this point, see Campany, *Strange Writing*, chapter 8.

42. NP 16/286 (Ware, 268).

43. Mircea Eliade, *The Forge and the Crucible: The Origins and Structures of Alchemy*, 2d ed., trans. Stephen Corrin (Chicago: University of Chicago Press, 1978; originally published in 1956 as *Forgerons et alchimistes*), 169.

44. The "pro-naturalist" might perhaps point to this as evidence of his own position, noting that for Ge Hong the highest and most potent writings are to be found, at their source, not in the cultural but in the natural world.

45. NP 18/323 (Ware, 302–3).

46. NP 18/324 (Ware, 303).

47. NP 19/336 (Ware, 314).

48. Thus *Taiping yulan* 187 and *Chuxue ji* 24. But *Taiping yulan* 663 reads as

follows, and *Xianyuan bianzhu* (HY 596) largely agrees: "[The words] were those—carved by someone in ancient times—of the method of [making] divine elixirs from the *Scripture from [the Realm] of Grand Clarity* (*Taiqing zhong jing*) as well as those of the *Celestial Writings in Large Characters of the Three Sovereigns* (*Sanhuang tianwen dazi*) and the *Charts of the True Forms of the Five Marchmounts* (*Wuyue zhenxing tu*). These were all manifest on the stone wall. Bo He recited all ten thousand words of them, but there were places where he did not understand the meaning. Lord Wang therefore bestowed oral instructions on him, [further] telling him to become an earthbound-transcendent on Linlu Mountain."

49. Thus the *Taiping yulan* version; another early version, in *Chuxue ji*, has "he chanted them." Whether it is a matter of "studying" (*du*) or "chanting" (*song*), the "reading" involved would have been in itself, in this period, a devotional act and would have involved vocalization accompanied by preparatory purifications, the lighting of incense, and so on; it was not simply the silent reading for content to which moderns are accustomed.

50. The translation of this particular passage of Bo He's *Traditions* hagiography is based primarily on *Taiping yulan* 187/4a and *Chuxue ji* 24/585. For other sources, see the listing in my *To Live as Long as Heaven and Earth*. One also thinks here of the legend of its own origins embedded in HY 388, a formative text in Ge Hong's patrimony; for a partial translation and discussion, see Stephen R. Bokenkamp, "The Peach Flower Font and the Grotto Passage," *Journal of the American Oriental Society* 106 (1986): 65–77.

51. NP 17/299 (Ware, 279–80).

52. NP 19/335 (Ware, 313). Ge Hong goes on to mention that talismans in his times were riddled with errors because few practitioners knew how to read them anymore; he also mentions, however, that the adept and eventual transcendent Jie Xiang possessed this ability (and this is documented also in Jie Xiang's hagiography as preserved in extant *Traditions* fragments).

53. See NP 17 passim; there are also passages in NP 11 (Ware, 179, 185–86) and 13 (Ware, 219) on talismans as prerequisites for entering mountains.

54. The need for teachers—good ones—is a main theme of NP 14 and 20. For a passage on the importance of oral instructions, see, for example, the one translated at Ware, 271.

55. Note, however, that the process of dying and one's spiritual fate after death are also described—by Ge Hong and many contemporaries—in anything but naturalistic terms; that is, they are imagined (and were so already in the early Han if not earlier) as utterly bureaucratic in character. And for this bureaucratic set of predicaments, Ge Hong and his cohorts have likewise bureaucratic solutions. See Campany, "Living Off the Books," and works cited there.

56. The first passage occurs in NP 18/326 (Ware, 307); the second, in NP 15/275 (Ware, 259–60).

57. Two examples of five-phase correlations with the five primary minerals used in alchemy and medicine appear in the two slightly different lists which Ge Hong gives in NP 4 and 17 (the latter list replaces magnetite with orpiment).

58. See especially Joseph Needham and Nathan Sivin, "The Theoretical Background of Elixir Alchemy," in *Science and Civilisation in China*, ed. Joseph

Needham, vol. 5, no. 4 (Cambridge: Cambridge University Press, 1980), 210–322.

59. NP 11 is peppered with examples in which certain herbs or minerals may be harvested only on particular days of the sexagenary cycle, and NP 17 with examples of times when it is auspicious or inauspicious to enter mountains or attempt certain operations there. These are additional ways in which the natural ingredients on which the practitioner depends are of *only limited access*.

60. And space and time were so closely interlocked in the five-phase correspondence scheme that almost any temporal designation implied a spatial one, and vice versa; for example, spring (or dawn) suggests east.

61. As noted above, Ge Hong does not usually understand these as absolute preconditions for success, but he does at more than one point outline the celestial-terrestrial systems by which natural endowment was determined.

62. Although I here use the masculine pronoun, women are on record as having successfully sought transcendence from early times; see my *To Live as Long as Heaven and Earth* for examples and discussion.

Sectional Discussion: What Ecological Themes Are Found in Daoist Texts?

JAMES MILLER, RICHARD G. WANG, and
EDWARD DAVIS

Introduction

The purpose of this group of essays has been, to quote Kristofer Schipper, to "introduce the evidence that there is more to Daoism and ecology than just a philosophy of nature." The purpose of this review is therefore to assess what this "more" comprises, by considering the ecological themes that the writers have discovered in the texts that they study. It should also be pointed out that these essays stand on their own as important textual studies. The reason for this is not only attributable to the scholarly credentials of the respective authors. By focusing on these texts with an ecological lens, the authors have also brought to light important, authentically Daoist themes. In no way have the authors had to struggle to scrape together a few ecological crumbs that have fallen from the Daoist table. Rather, the hypothesis advanced here is that there is a genuine ecological consciousness already embedded within the religious texts of the Daoist tradition. The nature of this latent ecological consciousness is made explicit by considering three themes: 1) the role of human beings in the cosmic ecology; 2) the spiritual ecology of the inner landscape; and 3) the Daoist religious problematic of the human transcendence of nature and our implication within it.

The Role of Human Beings in the Cosmic Ecology

It is well known that the universe of Chinese thought is predicated on the cosmological constant of the interrelationship of Heaven, Earth, and humankind, in which human beings are considered the supremely intelligent embodiment of the creative interaction of Heaven (yang) and Earth (yin). The famous phrase in Zhang Zai's *Western Inscription*, "I consider Heaven as my father and Earth as my mother and the ten thousand things as my brothers and sisters," is surely the noblest expression of this worldview, but it is an expression that places the Confucian heart-mind at the nexus of this cosmic relationship. Chi-tim Lai, Zhang Jiyu and Li Yanguo, and Kristofer Schipper, by contrast, examine the ecological implications of this Chinese worldview in mainstream Daoist movements. This worldview is not esoteric or psychologically complex, but identifies the practical obligations of a normative Daoist religious culture.

Chi-tim Lai's examination of the concept of centrality and harmony (*zhong-he*) in the *Taiping jing* demonstrates that the onus for the harmonious—that is, successful—functioning of this ecocosmic relationship falls squarely on the shoulders of humankind. Richard G. Wang observed in his response to Lai's paper that "in the *Taiping jing* it is a great human duty and a heavy responsibility to maintain the peaceful balance of the qi of Heaven and Earth and safeguard the natural growth and development of all living things." There can be nothing like a final resort, a *deus ex machina*. The ideal Confucian response to this absolute presupposition of the Chinese worldview might be to institute practical policies aimed at promoting the well-being of the natural environment and simultaneously to reform the moral character of the people. But the Daoist way is an intriguing alternative to the Confucian strategy of political and psychological reform and hinges on the notion of transparent communication (*tong*) within the ecology of the three dimensions of being. This is why Chi-tim Lai's criterion for harmony is the natural flourishing (*ziran*) of the ten thousand things and why Kristofer Schipper observes that the precepts he has studied are formulae of non-action (*wuwei*): we must refrain from whatever action obscures the transparency of communication within the cosmic ecology. This insistent and persuasive leitmotiv, that all things must be permitted to flourish naturally, we now recognize as a wholeheartedly ecological theme. But, as Richard G. Wang notes, this is not the same spontaneity and non-action that the *Daode jing* and the *Zhuangzi* attribute to the way of Heaven. The ecological theme present within these texts is more than a philosophy of nature, a policy, and a psychology: it is the religious orientation of human beings toward the harmony of their cosmic environment.

The Spiritual Ecology of the Inner Landscape

Zhang and Li's examination of the Yinfu jing goes a long way toward spelling out what this religious orientation consists of. The theory of "mutual stealing among the three powers" describes in concrete detail the workings of an ecosystem of cosmic power, in which "stealing" is to be viewed not so much as the exploitation of finite resources as the wise exercise of authority and power. The underlying premise is wholly ecological in that our action toward the environment is fundamentally to be construed as an action toward the self. There is no separation of powers in the constitution of the universe.

This presupposition lies at the heart of Kristofer Schipper's felicitous phrase "spiritual ecology" and is evident from the refrain to which he draws our attention: "the Dao is in ourselves and not in others." Thus, the second ecological theme that we note in Daoist texts is that of the landscape of the body. This means that the individual's material being is also a field that must be tended and nurtured from the overarching perspective of its being implicated in this universal ecosystem. The training of the body and the mind in the Daoist arts accomplishes a significant religious and therefore ecological goal. There cannot ultimately be any separation of the religious life of the individual from the life of his or her environment.

Transcendence and Implication

If this is the case, the question that comes immediately to mind is how to reconcile the natural science of ecology and the religious arts of Daoism. The answer to this question can be found in Robert Ford Campany's careful analysis of the concept of nature according to Ge Hong. The argument, basically, is that nature is unnatural, or "marvelous." This means that the possibility for transcendence is already present and implicated within the materiality of nature: "All of the tracks leading beyond this world are to be found in this world; the only way to get There is from Here." This is an ecological theme because it takes as its core the mutability of nature that cannot properly be comprehended in a static scheme of Linnaean taxonomy. If the very nature of nature is to be marvelous—to bear transcendence within itself—then this has a direct implication for the way in which we view our relationship to it. In fact, this relationship must be considered as an ecological relationship, but also as a spiritual relationship. The ecological theme to be found in Ge Hong's writings is not simply that we are implicated within our environment, but that this implication is precisely the means for our transcendence of it.

In his response to Robert Ford Campany's paper, Edward Davis spelled out the stages of dialectic of transcendence: first, "it is 'nature' that provides the basic stuff of salvation"; and second, "Ge Hong rejects both the natural (and the supernatural) in favor of a clear anthropocentrism." But the third movement is resolutely religious: The secrets that nature harbors within itself are *written texts*. Now it may be that these written texts have a natural origin in the energy produced by the basso profundo of Chinese cosmogenesis, but it is only when they take on a definite cultural form, as written Chinese characters, that they are useful and even intelligible to humans. Similarly, it may be that these texts have a natural existence in the mountain crevices that hide them, but it is only through the mediation of human culture— through the mediation of the teacher-student relationship, of ideas about moral worth, of written forms such as talismans—that we, or at least the deserving among us, can have access to them, can understand them, and can achieve the improvement of nature that is laid out in them. The dichotomy of nature/culture is here subverted and transcended, just as the Daoist adept attempts to transcend the world, which is now understood to include both nature and culture.

The serious issue that is at stake here is whether or not the transcendence of nature is a properly ecological theme. Ought it not be the case that ecology demands a Copernican revolution in our perception of our place within the natural order? Should we not exchange the Enlightenment mentality of the transcendence of humankind over nature for an ecological sensibility of ourselves as a fragment of nature construed as the interactive matrix of all life? The Daoist perception of this matter is resolutely religious and insists upon the concept of transcendence. But it is not the transcendence of divine beings over human beings or the transcendence of human beings over nature, but the innate capacity for self-transcendence that is pregnant within life itself. The Daoist vision of the process of evolution is recursive: nature gives birth to itself time and again. It is an ecological vision of nature but with transcendence built in. The marvelous nature of nature, finally, engenders in us an attitude of marvel and respect.

Conclusions

Ecological themes are central to Daoist religion, but these themes are far more than a philosophy of nature. They encompass: 1) the *cosmological interdependence* of heaven, earth and humankind; 2) the view of the body as a *spiritual ecosystem*; 3) a religious respect for the *marvel of nature*. None of these three themes is free from religious contamination. In fact, the ecological themes authentically present within Daoist texts are nothing less than the cornerstones of Daoist religious being.

III. Daoism and Ecology in a
Cultural Context

Flowering Apricot: Environmental Practice, Folk Religion, and Daoism

E. N. ANDERSON

Abstract

Daoism is closely related to the wider complex of belief that makes up Chinese traditional religion. Pragmatic management of land, water, and plant and animal resources was typically represented in spiritual or (broadly) "religious" terms. Fengshui, sacred trees, guardian spirits of the locality, and village festival cycles were inseparably connected to the ordinary practice of cultivation and resource management. Discourse on landscape often took a religious form while having a purely empirical content. Also, sanctions that an outsider might call "supernatural" were typically invoked to convince individuals to act in a conservationist manner—i.e., for a long-term, widely extended benefit as opposed to a short-term, narrow one. Managing the world required proper understanding and awareness of all these entities and their actions.

There are implications for the study of early Chinese philosophy. Explicit conservation injunctions are well known in the writings of Confucius, Mencius, and the (received) *Liji* (Book of Rituals). Daoism focuses on inner experience, and its cosmology is visionary and imaginative. However, by adopting a worldview in which cosmic forces are constantly playing and transforming all things, Daoism gave a distinctive flavor to Chinese traditional thought about the environment. For example, Daoism encouraged preexisting tendencies to think seriously about interaction with, and appreciation of, nonhuman entities (whether "natural" or "supernatural"). It also encouraged people to pay serious attention to the complexity of human behavioral and emotional involvement with these entities.

I

My flowering apricot (*Prunus mume*) arches over the door, lifting pale flowers into pale January sky. At dawn, its flowers catch exactly the pink of first light on gathering storm clouds. At night, the flowers bring the moon down: white rays direct to the earth. Petals fall, covering the herb garden and the walkway. After I sweep the doorstep, I shake the tree, to renew the litter of bloom.

The flowering apricot—the *meihua shu*—has been the symbol of Daoism for many centuries now. It flowers in the middle of winter, indifferent to such accidents as snow and storm. Testing the Chinese height of refinement, I have made tea with the snow that lies on the flowers. Their nectar gives the tea a faint, delicate carnation flavor.

In English, this tree is usually miscalled "plum." It is not a plum. Both in the West and in China, the plum is a rather ordinary, pedestrian tree.

Early Daoist thought took account of ecological truths. The sages were, however, concerned more with the nature within, the "interior landscape." They were also concerned with social life. Like other Chinese thinkers of the time, they were concerned with how a society can arise from a raw rout of individuals. They were concerned with *xing*, "nature," in the sense of basic human nature and its connections with the cosmos—not with "nature" in the Western senses of the word. ("Nature" is notoriously difficult to conceptualize in English, let alone translate into classical Chinese. Perhaps the nearest Chinese equivalents are "all under Heaven"—*tian xia*—and "the ten thousand things." But these are not really very close to what Western "nature writers" and "naturalists" are talking about.) Nonhuman things served as examples, but were rarely considered as ends in themselves.[1]

II

Mencius, in his story of Ox Mountain, tells us perhaps more about the Chinese environment and its problems than any other single passage (then or since).[2] Ox Mountain was once forested, but it was logged off.

Then sheep and goats ate the young herbs and seedlings. The mountain became barren. Mencius, in what must be the most perceptive single passage in all environmentalist literature, compares Ox Mountain with us poor ordinary humans—our nature good, but our lives denuded.[3] This was no mere analogy; Mencius's accurate observations of the causes and effects of mismanaging mountains were made parallel to equally and similarly destructive effects of mismanaging humans. The basic (even cosmic) principle is the same.

Every human group—certainly every group as large and agriculturally developed as the Chinese—has to confront the realities of environmental degradation. Therefore, every human group has developed—as part of its culture—an ecological science of considerable depth and sophistication. Such sciences are folk rather than formal sciences, but ethnography has revealed that they are strikingly elaborate and systematic in even the simplest societies.[4] Peoples such as the Chinese have very ancient traditions of managing soil, water, plants, and animals.

Such folk systems of thought often phrase ecological knowledge in terms that modern observers call "religious." Revelations about the gods and hardheaded, pragmatic knowledge of hydrology and forestry tend to be combined in cosmological systems. We can see this in such early Chinese works as the *Shan hai jing* (Classic of Mountains and Seas),[5] in which fragments of realistic natural history and ethnography are mixed with a great deal of lore that appears to have come from shamanistic visions.

Such cosmologies may be ancestral to philosophy and religion; at least, philosophies and religions often arise from broad substrates of folk theory. Whether they arose from folk cosmology or not, early Chinese philosophies—including the Daoist and Confucian traditions—encoded much knowledge of the nonhuman environment, and many comments on how humans could and should interact with it. The present paper sets the stage by considering some existential realities of the early Chinese environment and then examines certain aspects of the Daoist and Confucian approaches to it.

In Mencius's time, China was a smaller and wilder world than it was even a century afterward. The area that concerns us included the drainage basins of the Yellow and Yangzi Rivers and the adjacent mountains and coastal plains. The core, the rich area in which Chinese civilization most flourished, consisted of the North China Plain, the neighboring hills, the Wei River valley, and the Yangzi drainage from central Sichuan to the sea. This early Chinese world was by no means isolated. Ideas and goods flowed easily through these margins.

However, by this time, the settled core of China was farmed and urbanized. The wilderness had been pushed back: upriver, upslope, up the country. Pockets of marshland, river-bottom forest, and rough grassland persisted. Mencius's story shows (and archaeology confirms) that the hills had been substantially deforested. No one even remembered whether the plains had been forested or not. We still do not know how extensive forests had been on the North China Plain and the loess hills.[6] Extensive fires were set for hunting (as we see in the *Liezi*).[7] Game parks were important—animals had to be preserved for ritual hunts. Gamekeepers protected wildlife for hunting.[8]

The ordinary person lived by farming. Liezi called this "stealing from heaven";[9] in an ecologically revealing passage, he tells us that the farmer "steals" rain from the sky to irrigate his crops and "birds and animals from the land, fish and turtles from the water." Heaven tolerated such theft, and the wise robber confined his attentions to harvesting its bounty; Liezi's foolish thief soon found that society did not tolerate stealing of human-owned wealth. This story, and many like it, imply that agriculture was universal and was accepted as the most general, widespread human activity by all the philosophers. On the other hand, the story also shows that farming was still highly dependent on the unpredictable vicissitudes of natural rainfall and weather. The exquisite control of the environment that characterized later Chinese agriculture was not yet a reality for Daoist writers. They knew irrigation, and no doubt had heard of the complex systems of the rice-growing south, but they lived in a simple world. Contrary to some earlier claims, China's highly sophisticated irrigation agriculture was rather late. Key innovations were made in Qin and Han.[10] North Chinese agriculture in the Warring States period was not the incredibly intensive, fine-tuned system that amazed later observers such as F. H. King and J. L. Buck.[11] Fields were small, crops were limited, and yields were apparently low. Millets of many varieties, primarily foxtail millet (*Setaria italica*), were the main crops and staple foods in the Yellow River drainage. Southward in the Yangzi valley, agriculture appears to have been more sophisticated. Rice was dominant, and the landscape seems already to have been one of paddy fields and controlled irrigation. (Rice was domesticated in the Yangzi country by

8000 B.C.E., if recent finds are correctly dated. It was a staple of the south long before 400 B.C.E., though millet was an important supplement.)

Millet and rice were both brewed into *jiu*, a sake-like drink (technically a beer or ale, but usually translated "wine"). This substance is significantly ignored by the early Daoists (and still less noticed were marijuana and other drugs). The Chinese poets' love affair with alcohol came much later, in the period of unrest following Han. Staples important today, such as sorghum, maize, and potatoes, were not yet known.[12]

Yields of grain were low. Millet probably yielded about five hundred kilograms per hectare, to judge from the pride the Han government felt at raising the yields to two or three times that.[13] Rice must have yielded much more. A good irrigated paddy in the early twentieth century yielded about twenty-five hundred kilograms per hectare,[14] and a lower Yangzi paddy of 400 or 300 B.C.E. might already have yielded almost that much, since soil fertility was high and water control already established.

Climate fluctuations devastated crops and produced famines. Ecological changes that devastate predators and allow pest explosions were beginning, and we read of locust outbreaks.

Soy technology had not yet reached the pinnacles it was soon to achieve. Primitive fermentation products were used, but such genuine technological wonders as soy sauce and bean curd were known only at the very end of our period, or even later.[15] Beans were low-class, regarded as inferior to rice and millet.[16]

Bottle gourds, of course, were common and were of great symbolic significance in Daoism.[17] Mallow (*Malva parvifolia* complex) and sow thistle (*Sonchus oleraceus*) were common cultivated vegetables, but were considered rather low class; they have since declined still further on the prestige scale and become mere weeds. Domestic animals were essentially the same as today's.

The nobles, with their great hunting parks, had a shot at game (in the literal sense). Their major targets were deer—creatures of disturbed habitat, of field regrowth, of burned hills. The animals of old-growth landscapes—the tapirs, rhinoceroses, and elephants—were already gone. (Han poems mention them but treat them more or less as they treat phoenixes and dragons, in a context of romantic hyperbole.) The ordinary people could hope for little above the level of rabbits and pheasants. In a story Liezi tells about deer and dreams, a "man of

Cheng" was so delighted to get a deer that "[h]is joy overwhelmed him."[18]

Sacrifices and feasts described in the *Liji* run heavily to meat, especially pork, but also beef, various fowl, and sheep or goat. Millets were the major grains used in these rites, but wheat and rice were also important. Recent excavation of the tomb of Han Jing Di confirms these emphases. Countless pottery models of horses, cattle, sheep, goats, pigs, dogs, and chickens were found. Masses of grain were largely millet, but sorghum has been identified (a quite early record for this native of Africa), and there are other grains.[19]

Wild plants mentioned in the *Book of Songs* were, already, species that grow in human-influenced landscapes. Plants mentioned in the *Songs of Chu (Chuci)*[20] are even more weedy; the ones that are not cultivated are mostly those typical of fallow fields and hedgerows.

Beginning in the late Warring States period, markets, cities, armies, and often dramatic political-economic activities mushroomed. The Daoists (and the thinkers that A. C. Graham calls the Primitivists and the Tillers) found this no gain. The *Daode jing* idealized the isolated village community in which people overheard the sounds of their neighbors' farm animals but had no occasion to visit.[21] Such communities lasted until recently; I have visited some places almost as remote, and Philip Huang tells of visiting a North Chinese village in which few people had ever gone as far as the next village, which was indeed within earshot.[22]

Landholdings ranged from large estates to small private farms. The warring states experimented with different agrarian policies. All early sources agreed that people had once lived as hunters and gatherers in tree nests and had later held at least some lands in common. However, the ideal of the small yeoman farm, with an independent farm family producing for both subsistence and sale, was already widely established in the late Warring States period. Soon after, Qin and Han policy declared that "agriculture is the basis of the state." Of course, land reform did not work very well, but, alas, it almost never has in history.[23]

Ecological crises are not new. In China, they are very old, and, like ours today, they involved the entire world—that is, all the world the Chinese knew, their "all under Heaven." During the Han dynasty, cultivation had disrupted the drainage system, producing floods and droughts. Population was rising—the first census, in 2 C.E., showed about sixty million people in the Han realm. Population appears to have been substantial long before that; some of the late Warring States seem to have been very crowded places. Famines were well known. The relatively primitive cultivation system in most of the realm was simply not adequate to feed many mouths. Even the more productive system in the Yangzi valley was possibly stretched thin in some areas.

The state of Qin, located as it was in a dry area subject to violent climatic fluctuations, was one of the first to take agricultural policy seriously. When it conquered the other states and brought a true imperial regime to "all under Heaven," agriculture became a focal concern everywhere. The Han government sensibly built on Qin initiatives to develop agriculture. The Han dynasty launched a genuine green revolution, the world's first.[24] In the process, they virtually created agricultural development policy, as a concept and as a priority. They perfected the ever-normal granary system, and invented agricultural extension, case-control experimentation, and government research. They developed high-yield grains and highline production processes. The major period of innovation fell between 180 and 120 B.C.E. This was also the time period when most of our received texts of early Daoist and Confucian thought were being edited. Naturally, passages dealing with nature, cosmology, and conservation were preserved and emphasized, and texts were edited to highlight such concerns. The received texts were thus colored by the Han ecological crisis and by early attempts at what was to be its fairly successful resolution.

III

One can rather arbitrarily distinguish three ways of relating to the environment. First, there is direct observation and accurate description. Second, there is direct logical inference—often wrong, but based directly on experience, and especially on knowledge of cause and effect. People infer intervening variables according to their preconceived notions about the world. Thus, in Chinese folk logic, a particularly large old tree must have a particularly powerful spirit, and an undercut and failing slope must be the result of disturbing a dragon. This was a perfectly logical inference, given the animistic assumption that the world is inhabited by spirits. Animism, in turn, was the most economical and reasonable way to make sense of the world, given the

knowledge that the ancient Chinese had of it.[25] Insofar as we now disbelieve animism, we do so largely on the basis of observations made
with equipment not known to the ancients.

Third, there is environmental belief based on experiences that are
not directly connected with actual observation and experience. Such
beliefs can result from visions, hallucinations, and dreams; or they
can result from indirect and tenuous extensions (logical or otherwise)
of more directly grounded knowledge. Thus, once one has inferred the
existence of dragons, one can go on to infer more and more things
about them—assisted by dreams and visions. Soon, one is far from
any experiential reality. The less one interacts with nature, the more
one can construct fantasies about it. Farmers and fishers must base
their lives on actual experience, even if they do infer dragons in the
hills. Philosophers and divines, however, can construct without con-
straint their own phenomenological realities. (I am aware that I am
committing myself to a broadly "Aristotelian"—and, it may be said,
classical Chinese—view, as opposed to a "Platonic" one. So be it.)

"Daoist" writers of the time are labeled more or less in retrospect;
it was Han dynasty scholars, writing a few centuries later, who defined Daoism as a school. Yet, there are common threads among the
major writers and books involved. In particular, the leading works
later called "Daoist" were the *Zhuangzi* and the rather mysterious
Daode jing. Both are compilations rather than clearly single-authored
texts. There is a great deal of the *Zhuangzi* that is by a single hand,
presumably that of Zhuang Zhou ("zi" means "philosopher" here)
himself. This body of text is known as the *Inner Chapters*. Other
chapters, however, not only reflect different and later hands, but different philosophies too; they are only loosely "Daoist" in intellectual
affiliation. Writers of these chapters have been given labels such as
"the Primitivist" and "the Agrarianist." Their chapters include some
of the most interesting naturalistic observations. The *Daode jing* is a
more integrated work, but it is short and cryptic. Another work of interest is the *Liezi*, a collection of teaching stories compiled during the
Han dynasty. It is more or less in the *Zhuangzi* tradition, but reflects a
range of views.

From these and other roots, an actual Daoist religion, with its own
canon and rituals, emerged in the Han dynasty; it survives today, with
the expectable changes and reinterpretations. Its canonical writings

now run to thousands of volumes (see especially the papers in section two of this book).

It seems obvious that early Daoist thought about the nonhuman environment was built, to some extent, on existing folk belief. Moreover, the early Daoists wrote rather more about the supernatural aspects of the nonhuman world than about the natural ones. We hear not so much of trees as of sacred trees, not so much of real animals as of phoenixes and dragons, not so much of the real wind as of forces that the shamanic Daoist can ride or control. Even the *Zhuangzi*'s highly naturalistic story of the hunter who realizes he is himself hunted,[26] a fine exemplar of the "food chain" concept, includes a supernatural bird (a "magpie" in Graham's translation). Natural things, such as water, are used in a metaphoric or symbolic sense. We are definitely not in the universe of the *Shijing* (Book of Songs), whose superbly accurate observations of plants and animals reveal so much about Chinese nature in an earlier century.

Zhuangzi is the most sharp-eyed observer of the nonhuman world among Daoist writers known to me, but even his observations of animals and plants are rarely realistic. He gives us a gourd big enough to make a boat and a yak "as big as a cloud hanging from the sky,"[27] and countless trees too huge and strange to be used—unlike real trees, which bear fruit and other utilities.[28] Nature is a source of fantasy and imaginative symbol, not a reality to record. The dead become "a rat's liver" or "a fly's leg"[29]—dubious transformations in ecological terms; the point is that they get folded back into the great transforming flow of things. Realistic stories of gibbons and trees follow a wild tale of a dragon.[30] Zhuangzi's most famous stories, the tales of happy fish, dreaming butterflies,[31] and tortoises rejoicing to be alive, though ignored, are scarcely biological fact.[32] The frog in the well is realistic enough, but his encounter with the tortoise of the Eastern Sea is not.[33] The wishes of birds and fish are rather realistically described, but hardly for zoology's sake.[34]

Though he failed as a natural historian, he constantly used nature not just as metaphor but as real dynamic analogy. For early Chinese philosophers, analogies were not mere fortuitous or cute similarities, but were—sometimes—actual organic and systematic equivalences based on real relationships. A good philosopher knew which were fortuitous, as opposed to ones that were genuinely revealing—as in

Mencius's Ox Mountain story. Striking parallels or similarities showed the operations of the same qi (vital energy) or *li* (principle) or other cosmic forces.[35]

Liezi, too, tells us of animals unnaturally peaceful[36] or trained to fight for the emperor,[37] of a golem,[38] and, again, of wildly unrealistic transformations.[39] A fantastic set of transformations is ascribed to Liezi by Zhuangzi[40] and greatly extended in Liezi's own book.[41] Zhuangzi's realistic and stoical acceptance of death occasioned the most realistic naturalism in the books under discussion, but this was soon replaced by a focus on prolonging life, which led to extremely credulous passages.[42]

In general, the cosmology and "nature" reflected in early Daoist thought was a fantastical one, visionary and imaginative. Norman Girardot, Sarah Allan, and others introduce us to a world of primal chaos, giant gourds, the Queen Mother of the West with her peaches, and flying phoenixes.[43] In later centuries, the cults of deities and of immortality got still more extreme.[44] Much of genuine medical value resulted, but much dubious lore came with it.

A passage in which Robber Zhi rebukes Confucius has a reasonable scheme of human evolution, but it is not a very Daoist passage.[45] Later Daoists echoed it,[46] but it seems to have been a generally shared belief in old China, not a specifically Daoist point.

Joseph Needham may have been correct in his contention that Daoism was basic to Chinese science, but Daoism was not science. I am by no means persuaded by the recent revisionist attacks on Needham; he has certainly shown, beyond reasonable doubt, that Chinese science was extremely well developed, and that much of it arose from Daoist practice. Alchemy, for instance, most certainly begat serious chemistry and some serious medical practice. Sex, longevity, and drugs of all sorts were studied in detail, often by methods reasonably scientific (especially compared to the West of the same time periods). Botany nested in medical writings that had Daoist associations.[47] However, the vast majority of Daoist writings were typical religious productions. They told of saints and spirits, revelations and epiphanies.[48]

The same may be said of the arts. The brilliant, meticulously accurate, and mystically insightful landscapes and still-life paintings of later Chinese art are strongly and explicitly influenced by Daoism. However, both early Chinese art and Daoist religious art of later periods portray visionary scenes and beings far from everyday reality.[49] Mu Qi's persimmons are not just Daoist; they are the product of centuries of interaction between Daoism, Chan, and the Chinese folk beliefs and art traditions that underlie all Chinese teachings. Similarly, China's incomparable tradition of nature poetry goes back to the *Shijing*, not to specifically Daoist antecedents.

The Daoists said it themselves: the true judge of horses is indifferent to the appearance of the horse—he calls a black stallion a yellow mare, because he "goes right inside it and forgets the outside."[50]

Daoist mystical ecology grows within a wider universe of mystical geography and cosmology[51] from which it should not be isolated. The Chinese of the Warring States and Han periods looked behind the screen of reality to a visionary universe of soul travel (see the *Chuci*), fabulous beasts, gods and goddesses, and mysterious transformations. There is no evidence here of an environmentalist consciousness in the modern American sense—or in the Mencian sense. Rather, the desire is to transcend visible reality and put oneself in harmony with subtle flows and Ways—the inferred processes behind perceived facts. The *Daode jing* has virtually nothing to say about nature in the concrete; it merely tells the reader to stay in harmony with the Dao. Natural things, such as rhinos and tigers, are mere exemplars.[52] The description of the Daoist state, noted above, is the only real exception.

By contrast, the Confucian tradition, from Confucius's hunting rules to Mencius and the *Liji*, reveals a genuine knowledge of, understanding of, and desire to work with nature. The nearest equivalent word to "nature" is *xing*, but here I mean also nonhuman beings in general. Mencius's Ox Mountain story, analogizing nonhuman and human worlds, finds many later Daoist echoes,[53] but it is the Confucians that have the priority here. General beliefs about *xing* were evidently shared widely. Conservation was ritually represented and ritually sanctioned. Mencius is the major and pivotal source.[54] Mencius, perhaps more than any environmentalist before or since, understood both the practical and the spiritual reasons why conservation is necessary to individuals and states.

By early Han (second century B.C.E.), the probable time period when all these various texts were being put in order and taking their final form, environmental management was at the heart of *li* (usually translated "ritual," but really "social forms"). The *Liji* tells us: "The ruler of a state, in the spring hunting, will not surround a marshy thicket, nor will Great officers try to surprise a whole herd, nor with (other) officers take young animals or eggs."[55] Marshy thickets are favored refuges for animals, and the whole passage is unmistakably conservationist. These proscriptions follow the teachings of Confucius and Mencius—not yet sacred writ, but sacrosanct enough to guide ritualized and religiously represented practice. Conservation was a sacred duty.

The whole of Book Four, the *Yue Ling*,[56] is devoted to rules for the various months. This book, in turn, is taken from the *Spring and Autumn Annals*, with additions. (Whether or not Confucius really compiled the spring and autumn annals, the whole tradition is broadly Confucian, but it entered an eclectic-Daoist world too, being incorporated into the *Huainanzi*.)[57] Sacrifices, hunting activities, cultivating activities, and other actions are timed to go with the natural cycles of the North China Plain (an area roughly corresponding to the midwestern United States in geography). In spring, for instance: "Nests should not be thrown down; unformed insects should not be killed, nor creatures in the womb, nor very young creatures, nor birds just taking to the wing, nor fawns, nor should eggs be destroyed."[58] These are very reasonable rules, both ecologically and religiously—they accord with the order of nature as well as with hunters' common sense. If people carry out autumnal activities in spring, weeds take over the field—another bit of lore that accords with both practical experience and ritual correspondence. In late spring, nets are banned, and people are forbidden to cut down trees to get silkworm feed;[59] in winter, game wardens watch for poachers.[60] The emperor ate seasonal food—for instance, in autumn he ate marijuana seeds (for food, not as a drug) and dogs' flesh.[61] Seasonal tasks are allocated to game, fish, and forest wardens.[62]

In short, in the *Liji* we have an extremely detailed, explicit, and religiously sanctioned plan for resource management and conservation. The Han compilers of the *Liji* are quite explicit about the usefulness of religious belief in backing up common sense, and they share my Hong Kong friends' idea that the order of nature is divine and is something we humans should discover and follow. Insofar as the *Liji* is "Confucian," it is the Confucians, not the Daoists, who are the ecologists of East Asia. In fact, of course, the received text of the *Liji* reeks of Han syncretism. So does much of the Daoist material we have. Surely it was some Han environmentalist who had the brilliant idea of sticking the "Primitivist," *Nongjia* (Agrarianist) and Yangist writings onto the *Zhuangzi* text. The *Huainanzi* contains a summary of conservation material taken from the *Liji* or a closely related source.[63]

As noted above, the Han dynasty also created agricultural experiment stations and extension services.[64] These seem to have been inspired by yet a different philosophic tradition: the less extreme, more economically sophisticated version of Legalism (*fa jia*). This influenced, among other things, early agricultural extension manuals that taught conservation farming. The earliest one that survives is *Fan Shengzhi's Book*, dating from around the time of the Han emperor Wu.[65] It presents to us an agricultural science that is incredibly accurate, sophisticated, meticulous, and advanced in its ecological and environmental thinking.[66] Fan's book differs from the Confucian works (including the *Liji*) in that it does not talk about conserving game animals. (Perhaps lost sections of it did speak to that issue.) It does, however, have a great deal of highly sophisticated information about conserving and managing water, soil, and vegetation. Again, such ecological detail (driven by real farming concerns) is conspicuously absent in the Daoist works. These latter are more apt to speak against sophisticated new techniques, seen as detrimental to primal simplicity.

On the other hand, the Daoists eventually came to follow the Confucians in setting ecological rules. These first surface in the *"One Hundred and Eighty Precepts."*[67] These precepts for Daoist officiants include a comprehensive range of ecologically and environmentally sound rules. They later occur in various forms in a succession of Daoist works. Livia Kohn, for instance, quotes a Lingbao work in which we find the following: "[t]he sin to burn down fields or mountain forests. The sin to cut down trees or idly pick leaves and plants. . . . The sin to throw poisonous drugs into fresh water and thus harm living beings"[68] and "[t]he sin to capture and imprison wild birds or free animals."[69] These specific sins clearly resemble earlier Confucian and *Liji* rules more than Daoist traditions.

Thus, in time, Daoism did come to preserve nature. Daoist mountain retreats became nature sanctuaries. The Daoist temple at Castle Peak Bay—the Qing Shan temple—may be taken as an example of the best Daoist practice. It was situated on the ravaged slopes of Castle Peak (Qing Shan in Chinese). Castle Peak, like Ox Mountain, had been subjected to fire and woodcutting until it was a bare waste. The

Daoists redeemed a large part of this mountain. The temple directly managed a large area (tens of acres) as garden, grove, and fengshui woodland. This was allowed to grow to become forest. The nearer to the temple, the more carefully it was pruned into "garden" form; farther away, it was allowed to grow up as the mountain qi directed. Moreover, the temple's influence did not end there. Indirectly, it improved the environment of the whole mountain. No one would casually burn or ruin a mountain singled out as a Daoist site. Rather, pilgrims—some not much more than Sunday hikers, but still with religious feeling in their minds—climbed the peak from the temple and cared for the whole area.

IV

The age that gave us Fan's book also seems to have produced the versions we now have of the classic Daoist works and the *Liji*. All these share a broad and comprehensive concern with the cosmos and with humanity's place in it. All agree that humanity is part of a greater whole; that humans have an inner nature, related to outer natures; that this nature must be harmonized with the cosmos; and that the cosmic order must be preserved, with some degree of human help. All share at least some realistic, incisive observations of natural phenomena.

That said, the documents at hand range from hard-headed, practical advice for farmers to the soaring, visionary, imaginative tales of Liezi and the wildly lush and erotic nature poetry in the *Chuci*. There is a common core in folk ecological knowledge, folk religion, and traditional cosmology. Various philosophers and writers took their own ways in going outward from that common core. The Daoists went toward the skies and the remote and unknown regions of shamanic journeys. The Confucians, concerned with social responsibility as always, went toward conservation rules. The Mohists, concerned with universal love, also advocated the care of nature. The soft-path Legalists went toward agricultural manuals and experiments. The Han government accommodated all.

A Han individual with an environmentalist conscience—and there must have been many such—would have taken a general outlook on harmony with the cosmos from Daoism and from everyday religion and worldview; a conservationist view from Confucian traditions; and pragmatic knowledge from the agricultural manuals and economic writings.

However, the future association of nature and Daoism in poetry, art, and science was not without foundation. The Daoists, from the start, focused on human relations with the cosmos and on the inability of humans to achieve transcendence unless they understood and followed the "Way" of the great universe. The term *dao* was shared, of course, by all early schools; the special Daoist contribution was to concentrate on a mystical Way that was the dark, occult basis of all things.

In doing so, they directed human attention outward, toward the nonhuman cosmos. They also stressed the side of folk knowledge that saw the relationships and interactions of all things and their environments—in short, folk ecology.

John Patterson, in commenting on this paper in its original form (its "uncarved block" state?), pointed out that Daoist thought has been characterized by several foci that are now central to ecology.[70] He discussed three of these: cooperative adaptation, mutualism (mutual interactions, for good or ill), and benefit (the value of such mutual interactions when they are beneficial). Indeed, these are characteristic of Daoism—and also of ordinary Chinese folk morality, as I found out in Castle Peak Bay.[71] Maintaining mutually beneficial relationships with people, supernaturals, and the environment was the explicit goal of most formal activity and much, if not most, informal activity (whether religious or secular) at Castle Peak.

Thus, the Daoists provided an environment-oriented counterbalance to the Confucian focus on human life. Later, as China's growing population and economy impinged more and more on mountains and plains, Daoism became more ecologically aware, and such awareness became more necessary. The final decline of Daoism to near eclipse in the twentieth century was accompanied by a vast and devastating onslaught on China's fragile environment.

V

The more important question for us is the religious representation of this environment in that time.[72] There are many aspects of more recent Chinese folk religion that seem quite old. They are associated with a belief system that goes back to shamanism and with what Edward Tylor called "animism."[73] In particular, they seem implied in the early Daoist writings. In what lies below, I follow in the hallowed footsteps of those who have understood early Chinese thought through ethnographic analogy and through analysis of folk belief—the footsteps of Marcel Granet, of Edward Schafer and Wolfram Eberhard, and, in our own time, of Sarah Allan and many others.[74]

My introduction to Chinese traditional beliefs about the land came in the fall of 1965. I had just taken up residence at Castle Peak Bay, then an isolated, traditional, rural part of the New Territories of Hong Kong. One felt that one was living in a Qing dynasty village.

A first touch of modernization was coming in the form of a new hospital. Digging for the foundation of the hospital had undercut the slope of a hill. Older residents of the vicinity complained: "This is cutting the pulse of the dragon. Disaster will follow." Inquiring, I learned that every hill has within it a resident dragon. If his pulse is cut, some problem automatically occurs. I fear I thought of this as a quaint folk belief. But, when the next rains came, the undercut slope failed, and a landslide buried several houses. The old people said: "This is what happens when the dragon's pulse is cut."

I do not believe in the dragons, but I do know that slope failure is real, and that we moderns—like the ancient Chinese—have to infer the mechanism.[75]

At that time, I began to realize that fengshui was a science. Fengshui at the folk level centers on the pragmatic questions of siting buildings and graves so that they can be benefited rather than hurt by wind and water. Villages are situated on gentle slopes and on the lee side of hill ranges, so that they get gentle breezes rather than typhoons and reliable rains rather than floods. The ideal village is protected on all or most sides by hills. It lies at a point where streams come together and pool up. It is above the level of floodwaters. It is convenient to fields but does not occupy prime agricultural land. Graves are situated higher on remote and infertile hills, where they have beautiful views but do not interfere with cultivation. Groves of trees around villages and temples provide timber, firewood, fruit, and shade. Thus, the core of fengshui is purely practical. However, it extends into more metaphysical dimensions.[76]

The rural people of Castle Peak Bay knew their landscape in incredible detail. They had names and uses for every herb and grass.

They knew the birds, the winds, the sea currents. They knew every rock and every trail. I learned to pay attention to everything they said, for all of it was based on an encyclopedic understanding of the realities of local ecology. If they postulated something that seemed unreal to me, it was because they had reasoned logically from premises based on a cosmology I did not know.

As I worked on, I found that the traditional Chinese science of fengshui unified many such beliefs into a consistent picture.[77] I call fengshui a science because it was purportedly based on objective, factual principles, rather than on the will of supernatural beings. However, others have called it religious because it involves belief in a number of supernatural forces and presences.

Actually, the European opposition of "religion" and "science" is inadequate in talking about much of Chinese folk belief. "Religion" usually implies some sort of normative set of beliefs about supernaturals who act willfully and must be influenced by prayers, rituals, and offerings. "Science" implies some sort of open, flexible set of beliefs about the way the world works when left to run automatically, whether according to scientific laws or to more modest descriptive statements. Chinese folk ecology did not separate these matters, and I was repeated told so, explicitly, by people from all walks of life. Ecological practice[78] involved following natural rules, insofar as these were known, and also worshiping the appropriate beings. I was told that the gods must want the world this way, because they made it this way, and it is our job to understand that and act accordingly. Yet, also, a proverb said that "even the gods are subject to chance." Natural law is divine law, but perhaps both are less "law" and more "luck" than we might wish.[79]

Moreover, the people of Castle Peak Bay did not distinguish between "religion" and secular life. They made quite a different separation, and a revealing one: they contrasted ordinary life *and* folk religion with the organized bodies of teaching known as *jiao*. These last included Confucianism, sectarian Daoism, Buddhism, and Christianity. The ordinary folk of Castle Peak Bay also made no distinction between "supernatural" and "natural." They saw the cosmos as a shifting play of forces, in which gods, ghosts, spirits, and visible fleshly persons all interacted constantly. They did make a differentiation: some beings had to be contacted via incense and sacrificial offerings, while other beings, more visible and tangible, ate more material food and could be talked to directly. But they insisted that all were part of the same cosmos. There was no word similar to the English "nature," i.e., the real, tangible things of this world that are not human or immediately human-created. Chinese words like *xing* are, of course, vastly different from the English words so often used to translate them.

In Chinese folk ecology, the cosmos is held together by the flow of qi. This subtle "breath" circulates along channels similar to veins. It can rush, swirl, pool up, and otherwise act like a liquid as well as like a subtle force or energy. There are particular nodes where qi flow is especially pronounced; these more or less correspond to acupressure points on a human body.[80] There are places where one can feel the flow of qi. Often, one can feel the earth moving at such points. We in California call them earthquake faults. Qi flow was compared to the wind, which one could sense but could not see.

Dramatic or strangely shaped rocks ("qi stones") and crags show that they have a great deal of qi. So do old, gnarled trees. The knots and twists of an aged tree are evidence of the swirling, eddying flow of qi. Thus, ancient trees and striking crags were paid reverence. Incense sticks were burned in front of them at dawn and dusk.

Trees, rocks, and, indeed, all beings also had some sort of indwelling spirit or soul. Knowledge of the exact nature and structure of this spirit was vague. Human souls were divided into the familiar *hun* and *bo* components, but each of these had its own complex structure, about which no two people seemed to agree. The level of consciousness of tree, rock, and landscape spirits was also debatable. Rituals thus not only invoked the power of qi; they also called down, and opened communication with, animistic entities.

Rituals might directly influence the flow of qi. They certainly created a sacred space; in rural Cantonese thought, a sacred space was created and defined by the smoke of incense more often than by a formal boundary line. They also pleased the supernatural beings, who then were disposed to do what the worshiper wanted. Sacrifice could pay back a supernatural who had been helpful, or incur debts that a well-disposed supernatural would be happy to repay. Failure of offerings could bring trouble—either the automatic working of qi or the active anger of an offended god or dragon.

When rural Chinese looked out over a landscape, they saw a world they knew in meticulous and microscopic detail. But they saw more than details. They saw a play of invisible force, operating in a landscape inhabited by countless beings, tangible and intangible.

Intangible qualities other than qi inhered in the landscape. These included good and bad influences (sometimes identified with types of qi flow, sometimes seen as something else). Such influences were impersonal and quite separate from whatever benevolence or malevolence the gods would send. They were also different from simple luck. These flows of good and evil force were tapped by fengshui. A well-sited building or grave lay at a point where good influences came together but the flow of bad ones was blocked. Cold and hot winds, the invisible arrows of disease, and pure and impure forces were also part of the intangible scene. The conscious, agentive beings of the other realm—the gods, saints, spirits, immortals, ghosts, and ancestors—had to deal with such invisible currents, just as humans do. They sometimes had more control than we do, but not always.

The relation of humans and these intangible persons was mutualistic. Humans controlled the wealth goods of this world: food, gold, silver, clothes, houses, and much more. The spirits controlled some share of luck; they could influence destiny. They could help or harm by influencing disease, wealth accumulation, weather, and other important variables that do not seem to be under human control. Thus, sacrifices of food and "paper goods" (paper representations of wealth goods) were intended to initiate a trade, exactly as goods in this ordinary world create *ganqing* (good feeling, sympathy) and thus *guanxi* relationships (helpful reciprocity).[81]

Several of the beliefs current in Hong Kong in the 1960s already existed in the late Warring States and early Han periods, and are attested in the early Daoist texts. Notable among these are sacred trees, sacred rocks, sacred waters, and sacred mountains. The flow of qi seems implied. Transformations and conversions from one state or being to another were common. People had access to the supernatural by similar means: trance, meditation, and formal ritual, including sacrifice rites. Sarah Allan has shown that rather similar general beliefs and practices lay behind both Confucian and Daoist philosophy.[82] Zhuangzi, for instance, tells us a good deal about sacred trees. Good and bad fate were partially (but only partially) under the control of supernaturals who could be influenced (somewhat) by sacrificial rites. I am not aware of any discussion of fengshui per se, but sacred geography was well established.[83] By inference, good and bad influences were already postulated. Dragons writhed in the clouds and waters and, perhaps already, in the hills. By Han times, we have good evidence for a whole magical geography, in the mountain censers, mirrors, Mawangdui art, and much else.[84] This inner world of landscapes is different in detail from what I found in Hong Kong in the mid-twentieth century, but there is a close "family resemblance." Specifically, in addition to the reverence for rocks and trees, we can observe a sense that humans were ritually and spiritually bound closely to the landscape and that both were affected by the dynamic changes and fluxes that seem implied in Daoist concepts of the working of the Great Dao.

This is not to say that modern religion is identical with the folk-religious world known to early Daoist philosophers. There were many striking differences. For instance, soul travel was frequent in those days, but had become rare by the twentieth century. (I have never seen a case, but I have heard tales of shamans who still do it.) China's ancient religious practitioners included many shamans, persons who could send their souls traveling to the heavens and elsewhere (as we know from the *Chuci* and other early poems). Modern Chinese *wu shi* and *dang-ki* are sometimes called "shamans," but, in the vast majority of cases and regions, they do not send their souls out; rather, they become possessed by supernatural beings. They are thus spirit mediums rather than shamans, according to the standard anthropological usage of these terms. This is a fairly pedantic point, but it has significance here. The wind-riding, soul-traveling, and spirit-journeying of early Daoists has been seen (surely correctly) as shamanistic, as based on folk shamanic traditions. In Hong Kong in the 1960s, Daoist (and folk) religious officiants were possessed by the deities; they did not often engage in soul travel.

We have now moved fairly far from ecology, so let us return to Daoism and its actual relationship with nature.

VI

In summary: Ancient Chinese "philosophy" was not, or not only, what we now think of as philosophy; it was more like political science, even when it appears otherwise to our modern eyes. It was advice to rulers, and was intended as a direct guide for statecraft. The Mawangdui manuscripts have proved just that for Daoism[85]—indeed, the point had been made by many writers long before the Mawangdui discoveries. Their morality consisted largely of rules for a livable society—principles that a ruler might adopt.

The early Daoists saw that, to be decent and civil, society must take nature into account. However, the purpose of natural history for them was not to provide factual truth, but to provide symbols and analogies for teaching and to provide a general sense of the underlying principles of the cosmos.

By contrast, Confucianism made real and serious use of ecology. There is nothing in all the Daoist literature to parallel the sharp, accurate, and genuinely brilliant union of ecology and psychology in Mencius's Ox Mountain story. There is nothing to match the conservationist teachings of Confucius and the *Liji*. Daoist looseness about government would not be expected to find a place for the game, fish, and forest wardens seen as necessary by the Confucians (and the Han syncretists building on their teachings).

Early Daoism was neither environmentalism nor social glue. It was based on an ecological vision, but was not ecology. It expressed a social philosophy, but was not sociology. Daoism surely originated from the pragmatic, nature-managing, nature-aware aspect of Chinese folk religion, but it moved on into political and religious quietism. By contrast, early Confucianism stemmed from the social-ordering, social-structuring aspect of the same folk tradition. Beliefs about nature, whether human or nonhuman, came into all these traditions from the folk worldview.[86]

When these philosophies were often combined in the beliefs of individuals,[87] I think it was often less a "syncretism" than a logical pursuit of two different implicational streams that flowed from a common folk source. I believe the evidence suggests sources in the folk belief system for much of the ecological wisdom—facts, fantasies, and ideology—in all the early texts. It continued to be the repository of such valuable environmental concepts as fengshui.

VII

Many authors, including Baird Callicott,[88] have maintained that Daoism is the ideology we need if we are to save the environment. Their claim

is not without justice, but, if they knew more, they might perhaps prefer both Han syncretism and Chinese folk religion to organized Daoism.
Still better, they might combine all of these in a new syncretist wave.

Perhaps, in such a new syncretism, Daoism would contribute its
general sense of the environment, its clear sight, its mysticism. Traditional Asian folk religion would contribute its own, more individual,
more impassioned involvement with local environments. Confucianism would contribute its moral responsibility in conserving and managing resources and the economy. The Judeo-Christian tradition
would contribute its concepts of stewardship and mutual responsibility. Above all, I would make the ethics of Emmanuel Levinas central
in such an enterprise.[89] Levinas, teaching the literally infinite importance to the Self of the Other, and thus our infinite obligation to the
others around us, seems to me to take a Daoist spirit to new realms.
Levinas makes explicit an interactive approach to the world that might
be seen as implicit (or explicit—but in mystical terms) in the great
Daoist classics. I believe these interactive ethics afford real hope for
an ecological vision. This, obviously, is an agenda that lies beyond the
present paper.

Notes

I am deeply indebted to the organizers and conferees of the conference on Daoism and ecology for their many helpful comments, and especially to my commentator, John Patterson; for truly wonderful insights and help. I am also indebted to my colleagues Vivian-Lee Nyitray and Lisa Raphals and to Norman Girardot and anonymous commentators. I acknowledge special inspiration from the writings of Sarah Allan (her *The Way of Water and Sprouts of Virtue* [Albany: State University of New York Press, 1997] is the most direct inspiration for the present paper), and from discussions with her, all too brief and too long ago.

1. On general philosophical issues of Daoist history and ecology, see Sarah Allan, *The Shape of the Turtle: Myth, Art, and Cosmos in Early China* (Albany: State University of New York Press, 1991), and also her *Way of Water*; *Nature in Asian Traditions of Thought: Essays in Environmental Philosophy*, ed. J. Baird Callicott and Roger T. Ames (Albany: State University of New York Press, 1989); Isabelle Robinet, *Taoism: Growth of a Religion* , trans. Phyllis Brooks (Stanford: Stanford University Press, 1997); R. A. Stein, *The World in Miniature: Container Gardens and Dwellings in Far Eastern Religious Thought* (Stanford: Stanford University Press, 1990).

2. *Mencius*, trans. D. C. Lau (Harmondsworth, England: Penguin, 1970), 164–65.

3. Ibid., 164.

4. See reviews such as Daniel Bates, *Human Adaptive Strategies: Ecology, Culture, and Politics* (Boston: Allyn and Bacon, 1988); Fikert Berkes, *Sacred Ecology: Traditional Ecological Knowledge and Resource Management* (Philadelphia: Taylor and Francis, 1999); Brent Berlin, *Ethnobiological Classification* (Princeton: Princeton University Press, 1992).

5. *The Classic of Mountains and Seas,* trans. Anne Birrell (New York: Penguin, 1999).

6. For the record, the facts sustain Ping-ti Ho's claim that the drier loesslands of interior Shanxi and Shaanxi were grass and scrub. See Ping-ti Ho, *The Cradle of the East* (Hong Kong: Chinese University of Hong Kong Press; Chicago: University of Chicago Press, 1975); E. N. Anderson, *The Food of China* (New Haven: Yale University Press, 1988). Farther east and higher in the mountains, though, forest is known to have prevailed. We have no idea where the ecotone lay.

7. A. C. Graham, *The Book of Lieh-Tzu* (London: John Murray, 1960), 46.

8. A. C. Graham, *Chuang-Tzu: The Inner Chapters* (London: George Allen and Unwin, 1981), 118. All references herein to *Zhuangzi* are to Graham's translation, which in spite of massive rearrangements is the most accurate and useful edition. On the topic of gamekeepers, see also *Mencius*, trans. Lau, 106.

9. Graham, *Lieh-Tzu*, 30.

10. See Anderson, *Food*; see also Francesca Bray, *Science and Civilization in China: Agriculture* (Cambridge: Cambridge University Press, 1984).

11. F. H. King, *Farmers of Forty Centuries* (New York: Mrs. F. H. King, 1911); John L. Buck, *Land Utilization in China* (Chicago: University of Chicago Press, 1937).

12. Sorghum may have been known, but was certainly not important.

13. Anderson, *Food*; Bray, *Science.*

14. Anderson, *Food*; Buck, *Land Utilization.*

15. The claim that bean curd was a spin-off of Daoism, invented by the Prince of Huai Nan (*Huainanzi*) or his Daoist nutritionists, has never been substantiated and is almost certainly wrong (see Anderson, *Food*). On the other hand, some evidence for Han bean curd has emerged in the last ten years.

16. Graham, *Lieh-Tzu*, 124.

17. N. J. Girardot, *Myth and Meaning in Early Taoism* (Berkeley and Los Angeles: University of California Press, 1983).

18. Graham, *Lieh-Tzu*, 69.

19. Peanuts are reported—a very striking find, if confirmed, since peanuts are native to South America; see Wang Baoping, "Excavations of Emperor's Tombs near Xi'an, China" (lecture, University of California, Riverside, 11 May 1998).

20. *Ch'u Tz'u: The Songs of the South*, ed. and trans. David Hawkes (Oxford: Oxford University Press, 1959).

21. Cf. Robert Henricks, *Lao-Tzu Te-Tao Ching* (New York: Ballantine Books, 1989), 156 (citing the Mawangdui version; the Wang Bi version is similar here, as we know from so many translations). Ursula K. Le Guin has done an English version, *Tao Te Ching: A Book about the Way and the Power of the Way* (Boston: Shambhala, 1997); see also the present volume.

22. Philip Huang, personal communication.

23. Elias Tuma, *Twenty-Six Centuries of Agrarian Reform* (Berkeley and Los Angeles: University of California Press, 1965).

24. Anderson, *Food*; Bray, *Science*.

25. On the logic of animism see Edward Tylor, *Primitive Culture* (London: John Murray, 1871).

26. Graham, *Chuang-Tzu*, 118.

27. Ibid., 46–47.

28. Ibid., 73.

29. Ibid., 88.

30. Ibid., 120–121.

31. Ibid., 61.

32. Ibid., 122–23, etc.

33. Ibid., 155.

34. Ibid., 189; Zhuangzi's reputation for being in tune with nature is due in part to Primitivist and Agrarianist texts that got swept up into the received *Zhuangzi*. These other blocks are Daoist only by syncretist extension of the term.

35. See Allan, *Way of Water*.

36. Graham, *Lieh-Tzu*, 42–43.

37. Ibid., 54.

38. Ibid., 111.

39. Ibid., 98–99; one transformation on p. 99 that seems believable is that of the citrus that changes to the lowly dwarf orange when planted north of the Huai. This probably refers to the use of the dwarf orange for understock for grafting. North of the Huai, the scion would freeze, and the rootstock would grow; Joseph Needham et al., *Science and Civilization in China*, vol. 6, part 1: *Botany* (Cambridge: Cambridge University Press, 1986).

40. Graham, *Chuang-Tzu*, 184.

41. Graham, *Lieh-Tzu*, 21–22.

42. Graham, *Lieh-Tzu,* 35.

43. Girardot, *Myth and Meaning*; Allan, *Shape of the Turtle* and *Way of Water.*

44. As in *Huainanzi:* see Roger T. Ames, *The Art of Rulership* (Albany: State University of New York Press, 1994); Charles Le Blanc, *Huai Han Tzu: Philosophical Synthesis in Early Han Thought* (Hong Kong: Hong Kong University Press, 1985); John S. Major, *Heaven and Earth in Early Han Thought* (Albany: State University of New York Press, 1993). Cf. also Livia Kohn, *The Taoist Experience* (Albany: State University of New York Press, 1993); Robinet, *Taoism*; Kristofer Schipper, *The Taoist Body*, trans. Karen Duval (Berkeley and Los Angeles: University of California Press, 1993); Holmes Welch, *The Parting of the Way* (Boston: Beacon Press, 1957).

45. Graham, *Chuang-Tzu*, 237.

46. Ames, *Act of Rulership*, 16–19.

47. Needham et al., *Science and Civilization.*

48. See Judith Boltz, *A Survey of Taoist Literature: Tenth to Seventeenth Centuries*, China Research Monograph 32 (Berkeley: Institute of East Asian Studies, University of California, Berkeley, 1987).

49. Cf. Allan, *Way of Water*; Michael Loewe, *Ways to Paradise* (London: George Allen and Unwin, 1979); and Major, *Heaven and Earth.*

50. Graham, *Lieh-Tzu*, 170.

51. Allan, *Shape of the Turtle*; Loewe, *Ways to Paradise.*

52. See Henricks, *Te-Tao-Ching*, 122.

53. See, e.g., *Huainanzi*; quoted in Ames, *Art of Rulership,* 16–19.

54. *Mencius*, trans. Lau, 51–52, 61–62, 106, etc.

55. *Li Chi: Book of Rites*, trans. James Legge, new edition with new material, original edition 1885 (New Hyde Park, N.Y.: University Books, 1967), 106; see further measures on pp. 220–221.

56. Ibid., 249–310 in the recent edition.

57. Major, *Heaven and Earth*, 217–268.

58. *The Li Chi*, trans. Legge, 265.

59. Ibid., 265.

60. Ibid., 304.

61. Ibid., 287–289, 293.

62. Ibid., 433.

63. Ames, *Art of Rulership*, 163, 201; see his note on p. 239, tracing the attitudes back to Mencius.

64. Anderson, *Food.*

65. Shih Sheng-Han, *On "Fan Sheng-Chih Shu"* (Peking: Science Press, 1974).

66. Anderson, *Food*; Shih, *Fan Sheng-Chih.*

67. Kristofer Schipper, "Daoist Ecology: The Inner Transformation. A Study of the Precepts of the Early Daoist Ecclesia," in this volume.

68. Kohn, *Taoist Experience*, 103.

69. Ibid., 105.

70. See John Patterson and James Miller, section 3 discussion: "How Successfully Can We Apply the Concepts of Ecology to Daoist Cultural Contexts?" in this volume.

71. Mayfair Yang, *Gifts, Favors, and Banquets: The Art of Social Relationships in China* (Ithaca, N.Y.: Cornell University Press, 1994).

72. Anthropologists have always been slightly schizophrenic about religion and cosmology. The first person to hold a professorship in anthropology, Edward Tylor, in his *Primitive Culture* explained traditional religion as stemming from curiosity about the world—from questions wrongly answered because science had not advanced to the state where better answers could be given. Emile Durkheim, in *The Elementary Forms of the Religious Life*, trans. Karen E. Fields (New York: Free Press, 1995; French original, 1912), in contrast, saw religion as a collective representation of society. Religion and cosmology provided emotionally involving symbolic expressions of the social contract.

73. Tylor, *Primitive Culture*.

74. Marcel Granet, *Chinese Civilization*, trans. Kathleen Innes and Mabel Brailsford (London: Kegan Paul, Trench, Trubner and Co., 1930); Edward Schafer, *The Divine Woman* (Berkeley and Los Angeles: University of California Press, 1973); Allan, *Shape of the Turtle* and *Way of Water*.

75. As I have suggested in my *Ecologies of the Heart* (New York: Oxford University Press, 1996), this allows us to go beyond the foolish opposition of positivism and postmodernism. Our working knowledge—the knowledge that lies behind actual practice—is not absolute "facts," but not mere "texts" either. We have real factual knowledge of the world, but we have to construct stories—texts—about it. These stories are subject to change as we interact with each other and with the world.

76. See E. N. Anderson and Marja Anderson, *Mountains and Water: The Cultural Ecology of South Coastal China* (Taipei: Orient Cultural Service, 1973); and Anderson, *Ecologies*, for my full arguments on fengshui, with reviews of the literature.

77. Anderson and Anderson, *Mountains and Water*; Anderson, *Ecologies*.

78. In using this term, I follow A. Endre Nyerges, *The Ecology of Practice* (Amsterdam: Gordon and Breach, 1997), who in turn has been influenced by Pierre Bourdieu, as am I, especially by *Outline of a Theory of Practice* (Cambridge: Cambridge University Press, 1977) and *The Logic of Practice*, trans. Richard Nice (Stanford: Stanford University Press, 1990). I am indebted to Dr. Nyerges for very valuable exchanges of ideas on this issue. Practice theory has been brought to the study of Chinese religion, also.

79. Philosophic Daoism, of course, is subject to similar protean labeling. It is, at once, "science," "religion," "philosophy," and "political science"—but of course it is really all of the above, none of the above, both all the above and none of the above, and neither all the above nor none of the above. By contrast, the actual Daoist *religion*, the sectarian *Daojiao* that arose in the Han dynasty, is a perfectly straightforward religion in Western terms. It has its own sects, scriptures (the *Daozang*), rituals and practices, and beliefs.

80. The macrocosm/microcosm correspondence in Chinese folk thought is well known (Schipper, *Taoist Body*; Stein, *World in Miniature*), but its complexities have never been fully described. Obviously, there are only some ways in which the earth is like a body. Of these ways, only some were believed to be true correspondences (as opposed to fortuitous resemblances). What correspondences were recognized, and what ones were actually meaningful, is a matter that has not, to my knowledge, been systematically studied. The cited works of Allan and Loewe provide valuable beginnings.

81. Anderson and Anderson, *Mountains and Water*; Yan Yunxiang, *The Flow of Gifts* (Stanford: Stanford University Press, 1996); Yang, *Gifts*.

82. Allan, *Shape of the Turtle* and *Way of Water*.

83. See, e.g., Loewe, *Ways to Paradise*.

84. Allan, *Way of Water*; Loewe, *Ways to Paradise*; Major, *Heaven and Earth*.

85. Henricks, *Te-Tao-Ching*.

86. On the other hand, Needham is clearly correct in stressing the influence of at least some aspects of Daoism on the development of Chinese knowledge of the world. One need think only of the great polymath Tao Hongjing, one of the most brilliant and wide-ranging intellectual figures in all world history. Tao's career is long overdue for critical historical analysis.

One of the problems in any scientific tradition is that science has to balance two needs: the need for generalizing theories and the need for skeptical, critical stances. These two needs always get in each other's way. We have trouble with this today and so did the early Chinese. The conflict was clearest in Han times, when Wang Chong represented one pole and men like Dong Zhongshu represented the other. Wang's skepticism was too much for an early science to bear, and he had relatively little effect. Dong's (and others') excessive generalization and schematization had deadening effects on future thought. This matter, too, needs much more attention than it has received. One can only wish that more scholars would take Han thought seriously—as seriously as we take late Warring States thought. Adulating Zhuangzi and Laozi but totally ignoring Wang Chong and Jia Yi does not seem either fair or reasonable. It costs us the opportunity to benefit from some of the finest thinking in any early civilization.

87. Such eclectic drawing from various traditions to address real-world problems is precisely what we see in such Han debates as Pan Gu's *Po Hu T'ung* (trans. Tjan Tjoe Som [Leiden: E. J. Brill, 1949]) and the *Discourses on Salt and Iron* (trans. E. M. Gale, new edition [Taipei: Ch'eng-Wen, 1967]), both of which—the latter especially—would repay further study by ecologists. It has been suggested that the received form of the *Zhuangzi*, complete with the various additions, was put together at the court of Liu An, as was the *Huainanzi* (Major, *Heaven and Earth*, 9–10). This, of course, is difficult to prove, but if it was not the Liu An group, it was somebody very similar, and the intellectual world of early Han was not a huge one.

88. J. Baird Callicott, *Earth's Insights* (Berkeley and Los Angeles: University of California Press, 1994); see also *Nature in Asian Traditions of Thought*, ed. Callicott and Ames.

89. Emmanuel Levinas, *Ethics and Infinity*, trans. Richard A. Cohen (Pittsburgh: Duquesne University Press, 1985), and *Totality and Infinity*, trans. Alphonso Lingis (Pittsburgh: Duquesne University Press, 1969).

In Search of Dragons:
The Folk Ecology of Fengshui

STEPHEN L. FIELD

An argument could be made for characterizing the practice of fengshui as the longest-lived tradition of environmental planning in the world. With a formal history of some two thousand years and a pervasiveness in all levels of Chinese society that is exceeded perhaps only by the practice of Chinese medicine, fengshui has directed the location and orientation of human dwellings and tombs in such a way that the ecology in China was "miraculously" preserved.[1] That this preservation may have been just a favorable side effect of what many would call a pseudoscientific, superstitious body of folk remedies does not detract from its positive environmental results. On the contrary, due to the abandonment of this traditional worldview by the Chinese communists, "as much damage was done in fifty years as in the previous fifty centuries."[2] Certainly the current regime in China, and the rest of us for that matter, could learn something from the venerable practice.

While the current popularity of fengshui in the West is a sign that the public is receptive, it is the more superstitious and less "scientific" aspects of fengshui that have captured the popular imagination. Traditional fengshui masters are presumably well versed in the theories of both the Form and Compass Schools.[3] That is, they are able to take into account detailed aspects of the physical environment of the dwelling, as well as the astrology of the dweller. Practitioners in the West for the most part are concerned only with birth dates and construction times and are not trained in the skills of reading the terrain. This is not to say that such cosmological indications are without value. On the contrary, if the belief in the efficacy of fengshui is accompanied by a willingness to embrace the holistic principles that underlie the system, then there will certainly be a positive gain.

Much has been written on these principles, especially the philosophy of qi—that ineffable force affected by the pierce of the acupuncture needle. While the physiological characteristics of qi, as well as its philosophical and cosmological ramifications, have been the subject of numerous studies since classical times, very little research has been conducted on the concept of qi as understood by practitioners of fengshui. That will be the principle object of this essay. In a holistic view of the cosmos, the human anatomy is a microcosm of the earth, and the blood veins of one correspond to the rivers and streams of the other. When the ground is broken and the well is dug for a new house, or when the excavation for a tomb is conducted, such action taps the qi meridians of the earth—called "dragon veins"—just like an acupuncture needle. Regardless of the type of fengshui, all site orientation methods purport to locate and characterize qi in the physical plane.

While the Form School of fengshui seeks the dragon veins in the terrain of a site, the Compass School analyzes qi as a force that courses through productive and destructive orders of the five phases. These phases are correlated with the eight directional trigrams of the *Yijing*, and a person's year of birth corresponds to a particular trigram. When the natal trigram and its phase are correlated with the directional trigram and its phase, one can determine if a particular direction is productive or destructive to a particular individual. It is thus possible to avoid destructive qi by orienting dwellings or arranging rooms in productive directions. Such readings theoretically do not have to take into account the surrounding terrain, and Compass School fengshui therefore apparently has little to offer in our discussions of ecology.[4]

From an ecological perspective, it is Form School fengshui that can offer the most insight into the success of this long-lived tradition. It is this body of knowledge which teaches, for example, that "one should put one's house or village on a slight slope or rise" that is "surrounded on two or three sides by higher ridges and hills" with "a grove of trees upslope" and "a stream at hand."[5] Theoretically, such a locale will ensure that the flow of qi is concentrated and not dissipated, but in practice it protects against floods and storms and provides materials for building and water for drinking and waste removal. We might ask why it was not strictly ecological concerns that motivated the Chinese to orient their villages in such manner, for surely the experience of generations of farmers would have assured similar results, but more directly. The answer is that the practice of fengshui satisfied various social needs that a science of environmental planning alone would never have addressed. In a densely populated country like China, with no laws to protect the rights of individuals, a system was needed that motivated one family to maintain its environment in order not to adversely affect another's. In the words of the anthropologist, "If a peasant was tempted to disregard possible future welfare and choose his own present self-interest, his neighbors would immediately raise the alarm: he was wrecking the community's *feng-shui*."[6]

Fengshui was a powerful motivator. If the possibility of a natural disaster was not sufficient to preclude the founding of a village on a hundred-year flood plain with its fertile soil and level building plots, then the surety that *sha*, or baneful qi, would eventually destroy the fortunes of the village was enough to arrest the construction. Religion is a much more powerful motivator than science, it would seem, when it comes to preservation of the environment. Were the bald eagle the object of worship, its numbers would never have dwindled to near extinction in the continental United States. Fengshui, strictly speaking, is not a religion, of course.[7] Nor is it a science in the modern sense, save for its original foundation on empirical observation. But we can consider its methods an art and its individual practice a work of art, especially if we adopt Roger Ames's and David Hall's description of the Dao as an aesthetic order of nature.[8] For regardless of how delimiting are the various prescriptions and proscriptions of the art of fengshui, inevitably, its practice is largely intuitive. But the intuition of the fengshui master must be based on an ethos, what Ames calls "the expression of the character or disposition of an integrated natural environment that conduces most fully to the expression of the integrity of its constituent particulars."[9]

The Form School of fengshui teaches the practitioner how to search for qi as it is integrated in the geophysical environment. The remainder of this essay will analyze the origin and early development of this art as recorded in the *Zangshu*, or *Book of Burial*, attributed to Guo Pu (276–324) of the early third century.[10] In particular, I will analyze the means by which fengshui masters locate qi in the mountainous landscape, then provide a general picture of the ideal location of the tomb site, and close with a discussion of accumulated qi and its advantages.[11] Before proceeding to the *Zangshu*, I will begin by tracing the origin of geophysical notions of qi.

One of the earliest statements of the theory underlying Form School fengshui[12] occurs in the *Guanzi*: "Water is the root of all things and the source of all life. . . . Water is the blood and breath (*xueqi*) of the earth, functioning in similar fashion to the circulation of blood and breath in the sinews and veins."[13] The use of the term *xueqi* to metaphorically describe the function of terrestrial water presupposes its use at the time in human physiology. Sure enough, in the *Zuozhuan* the term is used at least once to indicate the "blood and breath" of the human body and once for the "pulse."[14] It also occurs in the *Analects* to mean the blood-qi of a man in his full vigor.[15] So the image in the *Guanzi* is of living water pulsing underground, emerging as springs, and coursing in rivers. But qi also has other manifestations in the *Zuozhuan*.[16] Perhaps the most primitive concept of qi refers to the six qi of Heaven—cold, heat, wind, rain, dark, and light—which "descend and produce the five tastes," but "produce the six diseases when they are in excess."[17] Yin and yang, at this stage in the evolution of Chinese cosmology, are merely two types of qi (cold and heat) and have yet to subsume all binary paradigms. Thus far, qi is discussed only in its differentiated aspects and has yet to become a "universal fluid."[18]

By the time the *Guoyu* was completed in the mid-Warring States period,[19] yin and yang had been elevated to the status of the primary manifestations of qi. The following passage from the *Zhouyu* records the words of Bo Yangfu, the Grand Historian of the state of Zhou:

> The qi of heaven and earth do not lose their proper order. If they go beyond their proper order, havoc will be wrought upon the people. If the yang remains submerged and cannot emerge, and the yin remains oppressive and unyielding, there will be earthquakes.

If the *Zhouyu* was compiled in 431 B.C.E., this passage probably records the earliest known cosmological theory of qi.[20] In another passage from the *Zhouyu*,[21] we see a characterization of qi that is closer to the geophysical notions that we seek to trace. Here, Prince Jin is trying to dissuade his father, King Ling of Zhou, from damming the rivers to protect the palace from floods. The prince approaches the problem from a cosmological and ecological point of view:

I have heard that those who ruled the people in ancient times did not tear down the mountains, nor did they raise the marshes, nor did they obstruct the rivers, nor did they drain the swamps. For mountains are accumulations of earth, and marshes are gathering places for creatures. Rivers are channels for qi, and swamps are concentrations of water. When heaven and earth became complete, they had accumulated the high (mountains) and gathered creatures in the low (marshes). They had cut through rivers and valleys, to channel their qi, and had dammed and diked stagnant and low-lying water, to concentrate their fertility. For this reason the accumulations (mountains) do not crumble and collapse, and so creatures have (marshes) in which to gather. Qi is not sunken and congealed, nor is it scattered and dissipated. Through this the people, when living, have wealth and useful things, and when dead have places to be buried.

In this passage it is clear that qi is distinct from water even though the two occupy the same space, a departure from the view of the *Guanzi* author who equated the two but separated them into the different realms of the human and the earth. In Prince Jin's persuasion the rivers that cut through the accumulated earth channel qi, while the swamps naturally balance the flowing rivers and concentrate qi. In this fashion, the potential energy of the rivers is stored and prevented from scattering and dissipating. The place where qi naturally concentrates (swamps and marshes) is where the fauna and flora are most abundant,[22] and any alteration of this terrain will threaten the environmental balance. The implication of the argument is this: when Heaven and Earth dam the rivers to store their potential, a balance is achieved, whereas a similar attempt by King Ling will risk the release of violent, destructive force. Incidentally, when Prince Jin says that the dead will "have places to be buried," he probably refers to a belief in something like *yinzhai*, or burial site fengshui,[23] the subject of the *Zangshu*. His recognition of this fact follows right after his statement that qi is not scattered or dissipated in those fertile locales, which is a requirement for proper burial.

The *Zhouyu* passage just quoted is one of the earliest codifications of the theory of *ju* conglomeration and *san* dispersal, although here it is the concentration of Earth and waters that is the main subject of discussion. This view is similar to the principles of accumulation and dispersal of qi that govern Form School fengshui. The concept was widespread by the early Han dynasty when it had expanded to govern the physiological qi. In chapter twenty-two of the *Zhuangzi* appears this passage: "Man's life is the assembling of qi. The assembling is deemed birth, the dispersal is deemed death."[24] So, by this time the concept of agglomeration and dispersal of qi had both macrocosmic and microcosmic applications. Such a view is the premise underlying the theories of fengshui as understood by the author of the *Zangshu*. From its opening chapters comes the following passage: "Truly, life is accumulated qi. It solidifies into bone, which alone remains after death. Burial returns qi to the bones, which is the way the living are endowed."[25] Here the physiological qi of Zhuangzi and the geophysical qi of Guanzi and Prince Jin merge, as the corpse is interred underground to receive the influence of the "blood and qi" of the body of the earth. Indeed, the metaphor of the body also appears in the *Zangshu*: "Earth is the body (*ti*) of qi—where there is earth there is qi. Qi is the mother of water—where there is qi there is water." So the *Book of Burial* also maintains the correspondence between water and qi that appears in the Warring States texts, except that the relationship is clarified in a very important way—qi gives birth to water. The relationship is an important one. As mother and offspring, qi and water exhibit a natural attraction. Obtaining one is the means of acquiring the other. This is the *sine qua non* for the practice of fengshui, and it is explained in this passage from the *Zangshu*, which is the first appearance in extant texts of the term, fengshui:

> The *Classic* says, qi rides the wind and scatters, but is retained when encountering water. The ancients collected it to prevent its dissipation, and guided it to assure its retention. Thus it was called fengshui (wind/ water). According to the laws of fengshui, the site that attracts water is optimal, followed by the site that catches wind.

However, the means by which wind can be collected and water can be guided depends entirely on the topography of the land.

The following passage from the *Zangshu* describes the relationship between topography, qi, and water:

> The *Classic* says, where the ground holds auspicious qi, the earth conforms and rises. When *zhi* ridges hold accumulated qi, water conforms and accompanies them.

Thus, the elevation of features above the level ground is the result of the presence underground of "auspicious qi." These topographical features are called *zhi*, a term also borrowed from human physiology, which means arterial branches. Such arteries are the conduits of qi, and when they are full of accumulated qi, water appears and follows their outward manifestations, the arterial ridges. Since qi is the mother of water, presumably the presence of qi will generate water. The question that remains is this: just how does qi accumulate in these arteries? Another passage clarifies:

> The *Classic* says, where the earth takes shape, qi flows accordingly; thereby things are born. For qi courses within the ground, its flow follows the contour of the ground, and its accumulation results from the halt of terrain. For burial, seek the source and ride it to its terminus.

The words "contour" and "terrain" translate the same Chinese term, *shi*, a word that in most contexts means "power" or "force." However, at least since the late Warring States period, in combination with the character *xing*, or "form," it has referred to the inherent strengths of topographical features.[26] Where these features run their course and come to an end is where the qi naturally accumulates. According to the *Zangshu*, "Qi collects where forms (*xing*) terminate; it transforms and gives birth to the myriad things. This is exalted ground." Burial should take place in this exalted ground.

It would not be an exaggeration to say that in the *Zangshu*, *xing* and *shi* are the most important technical terms.[27] The author goes to great lengths to explain their meaning, because the search for accumulated qi depends entirely on the ability of the fengshui master to perceive the subtleties of the topography. In the following passage, *xing* and *shi* are described in some detail:

> Arteries spring from [low] land terrain; bones spring from mountain terrain. They wind sinuously from east to west and from south to north. Thousands of feet [high] is [called] forces (*shi*); hundreds of feet [high] is [called] features (*xing*). Forces advance and finish in features. This is called integrated qi. On land of integrated qi, burial occurs in the terminus.

The passage begins with another borrowing from human physiology. What previously were called *zhi* branches are here specified even more clearly as "arteries" (*mo*).[28] We saw previously that the presence of underground qi caused the earth to protrude, and here those protrusions are clearly the bulging arteries of the skin of the earth. Bones, on the other hand, rise like ribs high above the earth as mountain ranges. Specifically, terrain thousands of feet high is called "forces," and terrain only hundreds of feet high is called "features." From a fengshui point of view, terrain originates in the forces of alpine heights, slowly winds around as it decreases in altitude, and finally runs its course and finishes in the hills and knolls of the lowlands. When such a terrain can be traced from its highland origin to its lowland terminus, this is the optimum topography. Again, according to the *Zangshu*, "When the forces are fluid and the features are dynamic, unwinding from terminus to source, according to the art of fengshui, if interment occurs here, good fortune is eternal and misfortune nil."

At this early stage in the development of a methodology to analyze the topography, a technical vocabulary for distinguishing auspicious and inauspicious terrain is still somewhat primitive. Instead, the author of the *Zangshu* resorts to metaphorical language from the human and mythical realms to represent topographical features. The following passage describes the optimum topography:

> The mountains of exalted ground descend from Heaven in succession as if doing obeisance. Like billowing waves, like galloping horses, they come in a rush and cease as if laid to rest. As if embracing 10,000 treasures, yet they rest peacefully. As if preparing 10,000 feasts, yet they fast in purity. They are like bulging bags, like brimming plates, like dragons and phoenixes soaring and circling. Birds hover and beast crouch, as if paying homage to (a noble of) 10,000 chariots.

The appearance in this passage of the image of dragons and phoenixes is important. Another passage concentrates the view from a panorama of *shi* forces to a focus on *xing* features and depicts the terrain as a recumbent dragon:

> The *Classic* says, where forces cease and features soar high, with a stream in front and a hill behind, here hides the head of the dragon. The snout and forehead are auspicious; the horns and eyes bring doom. The ears obtain princes and kings; the lips lead to death or injury from weapons. Where terrain winds about and collects at the center, this is called the belly of the dragon. Where the navel is deep and winding, descendants will have good fortune. If the chest and ribs are injured, burial in the morning will bring sobbing that night.

Here we see a picture of the terminus of *shi* forces, as if the winding terrain were a coiled dragon. At the point where the mountain range runs its course, terminating in soaring foothills, if the image of a dragon's head can be visualized, this is the ideal locale for burial. This is called the *xue*, or dragon's lair.

If the topography of a particular locale conforms to the descriptions outlined above, then qi will be generated along the flow of terrain, and the appearance of water at the terminus will be proof of the coalescence of that qi. Thus, in the passage just quoted, a stream flows in front of the dragon's head. This water is the means by which the qi generated by the dragon can be harnessed to revive the spirit of the interred bones. According to the *Zangshu*:

> External qi is that by which internal qi is collected. Water flowing cross-wise is the means for retaining advancing dragons. Lofty forces wind around and come to a rest. If the external has no means to accumulate the internal, qi dissipates within the ground. The *Classic* says, the lair that does not hoard will only harbor rotting bones.

The water that fronts the burial site must flow transversely across the axis of the advancing mountains. The *Zangshu* is quite specific in regard to the disposition of this water:

> Where water is the Vermilion Bird, decline and prosperity rely on the features of the terrain. Swift currents are taboo and are said to bring grief and lamentation. From a source in the Vermilion Bird vital qi will spring. (Waters that) diverge will not bring prosperity; pooling (waters) will accumulate great abundance; stagnant (water) brings decline. . . . The *Classic* says, where mountains advance and waters encircle, there is nobility, longevity and wealth. Where mountains imprison and waters flow (straight), the king is enslaved and the prince is destroyed.

Several important points are made here. Springs are auspicious, and the stream thus formed should meander and encircle the lair, pooling its accumulated qi. Swift-flowing streams and diverging waters dissipate qi, while in stagnant waters qi is "sunken and congealed," in the words of the *Zhouyu*. The Vermilion Bird is one of the *sishi*, or Four Forces, the other three being the Cerulean Dragon, the White Tiger, and the Dark Warrior. These originally referred to the four celestial palaces, the four great macro-constellations of the twenty-eight houses of the Chinese zodiac. Here they mark the cardinal directions—the Cerulean Dragon of the east, the Vermilion Bird of the south, the White Tiger of the west, and the Dark Warrior of the north. The ideal lair faces south, the direction of the Vermilion Bird.

Ideally, topographical features should surround the lair. This is to insure that the vital qi thus accumulated does not dissipate in the wind:

> Blowing qi has the ability to dissipate vital qi. The dragon and tiger are what protect the district of the lair. On a hill among folds of strata, if open to the left or vacant to the right, if empty in the front or hollow at the rear, vital qi will dissipate in the blowing wind. The *Classic* says, a lair with leakage will only harbor a decaying coffin.

In practice, however, the features that front the lair to the south should be minimal so that the encircling waters will have means of egress. Those to the east in the direction of the dragon should be the most imposing. The features toward the rear of the lair form a backdrop, and the image from the text is of a tomb "backing up to an ornamental screen."

Let us now review the main principles that have been discussed so far and generate a picture of the ideal location of the *yinzhai*, or tomb. First of all, for burial to return qi to the bones of the deceased, the ground must lie in the vicinity of accumulated qi. To locate accumulated qi, one must look for a landscape of integrated qi. Such terrain should be continuous from its highest reaches to its terminus, without any breaks in its progression. The vista should be undulating (like the rise and fall of a dragon), and each successive level should decrease in height from its successor (like descending ranks of subjects bowing to the emperor). The sequence of features should describe circular contours (like a coiling dragon). Where such terrain runs its course is where qi naturally accumulates.

Once accumulated qi is located, the lair can be selected. The most auspicious ground will be surrounded on all sides by rising terrain. On the east the terrain towers highest and should resemble a recumbent dragon. Opposite this principal feature, toward the west, is another high feature, which should resemble a crouching tiger. Behind the lair, toward the north, is a topographical form that backs up to the burial site like an ornamental screen. The terrain toward the south is lowest in elevation. In this direction there should be flowing water, since qi is the mother of water. The water should meander, embrace the site, and form pools in front of the lair. Therefore, landscape of integrated qi will insure that the natural flow of the topographical *shi* forces will funnel the qi and concentrate it in a single location. The presence of appropriate *xing* features will insure that the wind does not encounter the concentrated qi and cause it to dissipate. Since qi is the mother of water, the appearance of water in the vicinity of the lair with the proper flow and configuration is proof that qi has accumulated beneath the site. If qi is present, then the bones of the deceased—the solidified remains of life—will be immersed within that qi.

The *Zhuangzi* explained how the dispersal of qi brings death to the human body. But by the time of Wang Chong, the later Han dynasty skeptic, the concept had evolved somewhat: "As water turns into ice, so the qi crystallizes to form the human body." Furthermore: "That by which man is born are the two qi of the yin and the yang. The yin qi produces his bones and flesh; the yang qi his vital spirit (*jingshen*)."[29] From Wang Chong's elaboration we can deduce that the qi of bones is yin qi. So if interred bones are to be charged, like a dead battery, by the influence of the accumulated qi of the burial lair, it must be yang qi that has coalesced. The *Book of Burial* identifies accumulated qi in several places as *shengqi*, "vital" or "living" qi. The earliest reference to *shengqi* is in the *Lü shi chunqiu*, in a passage describing the cycle of the seasons. In the last month of spring, we are told, "*shengqi* flourishes, and yang qi flows forth; shoots emerge, and buds unfold."[30] *Shengqi* is therefore the precursor of yang qi, and it is yang qi that can energize the bones of the dead. The process whereby the bones are energized is *ganying* mutual resonance. According to the *Huainanzi*, "All things are the same as their qi; all things respond to their own class" (4.8.27), and "Things within the same class mutually move each other; root and twig mutually respond to each other" (3.2.27–28).[31] Since the qi of the interred corpse and the qi of the living descendants are identical, therefore, when the vital, life-giving qi of the burial site surrounds the bones, they are energized, and the lives of the descendants are thereby endowed.

We have seen how the lair located in the belly of the dragon acts as a conductor and focal point for the vital qi coursing through the terrain. The energized bones of the deceased ancestor resonate, and the qi of the living descendants naturally harmonizes. This process somehow enhances the qi of the living, thereby increasing the chance of good fortune. We might be tempted to speculate that it is the spirit of the deceased that consciously bestows rewards or punishments, solely based on the ability of the living to orient the tombs of the dead properly. But this would require that fengshui be a formal component of ancestral rites, something like the burning of paper money or the burial of objects in the tomb that will accompany the deceased to the spirit world. That does not seem to be the case in the formative stages of the art of fengshui when the classical rule of *zhaomu* was still followed, "according to which the positions of graves were to be alternated with each generation of the dead."[32] Besides, Chinese ancestor worship reveres the agnatic ascendants, yet in *yinzhai* fengshui it is just as important to orient the tomb of the wife and grandmother properly. In addition, tomb fengshui is only one manifestation of the art; residential fengshui has always been at least as important as the tomb variety, if not more so. Maurice Freedman is therefore correct when he says: "By geomancy then, men use their ancestors as media for the attainment of worldly desires. And in doing so they have ceased to worship them and begun to use them as things. . . ."[33] This could only be true if such use were a mechanical process independent of religious repercussions and the resulting need for apatropaic ritual; otherwise, the "spirit" might retaliate for being used in a disrespectful fashion.

While it may not be the spirit of the ancestor that endows the living, we cannot discount the possibility that the resonating qi reacts in something other than the physical plane. Wing-tsit Chan's definition of qi as a "psychophysiological power"[34] is an attempt to capture those aspects of the energy of qi that elude measurement. This would account for the purported intuitive powers of fengshui masters to locate qi in the environment, especially when the topographical features of a particular locale do not conform well to the requirements of the text of the *Zangshu*. This would also help to explain how the proper flow of qi benefits the living when there is no deceased to act as a "medium" between what we are accustomed to calling the physical and mental realms. In truth, the Chinese do not divide the world into a Cartesian duality, as the heart-mind is intrinsically connected to nature through qi. The American counterculture of the 1960s and 1970s developed a metaphorical language that attempted to describe something like this connection. Thus it was said that a place had "good vibes," that is, a person was "in tune" with the vibrations of a particular locale. Similarly, a Chinese might be prompted to remark that a location has "good fengshui."

Psychophysiological aspects of experience fall under the classification of paranormal phenomena in the West and are normally ignored by the scientific establishment. Yet, it is precisely this mode of thought that is enthusiastically embraced by the public and which has propelled the rise of so-called New Age religion. Such ideologies reflect a yearning for a new spiritual paradigm that will be neither hindered nor mollified by an indifferent academic community. On the other hand, the rigorous denouncement of New Age thought by the religious establishment serves only to demean those systems in the eyes of the questing public. It may appear that the intellectual is left in the uncomfortable position of either denying the efficacy of fengshui auspice because it cannot be proven rationally or, conversely, embracing the philosophy of qi whole cloth with all its related arts. Yet, the two views are not necessarily mutually exclusive. From the vantage of an aesthetic worldview we do not have to judge the reality of a phenomenon in order to derive gain from it. It is not principles of rational order that should solely determine the value of our vision, but the capacity of our vision to satisfy our responsibility for ethical order. The vision of integrated qi preserved the ecology of China for centuries, and a similar vision is needed in the West, where an anthropomorphic worldview has left the environment vulnerable to rapacious self-interests. The adoption of an ethos of holism would help to ensure that selfish interests would not take precedence over social good. One means of effecting such an integrated worldview is to translate the philosophy of qi into the American vocabulary.

Notes

1. "[C]ompared to the almost total failure of environmental planning in the United States—with its superior scientific establishment and law enforcement capabilities—*feng-shui* was nothing short of miraculous." See E. N. Anderson, *Ecologies of the Heart: Emotion, Belief, and the Environment* (Oxford: Oxford University Press, 1996), 26. While his comment may be somewhat exaggerated, Anderson's point is that the practice of fengshui across the great expanse of premodern China accomplished locally what a national policy could not achieve in the United States. A case in point is the devastation of the Great Plains caused by drought and poor land conservation in the 1930s. Had fengshui been practiced in the decades before the Dust Bowl, at the very least groves of trees would have been planted around every homestead. Moreover, in order to preserve the natural flow of water across the prairie, seasonal creeks and streams would not have been dammed, thus preserving springs and natural greenbelts throughout the environment.

2. Ibid. This is certainly true when one considers the ecological impact of such massive projects as the Three Gorges Dam on the middle reaches of the Yangzi River. Yet it is no less the case when one considers the thousands of factories emptying toxic waste into China's waterways.

3. The *xingshipai*, or "Forms and Terrain School," is also known as the *luantipa*, "Mountain Constitution School," the Jiangxi School (for its reputed province of origin), and the Intuitive School. The *liqipai*, or "Patterns of Qi School," better known as the Compass School, is also called the *fangweipai*, "Directions and Positions School," the Cosmological School, the Temple School, and the Fujian School.

4. Elsewhere, however, I have shown that even the theories of the Compass School "very likely [owe their origin] to the empirical observations of an agrarian people who saw the annual birth and death of the natural world around them and tried to structure their environment to mimic that world. The eight compass points of their residences were thus divided equally between east (for growth) and west (for decline) and simultaneously portioned equally between south (for sun) and north (for shade)." See "The Numerology of Nine Star Fengshui: A *Hetu*, *Luoshu* Resolution of the Mystery of Directional Auspice," *Journal of Chinese Religion* 27 (1999): 13–33.

5. Anderson, *Ecologies of the Heart,* 16, 19.

6. Ibid., 25.

7. Daoist priests were forbidden to perform fengshui. On the other hand, fengshui has traditionally been associated with the Daoist arts of divination, medicine, and inner alchemy, according to Eva Wong in her *Feng-shui: The Ancient Wisdom of Harmonious Living in Modern Times* (Boston: Shambhala, 1996), 4–6. For example, the most popular fengshui manual in use today is the *Bazhai mingjing* (Illuminating the eight halls) by the Qing dynasty sage, Ruoguan Daoren ("bamboo hat Daoist").

8. See David L. Hall, "On Seeking a Change of Environment," and Roger T. Ames, "Putting the *Te* Back into Taoism," both in J. Baird Callicott and Roger T. Ames, *Nature in Asian Traditions of Thought: Essays in Environmental Philosophy* (Albany: State University of New York Press, 1989), pages 99–112 and 112–44, respectively.

9. Ames, "Putting the Te Back into Taoism," 135.

10. Guo Pu based his work on an earlier text, purportedly the *Zangjing*, or *Classic of Burial*, attributed to a certain Qingwuzi, whose origin is unknown. The text con-

sulted for this essay is the *Dili zangshu jizhu* (The annotated geomantic Book of Burial) collected in the *Siku weishou shushulei guji daquan*, section 6, "Kanyu," vol. 4, 1889–1958 (Hefei: Huangshan shushe, 1995).

11. The remainder of this chapter discusses tomb site location because the primer of Form School fengshui, the *Zangshu*, chooses to analyze the *yinzhai* (house of the dead) variety. As far back as the tradition can be traced, *yinzhai* and *yangzhai* (house of the living) fengshui have been two sides of the same coin. Although tomb location had less impact on ecology, since essentially the same techniques were used for both types of site orientation, the theories outlined below are equally valid for residential fengshui.

12. According to Joseph Needham et al., *Science and Civilization in China*, 7 vols. (London: Cambridge University Press, 1954–), 2:42 n.

13. *Guanzi*, chap. 39. For the translation, see *Guanzi: Political, Economic, and Philosophical Essays from Early China*, trans. W. Allyn Rickett, vol. 2 (Princeton: Princeton University Press, 1998), 100–101.

14. *Zuozhuan*, Duke Xiang, 21st year (James Legge, trans., *The Chinese Classics*, 5 vols. [Oxford: Oxford University Press, 1892], 5:488, l.2), Duke Zhao, 10th year (Legge, *The Chinese Classics*, 5:627, l.13).

15. Benjamin I. Schwartz, *The World of Thought in Ancient China* (Cambridge, Mass.: Harvard University Press, Belknap Press, 1985), 184. *Analects*, book. 11, chap. 7.

16. In the *Zuozhuan*, qi is also used in its modern sense as states of emotion, such as *yongqi*, a "spirit of bravery," and *keqi*, "courtesy." Thus, at the time when the *Zuozhuan* was written, among other things, qi was something like the medieval Western concept of humors, the bodily fluids that determined a person's temperament.

17. *Zuozhuan*, Duke Zhao, 1st year (540 B.C.E.), Legge, *The Chinese Classics*, 5:580–81.

18. A. C. Graham describes qi as the "universal fluid, active as Yang and passive as Yin, out of which all things condense and into which they dissolve." *Disputers of the Tao: Philosophical Argument in Ancient China* (La Salle, Ill.: Open Court, 1989), 101.

19. The *Guoyu* was probably compiled beginning in the late fifth and continuing to the late fourth century B.C.E.. The earliest book of the compilation is the *Zhouyu*, which dates to 431 B.C.E.. See *Early Chinese Texts: A Bibliographical Guide*, ed. Michael Loewe (Berkeley: The Society of the Study of Early China, 1993), 264.

20. This is according to Zhang Dainian. See W. A. Wycoff's translation of Zhang's article, "On Heaven, Dao, Qi, Li, and Ze," *Chinese Studies in Philosophy* 19, no. 1 (fall 1987): 3–45.

21. The translation (slightly amended) is by James A. Hart, "The Speech of Prince Chin: A Study of Early Chinese Cosmology," in *Explorations in Early Chinese Cosmology*, ed. Henry Rosemont, Jr. (Chico, Calif: Scholars Press, 1984), 39.

22. "It may be inferred that the speaker sees there the transfer of energy from the motion of the rivers to the fertility of the swamps, though this point is never explicitly stated." See ibid, 45.

23. *Yinzhai* fengshui dictates the choice of burial sites, while *yangzhai* fengshui governs dwelling sites.

24. Graham, *Disputers of the Tao*, p. 328.

25. Or more specifically, according to the *Zangshu*, "If the ancestors' bones acquire qi, the descendants' bodies are endowed." We will return to a discussion of this process at the end of the analysis.

26. In book 16 of the *Xunzi*, the Marquis of Ying questioned Xun Qing about the resources of the state of Qin. Xunzi replied, "Its topographical features (*xingshi*) are inherently advantageous. Its mountains, forests, streams, and valleys are magnificent. The benefits of its natural resources are manifold. Such are the inherent strengths of its topography (*shi*)." See *Xunzi: A Translation and Study of the Complete Works*, trans. John Knoblock (Stanford: Stanford University Press, 1990), 246. The terms also appear as the title of a chapter of the *Guanzi*. While here they are interpreted to mean something like "condition and circumstance," it is interesting to note that several passages in this chapter could be construed as implying a fengshui context. For example, the opening passage of the chapter begins, "If a mountain rises high and never crumbles, sacrificial sheep will be presented to it. If a pool is deep and never dries up, sacrificial jade will be offered to it." See *Guanzi: Political, Economic, and Philosophical Essays from Early China*, trans. W. Allyn Rickett, vol. 1 (Princeton: Princeton University Press, 1985), 61–62.

27. *Shi* occurs thirty-one times, and *xing* occurs twenty-one times.

28. This view of the flow of qi is an old one. In the Qin dynasty, Meng Tian (d. 210 B.C.E.), the builder of the Great Wall, made the following confession: "I could not make the Great Wall without cutting through the veins of the earth" (*Shiji*, chap. 88, p. 5b). See Needham et al., *Science and Civilization*, 4:1, 240.

29. *Lun Heng*, chapter 62. See *Lun-Heng*, Part 2, *Miscellaneous Essays of Wang Ch'ung*, 2d ed., trans. Alfred Forke (New York: Paragon Book Gallery, 1962), 92.

30. *Lü shi chunqiu*, *Sibu congkan* edition, *juan* 3, 1b.

31. See John S. Major, *Heaven and Earth in Early Han Thought* (Albany: State University of New York Press, 1993), 65, 167.

32. John B. Henderson, *The Development and Decline of Chinese Cosmology* (New York: Columbia University Press, 1984), 196.

33. Maurice Freedman "Ancestor Worship: Two Facets of the Chinese Case," in *Social Organisation: Essays Presented to Raymond Firth*, ed. Maurice Freedman (London: Frank Cass, 1967), 88

34. *A Source Book in Chinese Philosophy*, trans. Wing-tsit Chan (Princeton: Princeton University Press, 1963), 784.

An Introductory Study on Daoist
Notions of Wilderness

THOMAS H. HAHN

In Lieu of a Preface: Wilderness (Re)emerging from Culture

Some time around the year 1622, the English diplomat James Howell traveled on various missions throughout Europe. He visited the Pyrenean mountains in southern France and roamed the alpine realms of Switzerland. The former he admonished for their "monstrous abruptness," whereas the latter he took to be altogether "hideous" and absolutely superfluous, like "ugly warts" on the surface of an otherwise well behaved and properly groomed surface.[1] Howell, and with him countless other travelers who came before or would follow after him, had a firm understanding, deeply rooted in Christian thinking, of what nature should mean to humankind and what relationship these two contending powers should engage in. Nature, it was presupposed, was to supply and completely surrender its assets so that humans could strive and take full advantage of whatever physical resources and social merits were to be gained from it.

Roughly 350 years after Howell's encounter with the wild and unordered (in 1971, to be precise) the geographer Edward Soja calculated that about eighty billion people had hitherto lived and died on this planet, 90 percent of them as hunters and gatherers, 6 percent as farmers and herdsmen, and only 4 percent as members of the industrial age.[2] It therefore strikes one as amazing that up to the end of the last century so many patches of land were in fact still left uncharted; many published maps sported numerous white blanks, in turn challenging newer generations of hunters and gatherers to annihilate these blind spots, appropriating the last patches of terra incognita. The modern wilderness debate emerged exactly around the time when nature as the unknown physical environment was flooded with this type of the contemporary scientist-explorer. It is perhaps overstating the point, but after the demise of the romantic period and, more important, with the deconstruction of nature to its core elements and raw data by modern science, we witness a debate (again) about responsibility, ethics, and the politics of ecological judgment.

To demonstrate where we stand regarding this ongoing debate, let me briefly summarize the main positions and introduce some of the key terminology. But before I do, it is perhaps in order to state that the discourse outlined below is purely American (some even say *Americanist*), both in nature and in locality, stemming from a deep individualistic passion that became anti-urban; but nevertheless, it owes its roots and rationalization to the Darwinist observation that humankind is, in fact, more involved with nature than physico-theologians,[3] for example, had hoped for. Accepting the evolutionary paradigm, the Western mind, undecided still, oscillates between anthropocentrism and utilitarianism. God the creator, expelled from nature by modern science's proof of its proprietary "stand-alone"–type organistic causality, yielded to the collected evidence and fled the countryside, seeking a secure place in enclosed churches and cathedrals. In his absence, modern human beings face two choices: to carry on exploiting the so-called ecomachine without a second thought, or to invent measures to assume what once was deemed God's responsibility, that is, to act as an invisible but effective lubricant in order to keep the process of organic growth going indefinitely. However, the material world, which, according to the influential Greek Christian philosopher Origen (185–254 C.E.), constitutes a "pernicious wilderness" in and of itself, is perceived as a highly controversial platform when it comes to the question of spiritual salvation.

Recently, however, the issue was reversed. It is not humans who needs to be salvaged from naturalistic instincts and materialist bonds, but nature that needs to be disassociated and protected from human intervention. Theological discourse, ill-prepared for entering modern ecological thinking, has produced such views as biocentrism, stressing that "all living things have intrinsic value because no thing that lacks intrinsic value—or a good of its own—can be the object of respect"[4] (thus easily garnering protection and resulting in what J. Baird Callicott calls bioempathy). Likewise, the notion of stewardship has been introduced, a managerial approach to environmental ethics and a notion not so much concerned with air pollution control as with wildlife protection and the reintroduction of cautious, long-term land tenure patterns. To finalize this short and necessarily incomplete overview, the whole debate about anthropocentrism, biocentrism, and deep ecology has been criticized as a kind of Third World critique, by Ramachandra Guha and others, as being "at best irrelevant and at worst a dangerous obfuscation"; it is denounced as a smokescreen employed by special interest groups that mistake wilderness areas for a "commodity" in a country that is overmilitarized and where every citizen is engaged in the fallacy of overconsumption. The global perspective of such groups, Guha suspects, "is nothing but the platform for a new form of imperialism."[5]

Militarization also accounts for much of China's ecological downfall. Mark Elvin presents the hypothesis that "the central driving force of environmental degradation has been the intensified exploitation of nature linked to the drive to acquire the means of political, economic and military power. . . ."[6] Using Donald Hughes's work *Pan's Travails: Environmental Problems of the Ancient Greeks and Romans*— in which Hughes laments that a "most damaging aspect of Greek and Roman social organization as it affected the environment was its direction toward war"[7]—as his point of departure, Elvin argues that in China wars were waged and new territories exploited and subjugated because of "competitive necessity." Thus, military effectiveness and agrarian-urban transformation went hand in hand. "Undone extremities of the globe," to borrow an expression from Thoreau, were invaded with full force, forests were cut down, and wildlife declared as "game" and ultimately extinguished. The "derivate effects," as Elvin calls them, of this competition among the (Warring) states resulted in the intensification of a "market-oriented economy that created pressures to cash in natural resources at a faster rate."[8] The construction of the Three Gorges Dam, according to Elvin a "demented" undertaking, then, can be interpreted as a diabolically consequential act of the Chinese, who, to put it mildly, are now—with Western technology imported and set into motion—caught between a rock and a hard place: they can either acknowledge what is perceived as sacrificing the notion of regional or even national growth or put up a technologically challenging fight to tame the highly unruly and wild Yangzi Jiang, thus seeking tighter spatial control over the commodity water. This decision-making process is paired with a marketable demonstration of skills and a stubborn perseverance, perhaps equal to Li Bing masterminding the Dujiang Yan irrigation system in Qin times (221–210 B.C.E.), only this time it is broadcast live via satellite TV and CNN to the Ba and Shu peoples' neighbors, who have been driven off their land by the millions.

I have not come across a Daoist pamphlet or scripture denouncing the Dujiang Yan project,[9] then or afterwards, on the grounds that it would destroy precious wildlife or that it would no doubt add to the spiral which, two thousand years in the future, would then inevitably lead to one of the most controversial engineering projects our world has ever witnessed. In fact, a huge and beautifully elaborate Daoist temple (the Erlang Miao) was constructed alongside the hub of the irrigation system, placing the masters of the Dao in the service of the master engineers of the water (Li Bing and his son), whom it meant to honor.

On Early Chinese Terms Designating "Nature" and on the Characters for Wilderness

The philosophical concept of nature, as well as nature as the perceived space in which all beings (or should I say ontological entities) operate and bear fruit, in China as elsewhere, has undergone fundamental changes throughout recorded history. Early Chinese notions of nature include the terms Heaven (*tian*) and Heaven and Earth (*tiandi*), and also—for the convenience of argumentation—designate that which is in between (*tiandi zhi jian*). In the Laozi and Zhuangzi "nature" can (with some caution) be read into phrases or terms like "that which is so by itself" (*ziran*) and into the inherent "nature" (for which Mengzi uses the word *xing*) of the ten thousand things (*wan wu zhi ziran*).[10] Modern Chinese language uses the compound *daziran* in order to differentiate between objective, physical nature and the specific virtue of one of its components.

Nature, or Heaven and Earth, as we can read again in chapter twenty-three of the *Laozi*, is responsible for thunderstorms (*piaofeng*) and other calamities. Chapter five cautions us that nature (*tiandi*) is "indifferent to human-heartedness" (*bu ren*):[11] it treats the ten thousand things like straw dogs that can be abandoned or sacrificed at will. Nature was recorded as turning chaotic and unpredictable at times, or it was described as untamed, uncivilized, and uninhabitable per se. This notion of nature's indifference to humankind's intentions and the human plight clearly shows the rupture already felt two thousand years ago when the psychic balance between the civil and the natural was at stake. The same insight translated into the modern setting of the wilderness debate means that Thoreau needed to climb Mount Ktaadn in order to overcome Emerson's transcendentalism and finally acknowledge the truth in Laozi's dictum (literally speaking—whether Thoreau ever read the *Laozi* is completely irrelevant): during his climb, "nature was losing its human countenance, turning from familiar friend to potentially hostile stranger."[12]

Expressions for territories that were found to be uninhabitable (for "civilized" people like the Han, at least) comprise the term *huang*, a term that Remi Mathieu in his analysis of the *Mountain and Water Scripture* (*Shanhai jing*) translates as "vast uncultivated territories."[13] Historical records like the *Records of the Historian Shiji* and classical literature also use "great expanse" (*da huang*), "wilderness expanse" (*huangye*), or "vast earth expanse" (*huangdi*).

The central term that concerns the notion of wilderness (*ye*) is written in pre-Han terms with two trees above the earth radical. Oracle bone inscriptions and texts extending well into the Warring States period and even into later times (thereby exhibiting great semantic stability) generally employ the two-tree version of the *ye* character to designate the territory that is beyond administered boundaries (*jiaowai*).[14] Venturing into such space (*ye you*) deserves special preparation and is usually associated with hunting trips or imperially sponsored voyages, with the intention of executing important nation- or community-building rituals. Daoists are notorious for venturing "out yonder." The *Zhen gao* (Declarations of the Perfected), compiled by the famous Daoist master Tao Hongjing between 490 and 516, include, for example, a stanza that reads:

> I roam the Sovereign's mountain.
> All is deeply perfect.
> The peaks are jagged and high.
> They are spirits in themselves.

The place in question here is Kunlun, the seat of the Queen Mother of the West, the starting point of all manifest physical space in Chinese thought, here visited by none else but a very comfortable Cao Cao, the famous Han dynasty general who had just accepted the surrender of the Daoist theocrat Zhang Lu and his forces around 215 C.E. and who may have had Daoist inclinations himself.[15] Despite General Cao Cao's involvement in the poetic mindscape quoted above, the notion of roaming the land (*ye you*) in early China is not linked with military missions or even with leisurely outings. People dwelling in such spaces are the like of the "ten sorcerers" (*shi wu*) or "wild" men (*yeren*) who cultivate special arts and usually are not very sociable. They are, however, knowledgeable and in charge of the "one hundred drugs" (*bai yao*) growing in these remote, perhaps semi-mythological areas.[16]

And yet, *ye* as space is still mentally integrated into other types of spaces. It is not necessarily that kind of land which is farthest removed from the center, from civilization, from early urban humanity. The original hierarchy of social and uncultivated space is related to an early commentary to a song in the *Shijing*, the classical *Book of Odes*. Commenting on the character *jiong*, the source states: "What is outside of the district city (*yi*) is called 'outskirts' (*jiao*). What is beyond these 'outskirts' is called Wilderness (*ye*). External to the wilderness are the forests (*lin*), and still further out is what is called 'arid land' (*tong*)."[17] The anthropogeocentric viewpoint (as opposed to the biocentric) is evident: the measurement designates the different qualities of earth usable to mankind; *ye* as a territory stretching from the outskirts to the wooded areas is still a place which can be negotiated for agriculture and pasture. It is far enough from a district city or marketplace not to allow its potentially evil or irritating influences to affect settlements, but not far enough for speculation, plain curiosity, and, as border movements and land utilization patterns[18] show all too clearly, utilitarian scheming.

A different matter altogether is the search for the equivalent to our modern expression "environment," or *shengtai huanjing* in Chinese. Physical environment, it was argued in the Chinese historical context, was made up of mountains and forests, of streams and marshland. Hence, compounds like *shanlin* (mountains and forests), *chuanze* (streams and pools), or *shanlin chuanze* were used in sources like the standard history of the Han dynasty (*Han shu*) to designate specific localities as a natural resource (. . . *shan ze suo chu*). Calendars like the *Si min yue ling* regulated the direct access to these resources.[19] Imperial policies commonly stated that famous mountains and large marshes should not be included in fiefs and bestowed regions because interests of a national scale in these resources were at stake—as were the symbolic protection and values associated with them.[20] The many, well-documented cases giving details about prohibitions to alpine areas, swamps, and forests that were enforced, then rescinded, demonstrate the great importance of the issue. Guardians of these regions, for example, were instructed to check violations or, as during the eleventh month of the lunar calendar, help and instruct the people on how to obtain food in the form of game or edible fruits. Incidentally, the *Book of Rites* (*Liji*) calls these guardians the foresters of the outer regions (or the wilderness, *ye yu*).[21]

The Impact of Religion on the Land and the Beginning of the Dismissal of Wilderness

In the past fifty years a new academic discipline has raised its head in various disguises, a crossover between cultural geography, religious studies, and the vast field of historical studies: the geography of religions. Its classic treatise, using this same name as its title, was compiled by David Sopher in 1967. Sopher—a geographer by training and a teacher at Syracuse University—postulated four fields that constitute the major interests of this new discipline:

1. The significance of the environmental setting for the evolution of religious systems and particular religious institutions;
2. The way religious systems and institutions modify their environment;
3. The different ways whereby religious systems occupy and organize segments of earth space;
4. The geographic distribution of religions and the way religious systems spread and interact with each other.

I suggest that issues two (how religions modify their environment) and three (a more specific query: the different ways whereby religions occupy and organize space) are closest to the subject under discussion here. Let us hear what Sopher, who is careful to avoid the traps of "environmentalist and functionalist generalizations," has to say about "ecology and religious institutions." Let us then decide whether the following statements are also applicable to Daoism as an "ethnic religion":

In the simple ethnic system, religion often seems to be almost entirely a ritualization of ecology. Religion is the medium whereby nature and natural processes are placated, cajoled, entreated, or manipulated in order to secure the best results for humans. Even at a primitive technological level, however, every culture operates selectively in taking its "sacred" resources from its ecological milieu. The religious behavior of such societies becomes an extended commentary on selected, usually dominant, features of their economies.[22]

Translated into a Chinese context, "sacred resources," as Sopher calls them, can be dubbed spaces that are ritually sealed off to avoid exploitation by unwanted persons (as described in the *Guanzi*), such as extensive, resource-rich mountainous regions or, on a much smaller scale, diminutive fungi (*zhi*) gardens cultivated for stimulative consumption by Daoists and other "biotic citizens"[23] of the time. The process of "selectivity" and the process of resource extraction itself from a generally unprotected, but nevertheless abstractly structured and highly complex environment, in premodern societies precludes a ritualized framework for economic behavior. Religion, Sopher argues, provides for this ritual dimension—which is especially strong in Daoism—as what he regards as an "extended commentary" on basic economic functions. However, religious behavior is, one could counterargue, far more than only a "commentary" or a process to concoct alibis to approach and quell unruly natural agents. It is a mediation process between the individual self and a much greater self, between the necessity of social or economic demands and the land that nourishes these demands. In many cases it actually argues against taking what is precious, against tilling what is fertile. Were it only a "commentary" on a pre-established archetypal theme, such as plain hunger or the survival of a group, its impact would not result in mountains being moved, in rivers being dammed, or in swamps being drained. Survival in the wilderness was—this must be clearly underlined here—an art form that Daoists, with their knowledge of the natural environment and within their confined spaces of sanctuary enclaves, had mastered to perfection. Using their own bodies as taxonometric instruments, the physiological extremes they endured, due to all kinds of environmental challenges, were measured and painstakingly recorded. Their resourcefulness in surviving under extreme conditions actually is a good indicator of how sparingly natural resources were used and of how necessary these resources were to continuing with one's spiritual agenda.

The most fundamental "ritualization of ecology" of which I can conceive in Daoist terms is related to the notion of qi or refined qi (*jing qi*)—qi as the life-giving breath and force of nature, administered by mountains, which in turn are administered by Daoists acting as keepers and custodians. In this capacity Daoists served the entire empire, symbolically generating sacred or prescribed patterns in transforming earth space that had already been laid down in antiquity in the *Yueling* calendrical corporae. This tradition of governing qi in turn resulted, on a macro-level, in sustaining agricultural society, through the conscious and copious maintenance of a high status of fertility (the first and most basic principle in agriculture). Agriculture, of course, was, until quite recently, the dominant feature of sustained traditional Chinese economy.[24]

In order to distribute qi selectively, however, at least place-names had to be established as semantic and geographical cornerstones to which refined qi can be allocated. The Chinese abundance of (or should I say, obsession with) toponyms structuring the land and pinning down its people, as well as all the "shifting" topological features (like floating mountains, *fu shan*) has no equal in other cultures. By the Tang dynasty, about one thousand years before the naming of peaks in the Alps in the eighteenth and nineteenth centuries, the Chinese had already bestowed at least one name on every single peak of the Zhongnan shan (a mountain range in present-day Shaanxi that constitutes a great and important divide between the northern and the central plains). Likewise, when Western painters, finally approaching the alpine wilderness around 1750,[25] experienced aesthetical values that sought to transcend the traditional Chinese nature-as-matter theology with a more "user-friendly" nature-as-value model, similar values had already been under discussion (in writing and in image) among Chinese painting schools from Tang times onward.

Finally, visual evidence like the *Soushen shan tu*, but also textual in the category of administrative geography, suggests that, beginning with the Song dynasty, wilderness, along with its specific, hitherto very misbehaved elements—humans included—which were not only *jiaowai*, that is, outside the urban outskirts, but also *li wai*, "outside" or "ignorant of" any form of rites and socially proper conduct (a veritable no-man's land), had evolved into anything but an empty meta-phor. It was used by Chinese writers and officials to designate a type of territoriality of inferior value, which nevertheless was toponym-ically codified and firmly attached to well-established spatial land patterns.[26]

What Did Daoists Know about the Country They Lived In?

A pristine physical environment serving as a platform for perpetual bliss and readily available nourishment may be the constituent feature of a Daoist *dongtian*, a grotto-heaven, or a so-called blessed and be-nevolent earth, the *fudi*. Spaces bestowed with the most exalted virtue attributes constitute traditional Chinese notions of paradise—and their Daoist versions thereof. As naturalists roaming the countryside, thereby experiencing what Henry David Thoreau must at least have hinted at in his famous essay *Life without Principles*,[27] yet well equi-pped with talismanic protection (as can be found in various chapters of the text *Baopuzi* and insights into the world of illusions and delu-sions, Daoists were instrumental in creating a vast corpus of geo-graphical literature on both local and remote regions, on the so-called "géographie imaginaire," as well as on the concrete physical environ-ment. And, as Aldo Leopold not too long ago commonsensically claimed, "individual awareness was the starting point for ecological sanity," it can be surmised with some degree of certainty that personal experiences and observations were incorporated into the decision-making process of establishing and arranging places of worship, which were combined with usually quite modest plots of cultivated land within a given landscape.

It is an acknowledged fact that Daoists, whether they organized themselves within an ecumenical framework like the twenty-four dio-ceses (*ershisi zhi*) or preferred to operate independently like the leg-endary Geng Sangchu,[28] were among the first to explore the natural environment in a systematic, protoscientific fashion. The contemporary Chinese scholar He Shengdi credits Daoist "fieldworkers" of old as that group within traditional Chinese society that had accumulated unsurpassed knowledge in the fields of geology, geography, and earth sciences (*dixue*) in general.[29] In fact, he identifies instances in texts like the *Short Record of the Southern Marchmount* (*Nanyue xiao lu*) where Daoists unmistakably announce their programmatic agenda and purpose in life as surveyors and transmitters of both local lore and their own knowledge about nature, setting out to "note down everything about the forms of the mountains and rivers and the one hundred things within the nine regions."[30] This unique type of knowledge, accumulated throughout the centuries and woven into ritual texts as well as into oral master-disciple traditions, accounts for what He calls a high degree of awareness when it came to geographical and topographical matters. Nothing strange or hitherto unknown was left unrecorded and undescribed, especially in the mountains.[31] By meticulously noting everything there is to know, by embarking on trips to discover everything there is to discover, He concludes that Daoists' intentions were to let people know of the "greatness of the four seas (*si hai zhi da*), the pervasiveness of the ten thousand things (*wan wu zhi guang*)." Grasping the "big picture" (*liaojie quan yu*) would lead to a better understanding of the human position in the transformative process of life.

I have read and browsed through much of the geographical corpus of pre-Tang texts myself, and my findings corroborate He's viewpoint. Remarkably free from artificial concepts of "landscape,"[32] the descriptions given within the various texts—such as the already cited *Shi zhou ji*, passages from the *Daoji jing*, the vast mythological heritage preserved in the *Shanhaijing* and the *Shen yi jing*, the *Mutianzi zhuan*, the *Bowu zhi*, quotations in the *Xunzi* that originally stem from Wenzi's hand, and then all the way down to the first preserved illustrated topographies (*tujing*) and mountain gazetteers (*shan zhi*)—do demonstrate a high degree of command over geographical, topological, hydraulic, botanical, and related issues. Other evidence, such as the long and illustrious alchemical tradition of Daoism, as well as the transmitted legends and biographies of founders of monasteries and sacred sites (so-called *kai shan zu*, literally, "patriarch that opened up a mountain"), point toward a very rich stock of wilderness experience and territorial control over nature's most remote and inaccessible regions.

It also transpires that for the Chinese seeking purification and spiritual refinement, the ultimate experience with the often cathartic forces of nature is tackling a mountain.[33] Within the parameters of Christian theology, this space would be the desert. Inhabited by ghosts and powerful spirits whose names and real appearances (*zhenxing*) the wanderer has familiarized himself with beforehand, mountainous regions nevertheless constitute areas where both spiritual and material profit rank highest. Leaving the economic factor aside for a moment, the *Wushang biyao*, a Daoist encyclopedia of the sixth century, wants us to believe that living in the mountains is a highly ritualized and regulated lifestyle. If every rite has been properly observed, though, the reward will be enormous:

> Before taking a walk in the mountains, the adept must clasp his teeth nine times, close his eyes and visualize five-colored clouds which rise all about him and completely cover the mountain; after a long while the immortal officials of the five sacred peaks (*wuyue xianguan*) and all the animals and plants of the mountain come to place themselves at his service. Then he lifts his head and pronounces an incantation.[34]

Going even further than Arcadian traditions of paradise, the Daoist practice described here results in the *active surrender* of all plants and animals living in the region. Indeed, they all rush forward to be at the disposal of the Daoist adept. The incantation of a prayer after the mutual introduction of each party involved can be interpreted as a non-interventionist approach, an outspoken recognition of every thing's function and intrinsic quality. Daoists, I am quite certain, would not easily surrender to modern notions of relativism.

Looking at cartography, the long tradition of the fabrication of "Images of the True Mountain" (*zhenshan tu*) or the—more down-to-earth—mountain-related illustrated records (*tujing*)[35] speaks for itself. Whereas the "Images" serve as a type of magic seal of sovereignty over a mountain and its deities, the latter type of material is connected with actual fieldwork, charting roads that could be traveled, thereby gradually turning wild nature into a subjective resource compendium that eventually, with the arrival of the pleasure-oriented literati class in the Song dynasty, would be integrated into preexisting aesthetical standards and downgraded to a string of sceneries (*jing*). Concrete control over mountains that were identified as meteorological hubs of regional impact was sought by addressing the dragon king, bribing him with small golden counterfeits and asking for rain (or water control in general), on jade tablets. Such elaborate measures (generally accompanied by a great celebration of Daoist rituals that usually would span three days but could last up to forty-nine days) to cope with the dangers of the wild were not available to the ordinary people. Nor were meditation practices, such as embracing the one (*shou yi*), meted out on a prescription by health insurance companies. Yet *Baopuzi*, under-standing the corpus mundi as one living organism, suggests exactly this method of "inner calm" in order to obviate calamities like tigers, wolves, snakes, vermin, or their respective demonic manifestations. More important still, he uses the term *bu tai* (meaning "do not be disrespectful or insolent") as a mental measure to survive and flourish in nature.[36]

It is exactly this attitude that is demanded of modern human beings by environmentalists today. The ethical imperative of Lynn White,[37] one of many who hold that nature is not yet "complete" or—following the metaphor of the book of nature—even fully revelatory of its cre-ator and its purpose, calls out for ecological harmony but still operates along the lines of a Cartesian-Newtonian mechanistic cause-and-ef-fect model. Even Lynn White's projection of nature's ethos does not allow for a (scientifically speaking, much more complicated) model wherein all components are subtly interrelated in highly complex con-glomerates of interdependencies, redundancies, and backup plans. Their enforced disassociation, which may be caused by massive inter-vention on the human side, would result in acts of desecration and destruction.[38]

Daoists, I presume, in creating and defining myriad natural enclaves of social spaces as sacred spaces, had little problem with the reada-bility of nature. Perhaps it is overstating the point, but one could say that nature was their own text, written in their own writing, readable to them only. To strive for mental and physical harmony with the exte-rior world and to balance it with the cultivated interior counterpart (there are no white or unexplored regions on the famous map of the inner landscape of the body, the *neijing tu*!) was not a sentimental journey, but meant taking the road home, toward the Dao.

The individualism of their actions was not meant, however, to re-sult in the wholesale preservation of an infinite environment made up of equally infinite features. Spatial infinity was not and is not an item for the Daoist agenda. As stated above, Kunlun was acknowledged as the starting fixture of all space, the source of the Yellow River, the mystic fountain of Chinese civilization. The analogous construction on a time axis demonstrates that immortality (a sort of indefinite ten-ure-track appointment for the Daoist adept) can only be formulated if the postulation of the infinity of time has been previously introduced and formulated. Daoists, trained in scriptures, lore, and literature as well as in medicine and meditation, of course regarded temporal infinity of the body as their highest achievement, possible only because the Dao itself was considered inexhaustible. To preserve neutral time mattered more to them than to preserve actively contested spaces. Having achieved immortality, it was argued, the at times rather awkward human bond with space and place could be transcended, and every successful adept would have a cloud or a dragon at his disposal to roam freely beyond the "Four Seas."[39] Thus, the wild and therefore considered useless *shu* tree described by Huizi in the closing paragraph of the second chapter of the *Zhuangzi*, and which he claims was actually growing furtively in his garden, is definitely living in the "wrong" place. *Zhuangzi* advises Huizi to "plant it in Not-Even-Anything Village, or [in] the field of Broad-and-Boundless." Reinstate the tree to its original, unspoiled, natural, and unnamed environment, *Zhuangzi* says, and "axes will never shorten its life, nothing can ever harm it."[40]

Wilderness, then, to close this issue for now, is nothing like home. Home is social space, but the hermit does not ask for it. Wilderness has the attributes of infinity and "exists outside the dimensions of time as measured by the clock, and even of ritual time." It is "pregnant with as yet unrealized possibilities," the place of the "becoming." Kunlun, a Chinese Eden that is administered by the illustrious female Queen Mother of the West, a potent maternal icon of traditional China, is space conceptualized, a noetic projection irreconcilable with artificial urban order, or the square and the circle. It has more to do with Plato's *Chora*, a Greek concept that perhaps comes as close to the notion of the Dao as Western thought goes and that also forms the "matrix space, nourishing, unnamable, anterior to the One, to God and, consequently, defying metaphysics."[41] Compare this definition of the underlying machinations to chapter one of the *Laozi*, where the Dao is explained in similar terms, and all of a sudden the Dao, as the Chora, "provides a ground for all that can come into being." It assumes spatiality as unprecedented presence before ever a place was named. The Dao and wilderness are not synonyms, however. Neither are wilderness and sacred space. As an American concept, the notion of wilderness is controlled by rationalized and commercialized argumentation, with the Department of Fishery and Wildlife on the one side and theorizing ecotheologicians on the other. The assets of the respective arguments of these two contesting forces are controlled and regulated in one of the most powerful of all modern spaces, considered wild by some and sacred by others: Wall Street.

Notes

1. Jacek Wozniakowski, *Die Wildnis: Zur Deutungsgeschichte des Berges in der europaischen Neuzeit,* translated from the Polish (Frankfurt am Main: Suhrkamp, 1987), 22. It is interesting to note in this context that the earliest Buddhist description of paradise (the land of Sukhavati) is absolutely devoid of mountains, too: it is flat on all sides, and it is explicit that no dark mountains, not even mineral or emerald mountains, exist in this realm. See Wolfgang Bauer's partial translation of the scripture *Da wuliang shoujing*, published in *China und die Hoffnung auf Glück* (Munich: Hanswer 1974), 224ff.

2. E. J. Soja, "The Political Organization of Space," *Commission on College Geography. Resource Paper* 8 (1971): 40.

3. Physico-theology is the viewpoint that final causes within nature are not determined by nature's own design, but still rest on the shoulders of an omnipotent creator. It seeks to reconcile theological arguments with the advancements of natural sciences, especially in the seventeenth and eighteenth centuries, and is often associated with Immanuel Kant's *Physische Geographie* of around 1800. For more on physico-theology, see, for example, chapter 8 of Clarence J. Glacken's seminal work *Traces on the Rhodian Shore* (Berkeley and Los Angeles: University of California Press, 1967).

4. See Tal Scriven, *Wrongness, Wisdom and Wilderness: Toward a Libertarian Theory of Ethics and the Environment* (Albany: State University of New York Press, 1997), 152, for a discussion of Paul Taylor's viewpoints on biocentrism.

5. Ramachandra Guha, "Radical American Environmentalism and Wilderness Preservation: A Third World Critique," *Environmentalist Ethics* 11 (spring 1989): 71–83.

6. Mark Elvin, "The Environmental Legacy of Imperial China," *China Quarterly* 156 (December 1998): 739.

7. J. Donald Hughes, *Pan's Travail: Environmental Problems of the Ancient Greeks and Romans* (Baltimore: John's Hopkins University Press), 198–99.

8. Elvin, "Environmental Legacy," 743.

9. The *Taiping jing*, or *Scripture of Great Peace*, which was serving as a base text for Daoist communities of the Heavenly Master order in the Chengdu region and was presumably compiled in the third or fourth century c.e. in that same area, has nothing on irrigation, although it is very outspoken when it comes to wielding the ax in the forests (as in chapter 118, for example). See the essay by Chi-tim Lai, "The Daoist Concept of Central Harmony in the *Scripture of Great Peace*: Human Responsibility for the Maladies of Nature," in this volume.

10. *Daode jing* 64. See also *Daode jing* 25: "Man models himself after the earth, the earth models itself after Heaven, Heaven models itself after the Dao, and the Dao models itself after Nature (*ziran*)." Hermann-Joseph Roellicke sees in this chapter the fundamental Chinese discussion on the nature of the "beginning or origin" and concludes his thesis with the remark that *ziran* was used by Laozi and Zhuangzi in a very different way than the proto-philosophy in Greece, which later acquired the name "metaphysics." See Hermann-Josef Röllicke, *Selbst-Erweisung: Der Ursprung des ziran-Gedankens in der chinesischen Philosophie des 4. und 3. Jhs. v. Chr.* (Frankfurt: Lang, 1996), 440.

11. *Ren* is a highly complex term. Another translation of the two words *bu ren* would be "not akin to man's mind," i.e., nature is entirely different in its treatment of life per se and thus perfectly indifferent to the human will and the human "cause."

12. Max Oelschlaeger, *The Idea of Wilderness* (New Haven: Yale University Press, 1991), 146f.

13. Remi Mathieu, *Étude sur la mythologie et l'ethnologie de la Chine Ancienne: Traduction annotée du* Shanhai jing (Paris: Collège de France, Institut des Hautes Études Chinoises, 1983), 1:524. Mathieu fails to differentiate (or refuses to, for undisclosed reasons) between the different qualities of space when switching from the "inner" circles of civilization to the distant reaches, i.e., the *huang*. The *Xiang* commentary on the *Zuozhuan*, in a philological note on *huang* (besides explaining *huang* as "great"), also stresses the aspect of being "disorderly, chaotic" (*miluan*). See *Xiang zhuan* 27.5; *Chunqiu zuozhuan cidian*, 571.

14. The latter style of writing only appears with the *xiaozhuan* type of writing, and is thus recorded by Xu Shen in his *Shuowen jiezi*. See also *Yu pian* chap. 6, *zhong* (Zhonghua shuju, 1984), 214.

15. Cited from Terence Russel, "Songs of the Immortals: The Poetry of the Chen-Kao" (Ph.D. thesis, China Centre, Faculty of Asian Studies, Australian National University 1985), 217.

16. *Shanhai jing* chap. 16 (Mathieu, *Étude sur la mythologie et l'ethnologie*, 571). *Shanhai jing* chap. 7 already introduces the notion of territories or countries ruled by sorcerers and conveys the attribute "shamanistic" (Mathieu, 572) as a denomination for specific territorial settings. According to the *Record of the Ten Continents* (*Shi zhou ji*), the best continent for living is (tellingly) called the "Continent of Life" (*Sheng zhou*), which is densely populated with tens of thousands of immortals. See the translation by Thomas E. Smith, "Record of the Ten Continents," *Taoist Resources* 2, no. 2 (November 1990), 87–119.

17. See *Shijing*, section *Lu Song, jiong*. There is more to be said about this attempt at environmental stratification. It is interesting to note, for example, that at least one of the terms (*jiao*) was also used in respect to certain sacrifices. Furthermore, *ye* (written with two trees and the earth radical), besides being used as a place name (from which a family name was consequently derived), is also viewed by some scholars as a type of ritual (*yi wei jiming*; see *Jinwen da cidian* 1416). More on the qualities and immediate effects of untended land can be found in the Zhou Li, where the rites (*li*), the evolutionary hallmark of a civilized society, do not apply and get destroyed in such "rough" country (chap. 5, sec. 58).

18. On the development of land use statistics throughout Chinese history, see He Pingdi, *Zhongguo lidai tudi shuzi kaoshi* (A Numerical Examination of Land in Former Chinese Dynasties) (Taibei: Lianjing Press, 1995).

19. Compare Christine Herzer, "Das Szu-Min Yueh-Ling des Ts'ui Shih" (Ph. D. thesis, Hamburg University, 1963).

20. See *Li Ji*, chap. 11.

21. On this whole issue of resource exploitation and protection as evidenced in the Chinese classics and standard histories, see the useful overview by Liu Cuirong, "Zhongguo chuantong dui shanlin chuanze de kanfa" (Traditional Chinese Attitudes to Natural Resources), in *Zhongyang yanjiuyuan xueshu zixun zonghui tongxun* (Gen-

eral report on scholarship of the central research institute) 7, no. 1 (1998): 87–101.

22. David E. Sopher, *Geography of Religions* (Englewood Cliffs, N.J.: Prentice-Hall, 1967), 17–18.

23. For this term, see Max Oelschlager's discussion of John Muir's nature theology (Oelschlager, *Idea of Wilderness*, 194–95). Quoting Neil Evernden, Oelschlager argues that "the life world is a seamless living whole," a statement that Daoists would ratify without further questioning.

24. In 1911 a book appeared which was introduced to the Western public as something of an agricultural "travelogue" of China. F. H. King, author of *Farmers of Forty Centuries*, argues convincingly that Chinese farmers were experts at reusing the same land over and over again, yielding good crops year after year, by applying basic principles like economizing on time and composting away from the growth areas. King, professor of agricultural physics at the University of Wisconsin, obviously was very impressed with what he had the good fortune to record in such great detail. His compatriot Walter Lowdermilk, looking at the same issue from a very different angle, however, and who was visiting China to investigate land use patterns not even twenty years later, fled the country in utter despair because of his discouraging observations.

25. Compare Jacek Wozniakowski's excellent study (*Die Wildnis*) on wilderness and mountains as an evolving motive in European art.

26. The question here is: when was wilderness needed again? In the West, John Muir (1838–1914) raised exactly this question to the American public, creating the first national parks when wilderness began to be in short supply (Oelschlager, *Idea of Wilderness*, 178 ff.). The present-day Chinese equivalent to the secularized national parks as embodying representations of specific cultural (or perhaps acultural) values for the Western public is probably the monastic mountain enclosure. In my personal experience I have not seen Chinese tourists embracing nature in national parks that are devoid or even intentionally stripped of traces of human history.

27. Published posthumously in October 1863. See Oelschlager, *Idea of Wilderness*, 410, note 84.

28. Geng Sangchu, an alleged disciple of Laozi, left the civilized world to retire to Mount Weilei (Mt. Zigzag, as Burton Watson renders it; see his translation of the *Zhuangzi, The Complete Works of Chuang Tzu*, trans. Burton Watson [New York: Columbia University Press, 1968], 248). After he had lived there for three years, the area experienced such agricultural growth that a great surplus was achieved (*Weilei da huai*). Local people attributed this development to Geng Sangchu personally and wanted to pray to him. This is one of the instances where a Daoist master is directly linked to bringing "ecological" (and thus economic) success to a region.

29. See He Shengdi, "Bian you shanchuan shuo yudi—daojiao dixue sixiang jianshu" (Traveling throughout mountains and streams discussing geography—A brief account of Daoist geographical thought), *Daojiao wenhua yanjiu* (Research on Daoist Culture) 7 (1995): 148: "The above mentioned thoughts of Daoist scholars . . . have had an impact on the history of Chinese earth sciences which cannot be overestimated."

30. Ibid., 141. The term "nine regions" (*jiu zhou*) refers basically to the whole empire.

31. Ibid., quoting from the *Luofu Mountain Gazetteer*.

32. Here I follow the dictum of "Landscapes are culture *before* they are nature." Simon Schama, *Landscape and Memory* (New York: A. A. Knopf, 1996), 61.

33. Part of this "mountaineering experience" also included the entering of underground worlds, the *dongfu* or *mingjie*. One could argue that the Chinese concept of the underworld or netherworld stems from exploratory research trips undertaken by Daoists, coupled with mythological materials handed down from shamanistic traditions. *Fengdu* as the "mysterious capital" (*you du*) is a case in point. The *Songs of Chu* (*Chuci*) and the *Shanhaijing* both refer to this "third" world besides the world of heaven and the world of man. See Zhang Chenjun's views on the contributions of Daoism to Chinese traditional fiction in his "*Lun daojiao dui Zhongguo chuantong xiaoshuo zhi gongxian*" (On the Contribution of Daoism to Traditional Chinese Fiction)," *Daojia wenhua yanjiu* (Research on Daoist Culture) 9 (1996): 336. In any case, speleological explorations were undertaken in China on an astonishing scale centuries before Western adventurer-scientists dared to address this realm.

34. John Lagerwey, *Wu-shang pi-yao*: *Somme taoïste du VI siècle* (Paris: Publications de l'École Française d'Extrême-Orient, 1987), 172–93.

35. Curiously, the *Wushang biyao* quotes a title with inverted characters: here it is not a *tujing*, but a *jingtu*, with the emphasis on the probably emblematic visual description of a site (in this case the *Xuanxian Renniaoshan tujing*). See Lagerwey, *Wu-Shang pi-yao*, 76, discussing page 8b).

36. *Baopuzi neipian* 18, 2a.

37. Lynn White, Jr., "The Historical Roots of our Ecological Crisis," *Science* 1551 (1967): 1203–07; reprinted in *This Sacred Earth: Religion, Nature, Environment*, ed. Roger S. Gottlieb (London: Routledge and Kegan, 1996), 184–93.

38. Even Mencius, when talking about the destruction of nature, was already aware of the time factor. In his famous Ox Mountain *Niushan* parable, he laments both how quickly an ax followed by hungry cattle can destroy a forest, and that future people would never guess that the area in question actually was constituted of healthy wooded slopes not too long before. The sometimes fatal imbalance between the speed of destruction and the slow evolution of reforestation or renaturalization led him, on a metaphorical scale, to amplify the need for human-heartedness and benevolence and to apply it to what we would call today his "ecological concerns" (Mencius, 6:1.8). Interestingly, Mencius uses the word *xi* (to breathe) to describe the process by which vegetation flourishes. *Xi*, of course, is related to qi (vital breath).

39. According to the *Book of Changes* (chapter on the trigram *kun*), *ye* is the territory where dragons can be seen fighting.

40. *The Complete Works of Chuang Tzu*, trans. Watson, 35.

41. Julia Kristeva, "Women's Time," *Signs* 7, no. 1 (1981): 16, esp. n. 4. On Plato's Chora, which is introduced in chapter 52 of his *Timaios*, see Thomas Kratzert, *Die Entdeckung des Raums: Vom hesiodischen "chaos" zum platonischen "chora"* (Amsterdam and Philadelphia: B. R. Grüner, 1998), 88ff.

Salvation in the Garden:
Daoism and Ecology

JEFFREY F. MEYER

I begin with the assumption that there is nothing that deals with ecological and environmental problems in ancient Daoist texts. This assertion is based on the conviction that today's ecological and environmental movements are the outcome of the entirely new awareness—the unique contemporary realization that human beings now have the power to seriously degrade, fundamentally change, or even totally destroy the natural environment. Traditional cultures certainly recorded their concern and care for the natural world that sustained them, and realized that their practices could harm or help it. But none could have conceived of the radical control humans now exert over it. If at one time mountains were considered enchanted places harboring passages to other worlds, modern analysis treats them as no more than geologically formed mounds of chemicals. It is no longer required that you have faith to move mountains and fill up valleys. Technology can do it. At the extreme of human meddling with nature, we now have the power to destroy the habitability of the earth and bring an end to life as we know it. I am aware of no traditional texts, in any culture, able to conceive that such power lay in the hands of human beings. Archaic and traditional cultures, China included, assume the opposite: that Heaven and Earth (nature) are revered powers which dominate human life, to which humans must submit. Our contemporary Faustian stance as masters of creation and destruction places us in an unprecedented position.

This is more than an academic point. We can certainly find relevant texts among the sacred writings of traditional cultures to use as a foundation for a responsible environmental ethic. But unless we see the radically different position we occupy today, we may not be too different from Ronald Reagan, who announced during one of his presidential campaigns: "It is an uncontrovertible fact that all the complex and horrendous questions confronting us at home and worldwide have their answer in that single book," he said, indicating the Bible.[1] Find an apposite text, this literalistic approach suggests, apply it to the present situation, and morals will change. If only it were so simple.

Certainly all the world's major religions have a rich store of resources that can address the issue of environmental responsibility. In fact, if change is to come, it is difficult to imagine any more potent stimulus toward that change than the inspiration religion can provide. But finding the right texts, beliefs, and practices is only half the task. The other half is interpretation, which has been variously described by participants in the conference on Daoism and ecology and authors in this volume as "the hermeneutics of retrieval," "confrontational hermeneutics," or, most mischievously, as "creative misinterpretation." These descriptions all recognize that the Daoist texts used are a product of a different era, concerned with different questions than our own. Eric Hobsbawm has called this process of interpretation "inventing tradition," which he defines as "a set of practices, normally governed by overtly or tacitly accepted rules and of a ritual or symbolic nature, which seek to inculcate certain values and norms of behavior by repetition, which automatically implies continuity with the past."[2] That is exactly what I am trying to do in this paper.

The "invented tradition" is not the simple maintenance of genuine traditions which are still alive. Nor is it merely a movement for the preservation of the past, which Hobsbawm sees as a common trend among intellectuals since the romantic period. It is a movement that, by making conscious connections with a past history "as legitimator of actions and cement of group cohesion," is able to "socialize, inculcate beliefs, value systems and conventions of behavior." He makes an important point in suggesting that inventing tradition is not just an intellectual exercise. To succeed, it must make connections with ritual, symbolism, actions, and behavior. Only by being thus acted upon in the social context can it create a usable past. I believe that such a possibility exists within the Daoist tradition.

The world's major religions, as the approach of this volume hopefully presumes, are surely among the most effective institutions for exerting moral influence and providing a stimulus for change. The major religions all have a rich compendium of resources that may be used to inculcate the values of ecological responsibility: beliefs, texts, ritual practices, moral traditions, and powerful symbols. By a process of selective remembering and forgetting, all these resources may be reclaimed and used to reshape the environmental ethic. If Christianity is often faulted by environmentalists for encouraging an attitude of dominating nature traceable to Genesis 1:28, "Increase, multiply, fill the earth and bring it into subjection," it may instead choose to emphasize the command to care for the garden, given in Genesis 2:15, and build an environmental ethic on the idea of stewardship.

If such an approach has the whiff of expediency about it, it need not be considered inauthentic. History has shown how certain seminal ideas may grow and develop far beyond the initial meaning given them by their creators and early interpreters. Sacred texts provide numerous examples. The Declaration of Independence and the Constitution as originally written countenanced slavery. Yet the Declaration of Independence contained Jefferson's seminal phrase proclaiming that "all men are created equal." Never mind that its author was a slave owner and himself thought that women were already equal. The powerful statement thus proclaimed took on a life of its own, gradually expanding its meaning to include white male nonproperty owners, black males, black and white females, and finally all over eighteen years of age. When Lincoln insisted that the phrase "all men are created equal" meant that slavery must end, he was "inventing tradition." Martin Luther King, Jr., in his famous "I have a Dream" speech, again invoked the phrase. In proclaiming that the hallowed words now meant that all people should be treated equally in all respects, he went far beyond Lincoln and again "invented tradition." It is a valid tactic.

Turning to Daoism, it is obvious that its more than two-thousand-year history offers a wealth of concepts, ritual practices, and important texts that may be authentically appropriated to foster a responsible environmental ethic. In fact, it may be better suited to the task than other religious traditions because it places such a high value on human harmony with the natural order. The one hundred eighty precepts described in Kristofer Schipper's essay are a prime example of this. Daoism's early texts place humans within the larger natural context, as part of the whole and not necessarily the most important part.

Since the heart of environmentalism lies in the relationship between humans and the nonhuman world, what we call "Nature," Daoism gives humans an important and responsible but modest place in the scheme of things. This is a promising beginning for creating an environmental ethic, implying that the natural world is in some sense sacred or at least has its own autonomy which must be respected by humans.

Since others have already outlined the general Daoist concepts that may be useful in deriving an environmental ethic, I will limit my proposal to a single item in the rich Daoist repository, the garden tradition. Yet, it will become clear that this rather minor strand in the "great tradition" has resonance in many other important areas, especially as it relates to the enduring Chinese love of mountains, the practice of mountain worship, and pilgrimage. I will not spend any time on the issue of just how "Daoist" the garden tradition is. Many of the values and aesthetic principles it evokes may be considered simply "Chinese." Yet, like landscape painting, poetry, and geomancy, it is at least strongly influenced and inspired by Daoist ideals.

Craig Clunas has written a study of the Chinese garden in the latter part of the Ming dynasty, focusing on texts reflecting the interests of wealthy families of Suzhou during the sixteenth century. He argues for a change in the meaning of private gardens between roughly 1500 and 1600 C.E. What were at first essentially commercially valuable properties, producing fruit, wood, and other commodities, became chiefly aesthetic expressions of conspicuous wealth and refined taste one hundred years later.[3] It is this later meaning of gardens, connected with cosmic/aesthetic principles, upon which I wish to draw in this article.

One objection to adopting the garden as locus for a discussion of ecological and environmental attitudes may be the charge that it is elitist. Most of the classic examples of the garden tradition are found in imperial parks and summer residences of emperors or in the lavish homes of high officials and wealthy merchants. As I hope to indicate, the respect for nature embodied in the garden tradition runs more deeply and broadly and is present in all levels of Chinese society. As Rolf Stein pointed out long ago, even the *penjing*, the miniature gardens in a bowl, carry the same symbolism as the large imperial gardens.[4] As this concept of the traditional garden was developing during the Ming, Huang Xingzeng, a Suzhou writer (1490–1540 C.E.), remarked that "even humble families in rural hamlets decorate small dishes with island [landscapes] as a pastime."[5] In today's democratized landscape in China, the gardens of emperors and wealthy merchants are now open to everyone.

Choosing the garden as topos for environmental reflection has a number of advantages. It certainly reveals human attitudes toward nature, but it also requires, in its creation, a dialectic pattern of active cooperation between humans and the natural world. Ji Cheng's *Yuan Ye*, the earliest Chinese garden design manual extant, says "Don't commit crimes against the hills and forests. A man of sensibility will never treat them irreverently." Elsewhere, the garden designer states that "making use of natural scenery is the most vital part of garden design . . . the attraction of natural objects, both the form perceptible to the eye and the essence which touches the heart, must be fully imagined."[6] While they seem "natural" in contrast to Renaissance and classical gardens in Europe, the finest Chinese gardens are carefully contrived products, manipulated and arranged with exquisite care. Yet at the same time, the regnant principle was what garden designers perceived as the order of nature. As the late Ming scholar and garden designer Zhang Chao said, in painting, the artist can make all the creative decisions independently (*keyi zizhu*), but in garden design, the builder must "respond appropriately to the character of the terrain and the nature of the stone."[7] Although the basic principle of Chinese garden making "has its foundation in nature and its summit in nature," the force of its planning "lies in the blending of the natural and the man-made."[8] Nature is therefore an active partner in the dialogue that results in the creation of a garden. The effects of the human hand and the results of human construction are obvious. Gardens are surrounded by walls, entered through gates, often segmented by intervening courtyards, marked by pavilions, traversed by walkways and bridges. Yet nature itself always remains "the external teacher," shaping the garden according to the pattern or model suggested by natural mountains and waters.[9]

As a collaboration between the natural and human, the garden tradition seems to me to offer an appropriate model for future environmentalism. There is no place left on Earth not affected by human actions. It is important for a realistic environmental ethic to acknowledge, however sadly, that we humans are not going to leave nature alone. While we can designate preserves and parks to remain relatively untouched, the real crux of the movement is how to treat the greater part of nature which *is* touched and changed by human hands. The garden, as a site where this interaction must necessarily take place, suggests itself as an appropriate locus to work out an environmental ethic and a kind of textbook situation where humans may learn to treat nature with respect.

Certain characteristics of the traditional Chinese garden lend themselves to environmental reflection. One important characteristic lies in the garden's "otherness." The realm of the garden is entirely different from other forms of social architecture, existing in a world set apart from normal everyday life. This characteristic makes the Chinese garden comparable to traditions of garden building in most other major cultures in which, as representations of the highest aspirations of those cultures, gardens are a human attempt to represent "paradise," the ultimate state they seek. Since gardens are not obviously utilitarian in their function, their creators may indulge in symbolic imagination, using garden design to create an ideal of perfection in the natural environment, its ideality often heightened as it contrasts with actual social reality, as Clunas has argued.[10] Medieval Christian and Islamic gardens were symbolic representations, sometimes quite literal, of the golden age before "the fall," the biblical garden of paradise, with a fountain welling up in the center and flowing in four directions to water the land. Pure Land Buddhist gardens, with their lotus ponds, gems, trees, and flowers, depict the Western Heaven of Amitabha. It is significant that the paradise evoked by Daoist gardens is not located in a former or a future world, but is created by a miniaturization of the natural structure of this world.

Another salient feature of the Daoist garden is its "eccentricity." I mean by this word that the Chinese garden is off axis. It is entirely different in structure from the axial courtyard architecture characteristic of most other built environments, like private homes, yamens, monasteries and temples, or even entire cities. The garden is a world set apart from the spaces of normal human social interaction. It is usually off to the side, in the rear, or wrapping around two sides of the square or rectangle that constitutes a traditional Chinese home. The garden is therefore "unfamilial" in the Chinese sense, since the hierarchical and locational symbolism of the home, whether private or imperial, does not apply in it. It is non-axial and asymmetrical in contrast to the axial structure of the living quarters to which it is usually attached. It offers a place for the experience of a different dimension of life where there is at least temporary freedom from the normal constraints of social hierarchy. The traditional axial home, with its successive courtyards, is a perfect metaphor for the patriarchal family system of Chinese culture. This system of hierarchical ordering, with the extended family as model, dominated not only Chinese society, but the traditional conceptual world. Even the genealogies for the "family" of orthodox schools of thought[11] or of the mythic first cultural heroes, are represented as "on axis," with each individual a single unit in the vertical "tree" connecting each to the previous and subsequent generation. Any deviation off that axis devalues or subordinates those so located. The garden, in contrast, presents a natural environment free of both conventional and judgmental social construction.[12]

The realm of the garden diverges from the values of the social world in many symbolic particulars. Rather than the normally dominant yang images, the garden presents a complex of yin symbols: the gnarled tree, the meandering path, the asymmetry of the bent or interrupted directional line of bridge or walkway. Instead of the single vista of the domestic or public courtyard, which implies human control, the garden offers multiple vistas. Instead of compressing human experience of surroundings into one single comprehensive glance, it divides it into discrete moments of comprehension and multiple perspective, so that the whole cannot be understood until all these sights have been individually experienced. Ideally, the garden should be a place of wonder, unpredictability, and surprise. As the author of the classic treatise on garden design wrote in the late Ming dynasty (1368–1644 C.E.), he wanted to plan a magnificent garden, "arranging the five sacred mountains in one single place, marshaling them in serried ranks, even down to the exquisite details of precious grasses, ancient trees, celestial birds, and special adornments, making great nature to be seen in a way that it had never been seen before."[13] In the old Chinese saying, "Heaven is round, Earth is square," the garden is the realm of the circle, the marvelous, boundless and undefined. The different reality experienced there is symbolically suggested by the frequent use of "moon gates" as ways of entry into walled garden courtyards.

The garden tradition presents an opportunity for grounding an environmental ethic for several reasons. It is first of all a place where the human and the natural worlds intersect and where human attitudes toward nature can be clarified. To repeat Lynn White's now famous dictum: "What people do about their ecology depends on what they think about themselves in relations to things around them. Human ecology is deeply conditioned by beliefs about our nature and destiny."[14] In the analysis of the theory of Chinese garden making, we can observe how the Chinese people thought about natural things and their relationship to them. Even Chinese domestic architecture, which is in most respects opposite to the world of the garden, argues for an integration of the human and natural realms. Some have in fact argued that the gardens of southern China are superior to the northern on just that basis: they are airy and feel spacious, despite their sometimes small size, because of a "profusion of doors and windows."[15]

Second, there is a strong element of human deliberation involved in making a garden, a series of conscious decisions that lead to the final form the garden will take. Gary Lease has pointed out that "humans and nature exist in dialectical relationship, each imagining the other."[16] I take his statement to mean that as we offer explanatory theories of nature, which fluctuate between "constructionism" (humans determining what counts as nature) and "essentialism" (the view that there is an independent, unchanging character of the nonhuman world), one or another of the two parties in dialogue will inevitably become dominant. The Daoist garden provides a privileged place to see this dialectic in action, with both sides having a part to play. Nature is perceived as autonomous, valuable in its own right, asserting its own subtle order as a dominant paradigm. As William Chambers wrote over two centuries ago in appreciation of the genius of the Chinese garden: "The art of laying out grounds, after the Chinese manner, is exceedingly difficult, . . . this method being fixed to no certain rule, but liable to as many variations as there are different arrangements in the works of creation."[17] This statement seems to be almost a paraphrase of Ji Cheng's axiom that there are principles but no fixed formulae and the designer must seek the essence lying behind the forms. On the other hand, the garden is at the same time a completely artificial product, its elements constructed by the artisans and owners according to their own creative or imitative abilities (with some gardens obviously more aesthetically successful than others). Garden making presupposes the mastery of a number of arts, including painting, calligraphy, literature, and engraving.[18] All of them, to some degree, have the same dialectical dynamic as garden design.

Third, the garden contains the primary cosmic symbols which express the universe in its unity, the mountains (yang) and waters (yin), the heavens, the earth, and the human. It also contains many second- ary symbols of the "ten thousand things," trees, flowers, fruits, birds, and fish. Like the *shanshui* landscape paintings, the garden is there- fore a microcosm of the universe, but in three dimensions instead of two. Mountains and waters are most essential, the natural elements which have elicited reverential feelings and a worshipful attitude from the Chinese from the ancient period to the present. In recreating these two elements in their gardens, the Chinese were creating a micro- cosm, a little Heaven and Earth (*xiao tiandi*), that expressed a certain control through miniaturization but at the same time embodied re- spect for the "many variations as there are different arrangements in the works of creation." The principles, according to Ji Cheng, imply the use of the natural setting already given, including the adaptation to local conditions and borrowing scenery.[19] The mountains and waters in some sense dictated, by their natural condition, the forms they should assume when recreated in the garden. In this sense, the Chinese prac- tice implied a certain submission to nature as dominant partner in the dialogue. Without some degree of submission of humans to the natu- ral order, it is difficult to imagine how a compelling environmental ethic can be devised.

On the other hand, I have serious doubts about the viability of what is called "deep ecology," if that is taken to mean a "biocentric egali- tarianism" that asserts that human and nonhuman beings are essen- tially equal and that individual human selves should seek their true identity in a great Self that includes everything which exists. As ap- pealing as this philosophy may be, there is little evidence that it could ever achieve widespread acceptance. Even in Asia, where scholars be- lieve that it would be more in harmony with indigenous modes of phi- losophy and religion,[20] its currency may be limited to a select few. No longer accepting archaic views of the natural world, humans now know that they can move mountains, divert rivers, build a Three Gorges dam, and, in the worst case scenario, destroy the world. The contem- porary realization of human power leads not to "biocentric egalitari- anism" but to anthropocentric domination. Even in a culture like Ja- pan, schooled for centuries in Buddhist and Shinto values, there is no evidence that a "deep ecology" perspective, if indeed it exists there, has produced any healthier attitudes or a better record toward the en- vironment than exists in the United States.[21] Finally, I think that the experience of oneness with nature and the universe is generally the result of a mystical state, obtained through long-term and rigorous discipline, and therefore cannot be used as the basis for an ethic applicable to everyone. I will return to this point later.

The Chinese garden, then, provides a model for an environmental ethic that stands somewhere between deep ecology and an anthropocentric ecology (saving nature for the benefit of humans). It proposes a cooperative relationship, wherein each of the partners, "in a dialectical relationship, each imagining the other," contributes to the final result. The human contribution to garden-making is fairly obvious. The very act of miniaturizing implies control. The garden designer is able to recreate nature within a delimited area. At the same time, the natural elements, once chosen and located, begin to dictate form. The most important of the garden elements is the mountain. It provides the connection between the minor tradition of garden-building and the greater and ancient tradition of reverence for mountains.

Anyone familiar with Chinese culture will know that mountains have been considered sacred from the earliest historic period. Shang kings offered sacrifices to the *yue* (a mountain, probably Mt. Song). The *Rites of Zhou* lists five sacred mountains, representing the five directions, requiring imperial worship: Mt. Tai (east), Mt. Hua (west), Mt. Heng (north), another mountain named Heng (south), and Mt. Song (center). The *Book of Rites* describes a tour of inspection made by Zhou rulers every five years in which they offered a sacrifice on top of one of four sacred mountains during the second month of each of the four seasons. The mountains in ancient China represented points of ritual and mystical contact with the powers of Heaven. Throughout dynastic history, the five sacred mountains, and later two other sets of mountains, received imperial worship. The feeling of awe for these sacred forces is well captured by the prayer, dated 645 C.E., of Tang Emperor Taizong:

> The sacred mountains contain fine soaring peaks and fields of wilderness with special markers where strange animals roar and dragons rise to heaven, and the spots where wind and rain are generated, rainbows stored and cranes beautifully dressed. These are the places where divine immortals keep moving in and out. The countless peaks overlap each other; thick vapor wraps green vegetation, layers of ranges move into each other, and thus sunlight is divided into numerous shining rays. The steep cliffs fall one thousand leagues into the bottom and the

lone peaks soar to ten thousand leagues high. The moon with laurel flowers blooming in it sheds veiled light. The clouds hang over the pine trees with clinging vines. The deep gorges sound loud in winter and the flying springs are cool in summer. . . . Its rugged mass is forever solid, together with the Heaven and the Earth. Its great energy is eternally potent, in the span from the ancient to the future. . . .[22]

Mountain worship is recorded in the *Shanhaijing*, which indicates that deities and marvelous creatures live in the high peaks. Some are benign, some harmful. There is ambiguity in the text as to whether the mountains are the dwellings of the deities or are themselves spirits. One chapter describes Mt. Kunlun, so famous in later Chinese literature, as the lower capital of the Heavenly emperor. The funerary art of excavated tombs at Mawangdui indicates that the sacred mountains were the dwelling place of immortals and places of passage through which the souls of the deceased might proceed to the other world. The cosmic mirrors of the Han dynasty had bosses for the five mountains, and the *boshan* censors symbolized their awesome qualities, the smoke of their incense creating "an image of mysterious clouds rising from a sacred mountain."[23]

The *Baopuzi* (Master Who Embraces Simplicity) text in the fourth century C.E. also records the presence of mountain deities. Ge Hong, the author, recommended going to the mountains to find spirits who would help the devotee in the alchemical quest. Whoever goes must prepare by observing a seven-day fast of purification. The sacred character of the mountains is reflected in the approach/avoidance feelings of those who wished to visit them. "Never enter lightly into the mountains," says Ge Hong.[24] In post–Han China both Daoism and Buddhism gravitated toward the mountains as places of retreat and religious discipline. Eventually, the great mountains became the sites for temples and monasteries as well as for hermitages for individuals and small groups of devotees. The mountains, still considered as sacred places of contact with other worlds, retained their awesome and numinous qualities. The ambiguity continued. They were places of entrance to the underworld and access to heavens. Mao Shan, for example, was "a recognized halfway house, by way of which the adept might expect to ascend from this world to the starry realms above." Yet, at the same time, it was only the adepts who could know the true secrets of the mountains. "Part of the education of a Taoist was learning to detect the normally invisible contours of the true mountain, and particularly the disguised entrances and adits which lead to the mystic world below."[25]

From the Tang dynasty onward, pilgrimage to the sacred mountains became a practice prevalent among all social classes. In fact, the classic form of pilgrimage in China "involves a journey to a temple on a mountain peak," a place endowed with *ling*, a numinous quality belonging to a place where the pilgrim might contact the spirits and secure an efficacious response.[26] The words used to describe the act of going on pilgrimage in Chinese are *chaoshan jinxiang*, literally, "paying respects to the mountain and presenting incense." The phrase is parallel to that describing an individual who goes to the imperial court and pays respect to the emperor, complete with kowtow, so it suggests coming into the presence of a sacred being. In the case of pilgrimage, the sacred being is the mountain. So interconnected were the ideas of pilgrimage and mountain that the famous Buddhist site was called Putoshan, Mt. Puto, although it was not a mountain at all but a rocky island off the eastern coast of China near Ningbo.[27] It is important for my thesis to stress that, as pilgrimage sites, the sacred mountains were not destinations reserved for the elite alone. Those who came included "emperors, Ch'an and ordinary Buddhist monks, professional Taoists, hermits, tourists, and lay pilgrims." Since most of the pilgrimages were open to all and frequented by all, their meaning must be sought in an "overarching unity of shared ideas and practice which developed and survived in the context of competing doctrines, specialists, classes and institutions."[28]

Traditional Chinese gardens, with their stones arranged to evoke sacred mountains, reflect at least some of this long history of Daoist religious practice. The unusual stone in the *penjing*, the larger configurations of boulders that re-create in the garden the mountain with its meandering paths, its caves and mysterious grottoes, are meant to create a feeling of awe in the observer. These miniaturized mountains may function as a kind of icon in that they not only symbolize but in some way participate in the reality they represent. Stephen Kellert speaks of the tendency of the Japanese to experience and enjoy nature in highly structured circumstances. "The objective, as one informant suggested, was to capture the presumed essence of a natural object by adhering to strict rules of seeing and experiencing intended to best express the centrally valued feature." The Chinese word for the essence or soul of a natural object would be *shenqi*, but it is hard to imagine a more accurate description of what is intended by the creators of Chinese gardens.[29]

The garden is therefore a deliberately created object that, like an icon, evokes realities far greater than itself. If we focus on the power of the mountain as an ancient symbol, we have an icon which opens the door to ritual. Building on this long history, Daoist gardens can recall and revitalize the past, connect with other Chinese traditions which revered mountains (Confucianism, Buddhism, fengshui, and the aesthetic appreciation of poets and painters), and, on the basis of reverence for mountains, inculcate a more general environmental ethic which extends this reverence to all of nature.

A conference in Taiwan on 7 June 1995 focused on the theme of environmentalism and Daoist thought. The keynote presentation was given by Professor Fu Weixun. Three substantial responses followed, along with an informal discussion. All of this material, plus two other major articles, were then published in the *Journal of Philosophy* (*Zhexue zazhi*) (volume 13, July 1995). The conference was very philosophical in approach, presenting a careful and thorough analysis of Laozi and Zhuangzi while at the same time showing an awareness of contemporary writings on environmentalism in the United States and elsewhere, including such major issues as deep ecology, ecofeminism, and social ecology. The keynote speaker asserted that China's ancient wisdom, if creatively interpreted, has the ability to speak to contemporary environmental problems. He grounded his assertion on three relevant ideas advocated in the Lao-Zhuang texts: the spirit of detachment, the spirit of equality of all creatures, and the spirit of environmentalism, interpreting them as the possible cornerstones of a contemporary environmental ethic.[30] The second major article is a nostalgic celebration of the traditional Chinese closeness to nature, a call to turn away from consumerist society and to return to a simple lifestyle.[31] It is difficult to see how, in the overpopulated and industrial world of contemporary Taiwan and China, such a return to simplicity could be feasible.

The third major article is essentially a religious interpretation of the Lao-Zhuang texts. It confirms my doubts that these texts can become the basis of an environmental ethic applicable to everyone. The author, Zhuang Qingxin, makes it abundantly clear that the very texts normally cited as relevant to an environmental ethic are really descriptions of a high level of spiritual attainment—what I would call mystical realization—-the result of serious, long-term discipline. To achieve a proper attitude toward the environment, says the author, the Lao-Zhuang writings teach that we must first seek Daoist perfection (*xiu Dao*). In simple words, we are going to have to save ourselves before we can save nature. As Professor Zhuang says in his summary:

> The attitudes Lao-Zhuang thought invites us to form toward nature are: respect for nature, following environmental laws (or natural laws) in dealing with nature, and finally reaching that realm where Heaven and humans are one—these would be the approximate principles of so-called Western environmental theory. But to realize these attitudes, we cannot just say fine words without getting down to practice. We must really work diligently to take on the discipline, beginning with transcending desires and (self-serving) knowledge, then gaining the freedom to liberate the self and the inner spirit; finally, after this, to enter the level of attainment where there is no self: where self is forgotten. Then only can we rise to that highest level where, in knowing the Dao, we understand that Heaven and humans are one. The process and steps of this discipline are what humankind today so desperately needs in order to protect the environment.[32]

While Daoist principles taken from the Lao-Zhuang texts have been approvingly quoted in Western environmental writings, I think few of the authors envision such an arduous spiritual task as a prerequisite for the endeavor. Professor Zhuang is asking people to undertake the effort to become sages (*shengren*) in order to acquire a proper relationship to nature. If that is what is required, then we are indeed in trouble. Still, the author's words are a salutary reminder that no ethical principle stands alone. To function effectively, each principle must be a part of an overall philosophy and way of life. Much Western ecological writing is theoretical and analytical, advocating ideals that are not grounded in the communal and ethical context of any actual society. Such thinking fails to stress sufficiently the need for practical application. The right words and a logical structured ethical theory are necessary but do not guarantee any commitment to put such principles into practice. As always, how to motivate is the central issue.

At this practical level, the tradition of the garden offers an opportunity, providing both theory and the promise of motivation within the context of Chinese society. Based on the fact of Chinese love for and awe toward mountains, it recalls the ancient practice of pilgrimage to such sites as sacred places. If Chinese reverence for mountains, as

reflected in garden-making, can be connected to a reinterpreted practice of pilgrimage, then we might have the beginnings of a new "invented tradition," as defined by Hobsbawm. Although individual pilgrim attitudes would run the gamut from the religious to the touristy
and secular, there is no reason why most could not feel some of the
awe of the mountain experience. Chinese pilgrims, like Chaucer's
folk going to Canterbury, have always been a motley group, on pilgrimage for a variety of motives, some religious and some secular.
With Daoist and Buddhist temples and monasteries being rebuilt and
with a resurgence of mountain pilgrimage, there is hope for a renewal
of a religious attitude toward nature that these practices evoke.

Still, an "invented tradition" is not just a theoretical enterprise; it
requires popularization through education. Here, too, all the elements
are already in place. The tradition of garden-making and the creation
of artificial mountains is alive and well in China. So many writers
have published articles and books on the subject in recent years that
one scholar speaks of a *yuanlin re*, a craze or fad about gardens, with
authors taking up the topic like a "swarm of bees."[33] There is scarcely
a park in all of China which does not have some attempt to construct,
whether artistically or clumsily, the strange stone configurations that
represent mountains. In the middle of Beijing, where I am now revis-
ing this essay, single, oddly shaped stones can be found placed in many
school yards, in small green areas along the street, or in open spaces
in front of apartment houses. The symbolic "infrastructure" for the
invented tradition, the mountain as icon, is already in place and wait-
ing.

The final requirement is interpretation through education. The simple
"presence" of the mountain is not enough. Though artificial mountains are found all over China, they can become merely commonplace,
an ubiquitous banality easily ignored. They need to become the object
of attention, their significance consciously recalled. This kind of seeing and remembering can only be accomplished by integrating it into
the course of Chinese education. Once again, most of the infrastructure is already in place. The Chinese, in both Taiwan and the People's
Republic, have a long-standing tradition of emphasizing the aesthetic
mode in moral education. Some years ago, when I was involved in
research on moral education on both sides of the straits, I found strong
evidence for the continued vitality of this tradition, the belief that
beauty, especially the beauty of nature, could be used to shape moral

character. As one professor at Sichuan Normal College wrote: "Beauty is the bridge which leads to morality . . . appreciation of the beauties of nature, artistic activities, or experiencing the beauty of human nature in a social context—all these lead to a purification of the spirit." And in the language textbooks used in Taiwan elementary schools, more than one-third of all lessons in the twelve volumes were aesthetic/ moral in purpose, seeking to "promote a love of the beauties of nature, the land, the sea, of animals, flowers and vegetation, the moods of the seasons, of night and day, rain and sun, and so forth."[34] The Chinese garden tradition, now receiving so much international acclaim, is an element of national pride and could become part of the foundation of an environmental ethic.

Given the long tradition of moral/aesthetic education going back to Confucius (who advocated using music and ritual to shape character and harmonize society), this approach would have a better chance of succeeding in China than elsewhere. Many of the lessons in the schools' language, morality, social studies, and geography textbooks already focus on the beauty and power of nature. The connection to ecology and the environment can be easily drawn. Although a limited aesthetic setting, the garden's chief components, the mountains and waters, can become iconic symbols of the beauty and power of the natural world as a whole. Even the lesser elements can have moral significance. As Zhang Chao said: "The plum tree leads a person to loftiness, the orchid to quietness, the chrysanthemum to unpolished simplicity, the lotus to contentment, . . . the peony to heroism, the canna to gracefulness, the pine to leisure, the phoenix tree to clarity, the willow to sensitivity."[35] If the aesthetic/moral tradition in Chinese education still thrives, there is no reason why the garden tradition could not be one element in the larger effort to recall the historic respect for nature and, at the same time, to develop an effective environmental ethic.

Notes

1. Quoted in Martin Marty, *Religion and Republic: The American Circumstance* (Boston: Beacon Press, 1987), 49.

2. Eric Hobsbawm, "Introduction: Inventing Traditions," in Eric Hobsbawm and Terence Ranger, *The Invention of Tradition* (Cambridge: Cambridge University Press, 1983), 1–14.

3. Craig Clunas, *Fruitful Sites: Garden Culture in Ming Dynasty China* (Durham, N.C.: Duke University Press, 1996).

4. Rolf Stein, *The World in Miniature: Container Gardens and Dwellings in Far Eastern Religious Thought*, trans. Phyllis Brooks (Stanford: Stanford University Press, 1990), 1–112.

5. Quoted in Clunas, *Fruitful Sites*, 101.

6. Ji Cheng, *The Craft of Gardens*, trans. Alison Hardie (New Haven: Yale University Press, 1988), 50, 121.

7. Quoted in Xia Xianfu, *Wanming shifeng yu wenxue* (Beijing: Zhongguo shehui kexue chubanshe, 1994), 63.

8. Peng Yigang, *Zhongguo gudian yuanlin fenxi* (Beijing: Zhongguo Jianshu gongye chubanshe, 1997; first published 1986), 9.

9. Ibid.

10. Clunas, *Fruitful Sites*, 91–103; 203–7.

11. See, for example, Thomas Wilson, *Genealogy of the Way* (Stanford: Stanford University Press, 1995).

12. But see some examples of social hierarchy in Ming paintings and lacquer work depicting humans in their gardens in Clunas, *Fruitful Sites*, 157–60.

13. Ji Cheng, *Yuan ye*, quoted in Xia, *Wanming shifeng yu wenxue*, 64.

14. Lynn White, Jr., "The Historic Roots of Our Ecologic Crisis," *Science* 183 (1967): 1205.

15. Chen Congzhou, *On Chinese Gardens* (Shanghai: Tongji University Press, 1984), 25.

16. Gary Lease, "Introduction: Nature Under Fire," *Reinventing Nature: Responses to Postmodern Deconstruction*, ed. Michael E. Soulé and Gary Lease (Washington, D.C.: Island Press, 1995), 7.

17. William Chambers, *Designs of Chinese Buildings, Furniture, Dresses, Machines and Utensils* (1757; reprint, New York: Benjamin Blom, 1968), 19.

18. Xia, *Wanming shifeng yu wenxue*, 63.

19. Quoted in Chen, *On Chinese Gardens*, 6.

20. See Michael E. Zimmerman, *Contesting Earth's Future: Radical Ecology and Postmodernity* (Berkeley and Los Angeles: University of California Press, 1994), especially chapter 1, "Deep Ecology's Wider Identification with Nature," 19–56.

21. See Stephen R. Kellert, "Concepts of Nature: East and West," in *Reinventing Nature*, ed. Soulé and Lease, 103–21.

22. Quoted in Kiyohiko Munakata, *Sacred Mountains in Chinese Art* (Urbana: University of Illinois Press, 1991), 2. My summary of mountain worship is based mostly on this source.

23. Ibid., 28.

24. Kristofer Schipper, *The Taoist Body* (Berkeley and Los Angeles: University of California Press, 1993), 171.

25. Edward H. Schafer, *Mao Shan in T'ang Times,* 2d ed., rev. (Boulder, Colo.: Society for the Study of Chinese Religions, 1989), 3–5.

26. *Pilgrims and Sacred Sites in China,* ed. Susan Naquin and Chun-fang Yu (Berkeley and Los Angeles: University of California Press, 1992), 11.

27. See Chun-fang Yu, "P'u-t'o Shan: Pilgrimage and the Creation of the Chinese Potalaka," in Naquin and Yu, *Pilgrims and Sacred Sites in China,* 190ff.

28. Susan Naquin and Chun-fang Yu, introduction to Naquin and Yu, *Pilgrims and Sacred Sites in China,* 23–25.

29. Kellert, "Concepts of Nature," 112–13. On *shenqi,* see Chen, *On Chinese Gardens,* 49.

30. Fu Weixun, "Daojia zhihui yu dangdai xinling" (Daoist Wisdom and the Contemporary Spirit), *Zhexue zazhi* 13 (July 1995): 1–16.

31. Wei Yuankui, "Lao-Zhuang zhexue di ziranguan yu huanjing zhexue" (The View of Nature and Philosophy of the Environment in Lao-Zhuang Philosophy), *Zhexue zazhi* 13 (July 1995): 36–54.

32. Zhuang Qingxin, "Daojia ziran kuanzhong di huanjing zhexue" (The Environmental Philosophy of the Daoist View of Nature), *Zhexue zazhi* 13 (July 1995): 70.

33. Peng, *Zhonggo gudian yuanlin fenxi,* 1.

34. "Moral Education in the People's Republic of China," *Moral Education Forum* 15, no. 2 (summer 1990): 6–7; "Teaching Morality in Taiwan Schools: The Message of the Textbooks," *The China Quarterly* 114 (June 1988): 269.

35. Xia, *Wanming shifeng yu wenxue,* 67.

Sectional Discussion:
How Successfully Can We Apply the Concepts of Ecology to Daoist Cultural Contexts?

JOHN PATTERSON and JAMES MILLER

Introduction

The response to the previous section argued that the essays, taken together, point toward the recognition of three fundamental ecospiritual themes present within Daoist religious texts. The purpose of this review essay is to see whether ecological themes are similarly present in the practical contexts of Chinese cultural life that have been the focus of the essays in this section. The ecological themes outlined here are drawn from a survey paper by Eugene Odum entitled "Great Ideas in Ecology for the 1990s"[1] and include: 1) survival through cooperative adaptation; 2) mutualism and identification; and 3) survival through benefiting our hosts.

Survival through Cooperative Adaptation

Odum states that, in order to survive, an organism does not compete with its environment as it might with another organism, but has to adapt to (or modify) its environment in a cooperative manner.[2] If it were to compete successfully with the ecosystem in which it lives, it would destroy the very system that supports it. The idea of adapting to the environment is prominent in the *Daode jing*. Chapter 76 tells of how, when we are living, we are supple and yielding, whereas when we are dead, we are strong and hard. Rather than try to overcome the world in which we live, we should take our lead from other living things, from plants in particular, and learn how to get what we want through flexible adaptation. The message here is that by yielding, not by competing, we can adapt to our environment, as contrasted with familiar aggressive ways of trying to master the nature world and make it conform to our will.

Jeffrey Meyer offers a cultural parallel to this in his discussion of Chinese garden design. Whereas the painter can afford a certain independence from his subject, the garden architect must "respond appropriately" to the materials with which he is working. The concept of appropriate response provides a third way between anthropocentrism and biocentrism, recognizing the reality of the privileged position of the human species, and its concomitant responsibilities. The deeper, religious point is that nature can become the space for human re-creation. The flexible stance required by the principle of cooperative adaptation does not entail a negative "yielding" or "compromise" by human beings, but in fact provides the means for spiritual enrichment and "compromise" in the wholly positive sense of a future held in common.

Mutualism and Identification

This second theme is simple but perhaps surprising: the evolution of mutualism increases when resources become scarce.[3] "Mutualism" applies to species which co-evolve so that each ends up being of benefit to the other. An example would be the small birds which pick annoying insects out of the skin of large mammals: the birds benefit by having a ready supply of food and the mammals benefit from regular grooming. The relation to scarcity of resources is easy enough to understand. If there was plenty of food for all birds, they might not bother to risk coming near large mammals. But when the going gets tougher, new risks are taken. The moral is clear: mutualism between humans and other creatures is the way to go. *Daode jing* 56 does have a parallel message when it urges us to identify with or assimilate to the mundane world:

> Block the passage
> Bolt the gate
> Blunt the sharp
> Untie the knot
> Blend with the light
> Become one with the dust—
> This is called original unity.[4]

This link is even stronger when we reflect on the way that mutualism breaks down the differences of interest between species. The goal of mutualism, however, demands an alternative to the ordinary scientific scheme of cause and effect. As Thomas Hahn points out, this mutual assimilation takes place through complex networks of correlation and interdependencies. Moreover, getting to know the "greatness of the four seas" and the "pervasiveness of the ten thousand things" is a psychologically transformative process for human beings. There is thus a real religious investment to be made in the imaginative pursuit of assimilation of nature and person, mapping the wilderness of the environment onto the mystery of the body.

Survival through Benefiting Our Hosts

Odum invites us to consider the human species as a parasite on our environment, and he goes on to point out that any such species that benefits its community has survival value greater than a species that does not.[5] While the image may not be welcome to all of us, the ecological message is important. Odum urges us to stop exploiting the planet by coupling this model of humans as parasites on the biosphere with the general principle that parasites stand a better chance of surviving if they are not too unwelcome. If we are to survive as parasites, the survival and well-being of our hosts is in our interest as well as theirs. So, as Odum says, we must reduce virulence and reward feedbacks that benefit our host.

Although, obviously, the foundation in evolutionary biology is lacking, a parallel idea emerges in the *Daode jing* when it advocates that we benefit all things without competing with them.[6] The Way of Heaven (*tian zhi dao*) is to benefit all, harm none,[7] to pare back abundance and add to the insufficient.[8] In the context of fengshui, the concept of "not too unwelcome" is clearly expressed in the language of qi that emerges as the means for reading the contours of the natural environment. To be sure, the purpose of this reading of nature is wholly directed toward the benefit of human beings through the propitiation of ancestral spirits, but this can only take place by means of a burial that does not disrupt the figurative contours of the landscape. We must not be deluded that fengshui has any tangible benefits for the environment, but inasmuch as it offers a model for "parasitic" human development that at least has a vocabulary for respecting the landscape of its "host," it deserves to be taken seriously.

Conclusion: Text and Context

The *Daode jing* presents a good model for any dominant being to follow. Maybe the intended audience were political rulers of ancient China, but the model is equally applicable to the human species and its domination over other species. This is clearly one reason why the text appeals so much to environmental philosophers. In the papers presented in this section, the connection is made not with Daoist texts, but with cultural contexts. The connection is evident in Thomas Hahn's exposition of the relationship between the Daoist and the mountain. By adopting an attitude of respect, the adept is able "to survive and flourish in nature." The principles by which this survival is possible are formulated rationally and scientifically by ecologists, but, as Stephen Field indicates, may best be appropriated aesthetically: "It is not the principles of rational order that should solely determine the value of our vision, but the capacity of our vision to satisfy our responsibility for ethical order." This suggests that although rational principles should be the substance of text and theory, from the perspective of practice and context, an alternative vision is required. Perhaps it is in this practical dimension, in the folk culture that E. N. Anderson delineates, that we will best be able to satisfy the demand for an ethical orientation to the environment.

Notes

1. Euguene Odum, "Great Ideas in Ecology for the 1990s," *Bioscience* 42 (1992): 542–45.

2. Ibid., 545.

3. Ibid.

4. *Daode jing* 56. From *Tao Te Ching,* trans. Stephen Addis and Stanley Lombardo (Indianapolis and Cambridge: Hackett Publishing Company, 1993).

5. Odum, "Great Ideas in Ecology," 543.

6. *Daode jing* 8.

7. *Daode jing* 81.

8. *Daode jing* 77.

IV. Toward a Daoist Environmental Philosophy

From Reference to Deference: Daoism and the Natural World

DAVID L. HALL

Daoism as *Philosophia Perennis*

As readers of this volume will certainly appreciate, the term "Daoism" represents a complex interweaving of both "philosophical" and "religious" doctrines and practices. Since I shall be primarily concerned with the philosophical resources of Daoism, I should begin by saying something about the relation of "philosophical Daoism" to the larger Daoist tradition.

Unlike Confucianism, which can be identified with a principal founder, Daoism is an *ex post facto* creation, first named as a school by the historian Sima Tan in the second century before the common era. Since Sima Tan's concern was classifying schools in terms of their contributions to the activity of governing a state, there was no recognition of a relationship between Zhuangzi and Laozi, later recognized as the two principal Daoist thinkers.

A further complication in explicating "Daoism" results from the fact that the "School of the Way," *Daojia*, is to be distinguished from *Daojiao*, what may be termed the the "Doctrine of the Way," the latter traditionally dated from 142 of the common era. *Daojiao* is the foundation of the organized Daoist religion that persists until this day. Until recently, scholars of Daoism, particularly those in the West, have typically drawn a rather sharp distinction between the *daojia* and *daojiao*. More recently, with the translations of Daoist "religious" texts of the medieval period (from the collapse of the Han dynasty in

220 C.E. to the beginning of the Tang dynasty in 618 C.E.) there have been attempts to identify in the *daojiao* a "mystical philosophy" that has its roots in the thought of Laozi and Zhuangzi.

Perhaps it is fitting that the historical roots of what has come to be identified as Daoist thinking should be obscure. In fact, it is in part due to the relatively fluid identity of the Daoist tradition that it has managed to have such a widespread influence beyond the borders of China. Both the texts of the *Daode jing* and the writings of Zhuangzi have become classics of world philosophical and spiritual literature, taking their place as profound expressions of the *philosophia perennis* that touches the deepest layers of human sensibility.

In the following discussion, I shall be concerned with connecting my interpretations with classical or contemporary Daoist religious practices, and with the further articulation of philosophical Daoism as *philosophia perennis*. I will focus upon the relevance of ideas, principally resourced within the *Laozi* and the *Zhuangzi,* to certain broad ecological concerns.

The Aesthetic Understanding of the Natural World

Some years ago, in a work entitled *Eros and Irony: A Prelude to Philosophical Anarchism,*[1] I borrowed some insights of philosophical Daoism to elaborate upon a contrast between aesthetic and logical understandings of order. Briefly, that contrast is between a characterization of things, by appeal to causal laws or pattern regularities, and one that, indifferent to such appeals, celebrates the insistent particularity of items comprising the totality of things. A. N. Whitehead had explicitly suggested that distinction in his late work *Modes of Thought*[2] but had not developed it in any detail. I found the Daoist vocabulary useful in attempting such a development. This was the beginning of my interest in the Daoist sensibility.

At about the same time, Roger Ames and I began a collaboration that focused upon Chinese and Western comparative thought. He and I further developed the logical/aesthetic distinction with respect, first, to classical Confucianism and, later, to a number of issues addressed in classical Daoist philosophical texts.[3]

In an essay entitled "On Seeking a Change of Environment: A *Quasi-*Daoist Proposal,"[4] I have argued that the advantage of philosophical

Daoism over certain mainstream Western approaches with respect to issues of ethics and ecology lay largely in the former's understanding of the natural world as *aesthetically* ordered.

In that essay I introduced three characteristics of an aesthetic understanding of nature, but I was only able to discuss the first two of the three. I will take the opportunity provided me here to consider the third characteristic of aesthetic thinking. Before doing that, however, I will briefly rehearse the other two.

The first characteristic of aesthetic understanding is *natural parity*—that is, the denial of ontological privilege to any object, event, or state of affairs. From the aesthetic perspective, the world is the noncoherent sum of all orders defined from the myriad perspectives taken up by each item in the totality of things. Natural parity urges that the whole of things be entertained as a noncoherent totality comprised by an indefinite number of "worlds" construed from the perspective of each item in the totality.

The second characteristic of an aesthetic understanding of nature is a specification of the first. Belief in natural parity *a fortiori* excludes anthropocentrism and with it any objectivist construal of the world that benefits the sorts of beings we human beings deem ourselves to be. Such construals characterize the natural world from the perspective of explicit or implicit theories of *human experience* usually parsed as feeling, knowledge, and action. Aesthetic understanding, therefore, promotes norm-less, nontheoretical characterizations of the modalities of human experience. In philosophical Daoism these characterizations are expressed in terms of the *wu*-forms of unprincipled knowing (*wuzhi*), nonassertive action (*wuwei*), and objectless desire (*wuyu*).

The third characteristic of aesthetic understanding, which will be the principal subject of my remaining remarks, concerns the character of the language employed to communicate aesthetic understanding. Aesthetic language is nonreferential. It is, rather, a language of deference.

A Daoist language of deference sidesteps both the assumptions that the reference of an expression is the object denoted by it (Frege) as well as that which only proper names denote (Russell). The philosophical Daoist wishes to replace any precise language of denotation and description with an allusive language of "deference."

Referential language characterizes an event, object, or state of affairs either through an act of naming meant to indicate a particular individual ("The author of the poem is Emily Dickinson") or by appeal to class membership meant to identify an individual or individuals as one of a *type* or *kind* ("Emily Dickinson is a poet"). The language of deference eschews proper names or classes and depends upon hints, suggestions, allusions. The language of deference accepts Dickinson's injunction:

> Tell all the truth
> But tell it slant
> Success in circuit lies.

The circuitous path required by the language of deference is a consequence of yielding to the perspective of the things we would appreciate. The path is paved with metaphor—provided that we do not think of metaphors as ultimately translatable by appeal to some literal ground. At its best, the allusive language of deference celebrates a processive, transitory world of myriad transformations that cannot be fixed through reference or bound by class enclosure.

> The long slender bars of cloud float like fishes in the crimson night. From the earth, as a shore, I look into that silent sea. I seem to partake its rapid transformations: the active enchantment reaches my dust, and I dilate and conspire with the morning wind.[5]

Dissatisfaction with referential language is hardly controversial. Assessed merely on these terms, philosophical Daoism is in the company of the behaviorists, pragmatists, the later Wittgenstein, and much of British and American analytic philosophy. Nonetheless, the Daoist critique of reference is a distinctive one.

The Daoist language of deference may be most productively contrasted with languages of *presence, absence,* and *difference.* In Jacques Derrida's quasi-Heideggerian formulation, the language of presence, or logocentric language, depends upon a contrast of Being and beings. Such language makes Being (essence, essential characteristics) present through the beings of the world. Logocentric language re-presents an otherwise absent object. In contrast to the language of presence are the mythopoetic or mystical uses of a language of *absence* in which discourse is employed to advertise the existence of a nonpresentable subject. In both the languages of presence and absence there is a referent—putative or real—existing beyond the linguistic expression.

In the language of presence, reference may be illustrated through the employment of nouns as names. Such names are abstract nouns serving to identify the designated item. "Pen" and "whistle" are abstract nouns, class nouns, which may be further delimited through pronominal description—*this* pen, *his* whistle. Proper names ("Maurice," "Glacier Park") serve to identify particulars directly. But, apart from immediate, ostensive contexts, further descriptions are usually necessary to identify even particular persons or places.

The Chinese language does not give the same centrality to abstract nouns as do Indo-European languages. (In the following section, I shall be arguing that the principal reason for this fact is to be found in the meaning of the "being/nonbeing" [*you/wu*] contrast in classical Chinese.) Thus, the referential function of language is not nearly as important as the descriptive function. No ontological referencing is possible without a copulative sense of being as existence that serves to underwrite abstract nouns.

Derrida's language of *différence* limits the referential act by deferring meaning with respect to both the structure and process of language. Language as syntactical, grammatical, or semantical structure abstracts from language in *use*. Focusing upon language in use places linguistic structure in the background. Thus, neither structure nor process considered by itself allows for complete discussion of meanings. Meaning is always *deferred*.

The first approach to the Daoist understanding of language may be made through an emendation of Derrida's notion of *différence*. Introducing the homonymic "defer" ("to yield," "give way to") supplements the Derridean view by adding an active sense of "deferring."

Put in a slightly less "gobbledygooky" manner: in her attempts to understand others (as opposed to that "self-understanding" that involves assaying her efforts to construe the world from her own perspective), the Daoist uses language in such a way as "to give way to" encountered items—that is, to let them be themselves.

In the following two sections I shall be concerned with highlighting two principal elements within the culture of Daoism that support the development of a language of deference. One involves the problematic of *you* and *wu*—usually translated "being" and "not-being." The second concerns the vagueness and allusiveness of the Daoist vocabulary.

Being and Not-Being

Western philosophers learn to understand *Being* as a universal property or common structure that qualifies all distinct beings. Being is seen as indeterminate ground guaranteeing the determination of things. The relation of being and beings as the relation of indeterminate and determinate is rendered somewhat more complex by the notion of "Not-Being." Classically, Not-Being is considered the negation of Being either in a simple, logical sense or as the Nihil, the Void.

Understanding Not-Being as the Nihil has colored much metaphysical speculation with a rather mystical tone. The presumed experience of Not-Being may be thought to evoke a sense of existential angst or dread. The reason for the existential negativity of the experience of Not-Being is associated with the early Neoplatonic speculations that equated Not-Being with evil. In order to defend the perfection of the Divine Being, thinkers from Plotinus through St. Augustine and St. Thomas contrasted perfect Good as the fullness of Being with pure Evil as absolute Not-Being. Insofar as something has Being, it is, to that extent, good. God as Pure Being is perfectly good. All of God's creatures are ordered in a hierarchical scale of Being, from the perfect good to the absolute evil of Not-Being. Thus the sense of *tremendum* associated with the possible experience of Not-Being as the Nihil, the Void.

The distinction between Chinese understandings of *you/wu* and the Being/Not-Being contrast of classical Western philosophical theology could not be greater. Chinese thought is not disposed toward any fixed form of asymmetrical relation such as is expressed by that of Being and Not-Being or Being and beings. The Chinese verb *you*, typically translated "being," suggests "having" rather than "existing." *You* means "to be present," "to be around." *Wu*, "not to be," means "not to be present," "not to be around." Thus the Being/Not-Being contrast in the West lacks an adequate translation in the classical Chinese language. *You* indicates *presence* and *wu* expresses *absence*. The *you/wu* relation may be expressed not in terms of negation but as mere contrast in the sense of either the presence or absence of *x*, rather than an assertion of the existence or nonexistence of *x*. The logical distinction between "*not*-x" and "*non*-x" is elided.

Chapter 40 of the *Daode jing* contains a phrase often translated somewhat as follows: "The things of the world originate in Being (*you*).

And Being originates in Not-Being (*wu*)." Understanding this passage by appeal to classical Western senses of Being and Not-Being would lead to confusion. Following the general preference of the *Daode jing* for reversing certain classical contrasts, *wu* appears to be given preference over *you*—in the same manner that *yin*, as "passive," is given priority over *yang*, as "active." Interpreting *wu* as "Not-Being" would, then, suggest a preference for Not-Being over Being. Such interpretations have led to some rather ridiculous conclusions suggesting that Daoist thinkers share the Western classical understanding of the Being/Not-Being contrast, but somehow prefer the Nihil or the Void, as Nonexistence, to Being-Itself.

In his translation of the *Daode jing,* D. C. Lau has sought to overcome this problem by translating *you* as "something" and *wu* as "nothing." This translates the claim of chapter 40 into the judgment that nothing is superior to something. A suitable alternative might be "not-having" is superior to "having."

The absence in classical Chinese of any notion of Being as existence per se means that there is no notion of Being as ontological ground and no need for a metaphysical contrast between Being and beings. In the language of postmodernism, one need not seek to overcome the logocentrism of a language of presence grounded in ontological difference if no distinction between Being and beings, or beings and their ground, obtains. For the Chinese, a language of presence is a language of making present not the essence of a thing or event but the thing or event itself. Classical Chinese philosophical discourse, then, is in no need of deconstruction since the senses of *you* and *wu* within the Chinese sensibility do not lead to the creation of texts that could legitimately be targets of the deconstructor.

For the Chinese, and this is a particularly fruitful fact concerning the Daoist sensibility, language that does not lead one to posit ontological difference between Being and beings, but only a difference between one being and another, suggests a world without ultimate ontological foundations.

The absence of an ontological ground is directly related to the relative unimportance of abstract nouns in the Chinese language and tradition, for such nouns require the articulation and formal definition of concepts expressing the sense of the noun. Thus, a concept must be defined both denotatively and connotatively.

For example, the denotation of a concept such as "trustworthiness" is all of the instances of trustworthy behavior—past, present, and future—to which the term might be said to refer. The strict connotation of the concept is all of the properties or characteristics of the term "trustworthiness" that permit a complete characterization of the concept. Apart from ostensive uses of language, a language of reference depends upon the existence of formally defined concepts allowing for the identification of items as members of classes of such items. In the absence of strict connotative and denotative procedures, a language of reference would not be possible.

The famous first lines of the Daoist classic will afford us greater insight into the understanding of the *you/wu* relation in Chinese thought.

> The Way that can be spoken of is not the constant Way. The name that can be named is not the constant name. The nameless is the beginning of the ten thousand things.[6]

The term "nameless Dao" is best construed here not as ontological ground but merely as the noncoherent sum of all possible orders. The "acosmotic" cosmology of the Daoist entails a decentered world comprised of a myriad of unique particulars—"the ten thousand things" (*wanwu*). These things have no single-ordered coherence and, therefore, do not add up to a *cosmos*.

In Daoism, and in Chinese cosmology generally, the relevant contrast is not between the cosmological whatness of things and the ontological thatness of things, but between the acosmotic world as the sum of all orders and that same world as construed from a particular perspective that envisions but one of the myriad orders ingredients in "the ten thousand things."

The Daoist world is not to be seen as *a* Whole but as *many* "wholes." Because there is no sense of being as a common property or a relational structure, the world lacks a single coherent pattern characterizing its myriad processes. The order of the world is, thus, neither rational nor logical, but *aesthetic*. This is the case since there is no transcendent pattern determining the existence or efficacy of the order. Natural order is *ziran* (self-so). That is, order is the spontaneous expression of the particulars comprising the totality of existing things. Each item in the world construes an order from its own perspective.

The shifting character of orders in the Daoist tradition leads to the notion of *deference*. I will later address the issue of deference in terms of the principal Daoist *wu*-forms (*wuzhi, wuwei, wuyu*). For the present, I shall merely note that when a world is constituted from a single construing perspective, all other particulars are invited to defer to that construing item. The necessity of deferring and being deferred to is an implication of the natural parity of all things.

The *you/wu* problematic establishes one of the conditions under which a language of deference may replace a language of reference. In the absence of Being as ground, the relations of things are constituted by the things themselves rather than by *Being* as common property or relational structure. The character of these relationships is established not by *reference* to a common standard, but by mutual *deference*.

Vagueness and Allusive Metaphor

A second condition permitting the construction of a language of deference may be characterized in terms of the notions of *vagueness* and *allusive metaphor*. It is with respect to these notions that philosophical Daoism makes its most explicit contribution to aesthetic understanding.

> Therefore, the sage is like a mirror—
> He neither sees things off nor goes out to meet them. He responds to everything without storing anything up.[7]

The absence of the notion of Being as existence in the Chinese tradition entails the absence of notions of *essence* as well. Concepts, principles, or laws all require us to deal with the world in terms of essential features. If we store past experiences and organize them in terms of fixed standards or principles, we anticipate, celebrate, and recall a world patterned by these discriminations. The Daoist sage, however, mirrors the world, and "neither sees things off nor goes out to meet them." As such, he "responds to everything without storing anything up." This means that he mirrors the world *at the moment*, without overwriting it with the shape of a world passed away or with anticipations of a world yet to come.

There is a significant contrast in the employment of the mirroring metaphor in Chinese and Western cultures. In the West the mind mirrors nature by reflecting the world as a pattern of unchanging essences. Mind and nature mirror one another by virtue of their sharing a common structure. Indeed, Plato's doctrine of "recollection" expresses a doctrine of rationality dependent upon the fact that the mind is a storehouse of the essential patterns resourced in an unchanging world of forms. For this reason, the things of the world are pale replicas of eternal forms. True knowledge moves beyond shadows and reflections to the mathematical patterns constituting reality itself.

The mirroring activity associated with Daoism engages the world in its transitoriness and particularity. The events and processes themselves, and the orders construed from their particular perspectives, are reflected in the mirroring process. This mirroring activity is best explicated in terms of *wuzhi*—unprincipled knowing.

Wuzhi, as "no-knowledge," means the absence of a certain kind of knowledge, the sort dependent upon ontological presence. Knowledge grounded in a denial of ontological presence involves the sort of acosmotic thinking that does not presuppose a single-ordered world and its intellectual accoutrements. It is, therefore, unprincipled knowing. Such knowing does not appeal to rules or principles determining the existence, meaning, or activity of a phenomenon. *Wuzhi* provides one with a sense of the *de* of a thing, its particular focus, rather than knowledge of that thing in relation to some concept or universal. Ultimately, *wuzhi* is expressed as a grasp of the *daode* relationship of each encountered item that permits an understanding of the totality construed from the particular focus (*de*) of that item. In order to do justice to the insights permitted by the exercise of *wuzhi,* one must give up any pretense to clarity and celebrate the vagueness and allusiveness of a language of deference.

Clarity is one of the most romantic of the ideals of Reason. Clarity of the logical and semantic sort is associated with univocal definition guaranteeing unambiguous usage. Normally, clarity may be contrasted with the state of unarticulated ideas or feelings we call confusion, but the proponent of *wuzhi* contrasts clarity, not only with confused or muddled ideas, but with *vague* ones.

Muddled ideas may not be further analyzed because potential senses of the notion cannot, for whatever reason, be sorted into meaningful units. Vague ideas, on the other hand, are richly determinable in the sense that a variety of meanings are associated with them. Leaving muddled thinking aside, the exercise of *wuzhi* celebrates vagueness in what may be seen as two primary senses: vagueness may be of the *semantic* or *pragmatic* variety.

A concept is *semantically* vague by virtue of its possessing a number of actual or potential interpretations. Both concepts and actions may be *pragmatically* vague. A concept is said to be pragmatically vague when a number of discrete actions are occasioned by it. The vagueness of an action is a function of the complexity both of the world envisioned from the perspective of the nonassertive "agent" and of the number of perspectives from which the nonassertive action might be entertained.

Semantic vagueness may be either "literal" or "metaphorical." Literal vagueness is the consequence of the semantic ambiguity that attaches to our philosophical vocabulary. Notions such as "freedom" and "love" are literally vague in the sense that a number of stipulatable senses, derived from a variety of theoretical contexts, may be unpacked from them. Metaphorical vagueness entails the sort of richness associated with the new uses of words. It is from metaphors that new meanings emerge.

In Western cultures, metaphors have been perceived as parasitical upon literal significances. Thus, rhetoric, insofar as it employs the trope, metaphor, is ultimately tied to logic as ground. This serves to discipline intellectual and aesthetic culture, precluding untrammeled flights of the imagination.

"Expressive metaphors" (metaphors expressing in some extended or alternate context the literal sense of a term) must be supplemented by more radical tropes that I shall designate "allusive metaphors." Expressive metaphors extend the sense of terms that could be understood "literally" in terms of (objective) connotative properties. Allusive metaphors are interestingly different in that they are *groundless*. Allusive metaphors are hints and suggestions, and the "referents" of allusive metaphors are other allusive metaphors, other things that hint or suggest.

Here we see that referential language may be replaced by a language of deference through appeal to the allusiveness of metaphor. Properly understood, allusive metaphors do not refer, they defer. The Daoist world is referent-less. In place of a world of potential referents, the Daoist requires a world of particular things capable of serv-

ing as "deferents." The appropriate question is not "What is the referent of this term?" but "What is the *deferent* of this allusive expression?" In a Daoist world, deferents are functional equivalents of referents.

The Daoist *Wu*-Forms and the Language of Deference

The relationship of deference is not only a significant element of the
Daoist sensibility; it is central to Confucianism as well. Indeed,
Confucius said that the single thread that tied his thought together was
that of *shu* (reciprocity). In Confucianism, *shu* operates within ritual
patterns (*li*) shaped by familial relationships extending outward into
society and culture.

In contrast with Confucianism, Daoism expresses its deferential
activity through what I term the *wu*-forms: *wuzhi, wuwei, wuyu*. I
have alluded to *wuzhi* already as a sort of knowing that has no resort to
rules or principles. *Wuwei*, or nonassertive action, involves acting in
accordance with the *de* (particular focus) of things. *Wuyu*, or "no desire," is to be characterized as desiring that does not seek to own or
control its "object" (its deferent). Such desire, as I shall argue presently, is, in effect, "objectless desire." In each of the *wu*-forms, as in
the case of Confucian *shu*, it is necessary to put oneself in the place of
what is to be known, to be acted in accordance with, or to be desired.

Like *wuzhi*, which does not proceed from stored knowledge or in-
grained habits, *wuwei* involves activity that proceeds without reference to principles. Such action seeks harmony with the particular
focus (*de*) of those things within one's field of influence. *Wuwei* involves unmediated, unstructured, unprincipled, and spontaneous actions. These actions can only be successful if they are the consequences of deferential responses to the item or event in accordance
with which, or in relation to which, one is acting. *Wuwei* actions are
ziran—that is, "spontaneous," "self-so-ing." They are nonassertive
actions.

The greatest confusion in the interpretation of Daoism derives from
those who see this sensibility as essentially quietistic. Such an interpretation receives some justification, of course, from the fact that the
Daoism of Zhuangzi is, as Angus Graham has shown,[8] a collocation of
strands that includes a number of utopian and reclusive texts along-

side the standard philosophical core. My concern is not to separate out "authentic" Daoist elements from pseudo-Daoist accoutrements. Attempting to render the Daoist sensibility coherent in some final form is to miss the fact that the Daoist texts themselves, like the world of the Daoist thinker, is a complex of "thises and thats" passive to any number of interpretations. My concern is to employ notions generally associated with philosophical Daoism as a means of developing a language of deference. The interpretation I am offering cannot be sustained if the modes of disposition named by the *wu*-forms are understood as passive.

The calling forth of deference by those items that stand to be deferred to—as well as the exercise of deference with respect to recognized deferents—are activity shaped by the intrinsic excellences of both deferrers and deferrees. Deference is deference to recognized excellence.

The Daoist sage sees beneath the layers of artifice that mask the naturalness of persons and things and responds to the excellence so advertised. And deference is a two-way street. The excellence of the realized Daoist calls forth deference from others. The *wu*-forms operate within a context of yielding and being yielded to.

Deferential activity associated with *wuwei* and *ziran* is best illustrated by the consummate ruler:

> The most excellent ruler: the people do not even know that there is a ruler.
> The second best: they love and praise him.
> The next: they stand in awe of him.
> And the worst: they look on him with contempt.
> Inadequate integrity in government
> Will result in people not trusting those who govern.
> So hesitant, the ruler does not speak thoughtlessly.
> His job done and the affairs of state in order,
> The people all say: "We are naturally like this (*ziran*)."[9]

Like *wuzhi*, *wuwei* involves a mirroring response. As such, it is action that, by taking the other on its own terms, defers to what it actually is. *Wuwei* involves recognizing the continuity between oneself and the other, and responding in such a way that one's own actions promote the well-being both of oneself and the other. This does not lead to imitation but to complementarity and coordination. Handshakes and embraces require one to take up the stance of the other, and so to complete that stance. When things are going well, a couple on the dance floor constitutes a dyadic harmony of nonassertive actions.

Taijiquan, a Chinese exercise form of Daoist origins, provides excellent training in the achievement of such deferential activity. A basic exercise allied to *taijiquan* training is that of the so-called "push-hands" exercise. Two individuals facing one another perform various circular movements of the arms while maintaining minimal hand contact. The movement of each individual mirrors that of the other. *Wu-wei* is realized when the movements of each are sensed, by both parties, to be uninitiated and effortless.

Wuyu is certainly the most problematic of the three principal *wu*-forms. Translating that term as "no-desire" or "desirelessness" would definitely be misleading. The best manner of understanding the term is as "objectless desire." This translation is consistent with the senses of *wuzhi* and *wuwei* I have been developing. Neither unprincipled knowing nor nonassertive action leads to the objectification of the world or its elements. Likewise, the desires associated with the condition of *wuyu* are experienced without the need to objectify, through possession or control, that which is "desired."

Oscar Wilde's *bon mot* to the effect that the only way to avoid temptation is to yield suggests the essential Platonic claim that desire (*eros*) could only be cancelled through its final consummation. In the case of *wuyu*, however, desire is not "overcome" through possession and consummation; rather, desire is experienced deferentially. The sort of desire that presupposes the mirroring activity of *wuzhi* and the nonassertive relationships set up by *wuwei* is shaped solely by the wish to celebrate and to enjoy.

Desire is directed to things desirable. Desirable things *stand to be desired*. They are, therefore, objects of deference. Moreover, these things that stand to be desired are themselves deferential. This is but to say that they do not demand to be desired. Such things are shaped neither by the act of being desired nor by the desire to be desired. In a world of transient processes, all discriminations are conventional and themselves transitory. Both desiring and being desired are predicated upon the abilities to "let be" and "let go." This is the meaning of *wuyu* as "objectless desire."

The world is a complex set of processes of transformation, never at rest. Attempting to objectify things results in interruptions of the processive character of the events and processes of the world. *Wuyu* permits one to hang loose to the processive character of things; it allows for both letting be and letting go. In the Daoist perspective, there is ontological parity of all things in the sense that no-thing is finally privileged with respect to others. Thus, w*uyu* does not concern *what* is desired but the activity of desiring.

For Plato, *eros* is the desire for completeness of understanding. This is the only thing that can serve as the aim of both embodied and disembodied existence. Thus, *eros* is the only permanent, eternal desire. For Daoism, *wuyu* is transient desire. As such, it lets things be. Enjoyment for the Daoist is realized not in spite of the fact that one might lose what is desired, but because of this fact. *Wuhua*, "the transformation of things," means that one can never pretend that what we seek to hold on to has any permanent status. By refusing to take up a privileged perspective, it does not construe the world in an objective manner and so does not render static the world of changing things.

There is nothing which is not a "that," and nothing which is not a "this." Because we cannot see from a "that" perspective but can only know from our own perspective, it is said that "that" arises out of "this" and "this" further accommodates "that." This is the notion that "this" and "that" are born simultaneously.[10]

The Daoist world is a world of thises and thats. The locution "this and that" is most instructive in understanding the acosmotic character of Daoism. Clearly, Zhuangzi drew different conclusions from the special character of "thises" than did Hegel at the beginning of *The Phenomenology of Spirit*.[11]

Hegel's argument is that, upon reflection, one must realize that "this," which seems to name the particularity of a thing, names a class of all items capable of being characterized in terms of *this*-ness— namely, all particular things. The suppressed assumption for Hegel is that Being stands behind or beneath the beings of the world, requiring that all sense-claims be assessed as existential statements. Thises both exist and own properties, which means that all thises, all particulars, are conditioned by universality.

Hegel presumes that a world of objects requires media of objectification. These are universals. In a Daoist world, however, there are no media of objectification. As I have continually stressed, there is no Being behind the beings of the world, and thus there are only *these particulars*.

A Daoist language of deference shares some similarities with philosophical nominalism. But there is no easy sense in which one-to-one correlations could be set up between the two positions. In the first place, there are two principal forms of nominalism in the Western tradition—physicalist and tropic.

Physicalist nominalism presumes that the world is comprised of primitive individuals. Tropic nominalism of the sort introduced by the ancient Greek Sophists presumes a world always already interpreted by language and other forms of cultural expression. In both sorts of nominalism, discriminations—including those that presume universal properties—are merely conventional and arbitrary. But physicalist nominalism presumes a real world of individual objects as fundamental, while tropic nominalism is conventionalist to the core.

Daoism, as I understand it, shares with physicalist nominalism the idea that there is a real world of particulars that the sage may entertain through proper exercise of the *wu*-forms. Daoist "realism" is peculiar indeed, since the term "real world" is ambiguous, carrying both the sense of a noncoherent congeries of thises and thats, as well as the sense of any particular world envisioned from the perspective of a single member of the "ten thousand things."

If we are to begin to grasp the Daoist insight concerning the *wu*-forms generally, we must attempt to distinguish between "objects" and "objectivity." Both the *Zhuangzi* and the *Daode jing* presuppose what mainstream Western epistemology would term a *realist* vision. In spite of our distorted perceptions, and the confusions introduced by language, we must presume that there is an "objectively" real world.

We make a mistake, however, if we presume that the objective world is a world of *objects*. For the Daoist, there are finally no concrete, unchangeable things which we engage in tension with us— things that proclaim their identities by saying "I object!" This is not the sense in which the objective world is objective. The objective world—the world as really around us—is a constantly transforming set of processes that precludes the sorts of discriminations that would permit a final inventory of its contents.

> Maojiang and Lady Li were beauties for human beings, but fish upon seeing them would seek the deeps, birds on seeing them would fly high, and deer upon seeing them would dash off. Which of these four understands what is really handsome in this world![12]

We cannot discount any of the evaluations of Lady Li and Maojiang without taking away from the richly vague, allusively complex senses of what is *really* beautiful. By insisting upon the exclusivity of our own particular understanding of what is beautiful and what is ugly, we create a world of fixed objects, but we altogether miss the objective reality of the world in all of its vague complexity. By the time we have repeated this process over and over again by appealing to standards of truth, of justice, of goodness, and so forth, we shall have lost the world completely.

The real world is objectless. Only by envisioning the world as a complex congeries of transforming events—events that may, for whatever reason, be frozen momentarily into a pattern of discriminations but which are finally beyond the reach of such discriminations—will the world be seen in its reality.

The *wu*-forms offer means of engaging the ten thousand deferents in the manner in which they offer themselves for enjoyment—namely, as non-objectified processes, each of which focuses on a world from its perspective. There are crucial implications here for the Daoist understanding of the self or person. Knowing, acting, and desiring that objectify the items of the world ultimately lead to the objectification of the self.

Both idealist and materialist understandings envision the self as an object among other objects. The reflexivity of the Cartesian "I" references the mind in the same manner as all things are referenced. Likewise, the materialist vision of the person renders the self, as *body,* a material object amid other material objects. In dominant Western epistemologies, the self exists only through tension with the world and other selves. Feeling ourselves in tension with objectified others leads us to act in an aggressive or defensive manner to effect our will. Selves (including one's own self) become instruments for the achievement of an end. In Daoism, the self is forgotten to the extent that discriminated objects, or persons-as-objects, no longer constitute the world of the self.

In typical Western psychologies, desire is recognized as a need to possess that which is desired. Such a motive leads to the significance of the desired object being understood solely in terms of the needs it can potentially meet. This instrumentalizing aspect of desire creates a thin and narrow world of personal inventory.

Wuzhi, wuwei, and *wuyu* all have this in common: They seek to allow the processes of the world to unfold spontaneously, each on its own terms. Further, as processes on a par with all others, the practitioners of the *wu*-forms themselves stand to be known, engaged, and desired in a deferential manner.

When Wittgenstein suggested that the philosopher should leave the world as it is, he may not have been thinking of the world as does the Daoist—namely, as myriad spontaneous transactions characterized by emerging patterns of deference to recognized excellences. But, *mutatis mutandis,* the Daoist shares Wittgenstein's aim.

The Daoist world, on the one hand, is a noncoherent totality of things characterized by natural parity. On the other hand, as seen from the focal perspective of each item of the world, there are many specific worlds, conventional and transitory. Understanding the world in this fashion enjoins that we use a language of deference rather than of reference. For only if we are successful at avoiding denotation, classification, stipulation, and so on, can we engage the world as it really is. Further, since any particular meaning is contextualized by an indefinite sum of meanings defined by all other orders, language must be *vague*, in the sense of open to an indefinite number of interpretations, and *allusive*, in the sense of actively suggesting those alternate interpretations.

The consequence of mirroring the world in terms of the *wu*-forms is that the ten thousand things are seen, not as referents, but as *deferents*. Deferents are the non-objectified termini of deferential engagements. Obviously one cannot successfully substitute a language of deference for that of reference merely by fiat. The testimony of the Daoist is that practice, long and persistent practice, is required. The principal practices leading to a language of deference involve continued recourse to the *wu*-forms whenever possible.

The relevance of the preceding, admittedly arcane, discussion for ecological concerns is direct and immediate. The distancing language of reference in assaying the relationships between organisms and their environments is grounded in the presumption of a Superordinate Organism who gets to set the agenda in managing the ecosystem. This presumption is doubly disastrous: It is responsible for both the exploitative motives of those who claim Lordship over Nature as their birthright and the cloying sentimentality of the Soulful Caretakers whose futile, finger-wagging gestures only advertise their impotence with respect to environmental issues.

The promotion of deferential relations with the natural world and the creation of a language of deference that both embodies and articulates such relations tackles the problems at the source—namely, with respect to the manner in which relationships within and with the natural world are best formed.

Notes

1. David Hall, *Eros and Irony: A Prelude to Philosophical Anarchism* (Albany: State University of New York Press, 1982).

2. A.N. Whitehead, *Modes of Thought* (New York: Capricorn Books, 1938), chap. 2, "Understanding."

3. In 1973, an early Han tomb was excavated at Mawangdui, and two silk copies of the *Laozi* were discovered, along with four texts attributed to the Yellow Emperor that had a distinctly Legalist flavor. The bundling of these texts together suggests an attempt to borrow the authority of Emperor Huangdi, the legendary founder of the Chinese state, for a Legalist interpretation of Laozi's doctrines.

4. *Philosophy East and West* 37, no. 2 (April 1987): 160–71. Reprinted in *Nature in Asian Traditions of Thought*, ed. J. Baird Callicott and Roger T. Ames (Albany: State University of New York Press, 1989).

5. Ralph Waldo Emerson, *"Nature"* in *Ralph Waldo Emerson: Essays and Journals*, ed. Lewis Mumford (Garden City, N.Y.: International Collectors Library, 1968) 83.

6. *Daode jing* 1.

7. *Zhuangzi* 21/7/33.

8. See the methodological discussions in Graham's translation of the Chuang-tzu: *Chuang-tzu: The Seven Inner Chapters and Other Writings from the Book* Chuang-tzu, trans. A. C. Graham (London: Allen and Unwin, 1981).

9. *Daode jing* 17.

10. *Zhuangzi* 4/2/31.

11. G. W. F. Hegel, *Phenomenology of Spirit*, chap. 1. English translation available in *Phenomenology of Spirit*, trans. A. V. Miller (Oxford: Clarendon Press, 1977).

12. *Zhuangzi* 6/2/69.

The Local and the Focal in Realizing
a Daoist World

ROGER T. AMES

The etymology of "ecology" is *ec* (from Greek *oik-*) "household, habitat" + (from Greek *logia*) "the study of." Ecology is thus usually taken to be the science of the relationships that obtain between living things and their natural environments. At the heart of ecology, then, is an understanding of what it means to be defined relationally. There are identifiable and persistent themes in the Daoist literature that challenge familiar ways of construing "relatedness" and provide us with an alternative vocabulary for exploring an ecological sensibility.

David Hall and I have argued that a "focus-field" language is helpful in articulating the notion of "relatedness" as it functions in a Daoist world. In fact, we have suggested that the *Daode jing* might well be translated "The Classic of Foci and Their Fields."[1] After all, the etymology of "focus" is (from Latin) "fireplace, hearth," resonating rather nicely with "habitat." "Focus-field," then, denotes the relationship not only between one's "hearth and home"—one's insistent particularity—and one's natural environment, but also with one's social and cultural environments as well.

In working through the Daoist texts with students over the years, I have found a vocabulary and a set of recurrent themes that are helpful in making sense of the Daoist sensibilities. These Daoist assumptions about a properly ordered world have too often been betrayed by familiar ways of thinking sedimented into the semantics and syntax of the English language that we must appeal to in reporting on Daoism. At the same time, I have looked at the early work of the late Tang Junyi on Chinese cosmology and have found some considerable corroboration. His vocabulary is most helpful in presenting many of these same and related propositions from yet another perspective and reinforces what I take to be defining features of the classical Daoist worldview.[2]

In offering textual illustrations of these seminal propositions, I am going to reference the *Daode jing* as my primary source because of its canonical status as the core text in classical Daoist philosophy and in Daoist religious practices. Even so, my contention is that such characterizations are broadly applicable and would be relevant to an understanding of collateral texts, such as the *Zhuangzi* and *Huainanzi*, and to the kinds of religious practices that sought to give expression to these ideas.

My first summary proposition characterizing the Daoist ecological sensibility, then, is this: *there is a priority of process and change over form and stasis.* In the received *Daode jing* 25, the four-character phrase "*du li bu gai*" that describes Dao has been almost uniformly translated into English as some variation on "it stands solitary and does not change."[3] Yet, we must ask: what could it mean to assert that Dao "does not change" in the dynamic and "eventful" world of classical Daoism? Contradicting such an assertion, *Daode jing* 40 claims explicitly: "returning is the movement of Dao." The assertion that Dao does not change is not only hard to square with the line which immediately follows this one in the received text[4]—"Pervading everything and everywhere it does not pause"—but not insignificantly, it contradicts everything else that is said about Dao in the early Daoist literature. In fact, on reflection we must allow that "change" is so real in the Daoist worldview that it is expressed in many different ways—*gai*, the character used in the phrase "it stands solitary and does not change," being only one among them. A translation of this passage might need to distinguish among several of these different senses of "change," some of which are: 1) *bian*: to change gradually across time; 2) *yi*: to change one thing for another; 3) *hua*: to transform utterly, where *A* becomes *B*; 4) *qian*: to change from one place to another; and 5) *gai*: to correct or reform or improve upon *X* on the basis of some external and independent standard or model, *Y*.

While "does not change" might fall within the semantic tolerance of *gai*, a failure to distinguish among these several alternative understandings of "change" and to make this "reforming" sense of *gai* clear has led to some classic misunderstandings in translating important texts, one of the most obvious being this particular passage of the *Daode jing*.[5]

The meaning of this passage is not that Dao "does not change," but being the *sui generis* and autopoietic totality of all that is becoming (*wanwu*), Dao is not open to reform by appeal to something other than itself. The Wang Bi commentary on this passage observes that Dao has no counterpart (that is, there is nothing beyond it), and hence it depends upon nothing. Wang Bi also states explicitly that Dao is a dynamic process: "Returning and transforming from beginning to end, it does not lose its regularity."

As we read further on in this same chapter of the *Daode jing*, we are told that one kind of thing correlates its conduct with something other than itself—that is, "human beings emulate the earth, and the earth emulates the heavens." But because Dao is everything, and everything is Dao, we must ask, what could exist beyond it that it could take as its object of emulatation? The answer being "nothing," it is for this reason that the text breaks the parallel structure to state: "And Dao emulates itself" (*dao fa ziran*).

We might also interpret this passage as saying that Dao in the process of "self-so-ing" (*ziran*) entails the relentless emergence of spontaneous novelty and, while always "great" (*da*), is never complete. There is no finality or closure to "Dao-ing." Likely, the text is making both points at the same time: there is nothing beyond Dao that can either change it or provide it with a model for change; and, as a relentless process, it is never complete or final.

Tang Junyi's version of this proposition that privileges process and change over form and stasis is that there is no fixed substratum (*wu ding ti guan*).[6] The world is an endless flow. "Things" are a processional flux, with nothing categorical or foundational posited beyond this flux and flow on which the process would depend.

One prominent theme in classical Greek thought that has no counterpart in classical China is the presence of elemental theories "which claim that things are made up of minute ultimate parts that usually do not look like the parts that are big enough for us to see."[7] Corollary to this path not taken, observes Nathan Sivin, is the marked absence in the Chinese worldview of the familiar reality/appearance distinction, an absence that has enormous philosophical implications that effectively preclude the concerns of metaphysics and epistemology from

center stage. There is no putative Being behind the beings, no unchanging formal aspect behind the changing world, no One behind the
many, no atomic level where unchanging atoms rearrange themselves
to constitute an apparent world.

The Chinese counterpart to Greek "elemental" theories that explain
the ultimate static nature of reality are the dynamic "five phases"
(*wuxing*). These five phases are quite literally a "functional" equivalent of the Greek elements in that, rather than referring to ultimate
parts, these phases reference both the structure and function of the
various phases of the changing world itself as captured in the metaphorical language of metal, wood, water, fire, and earth.

Another expression of this summary proposition is the "vital energies" (*qi*) cosmology presumed by the authors of the early Daoist texts
as a kind of common sense. That is, the often unexpressed assumptions about qi would have the place of the genetic and molecular understanding we have of our world today. In *Daode jing* 42 we have
this language of "vital energies," a vocabulary of fluidity and process
that is pervasive in the language and culture in the explanation of everything from cosmology and health to weather and politics:[8]

> The myriad things shouldering yin embrace yang,
> And blending the qi together make it harmonious.

Qi is an image that, like water, is deliberate and provocative, defying the Aristotelian categories that structure and discipline our language and our thinking. That is, qi is at once one and many. When it
separates out into "things," it has a formal coherence that is at once
persistent and changing. Axiologically, qi is both noble and base. It is
conceived of as being like water, where water can be a "thing" (water)
and an "action" (to water) and an "attribute" (watery) and a "modal-
ity" (fluid).

Reiterating this point that Dao, like qi and water, is relentlessly
fluid and changing is *Daode jing* 32:

> As a metaphor for Dao in the world,
> It is like the streams and creeks flowing to the rivers and seas.

My second Daoist proposition, then, is: *there is a priority of situation over agency*. This proposition speaks directly to the nature of
relationships. When we speak of the "relationship" between the "in-

sistent particularity" (*de*) of things and their various "fields" or environments (*dao*), we must avoid the very real possibility of equivocation in what we mean by "relations." In a world of substances—of essences and accidents—people and things are related extrinsically, so that when the relationship between them is dissolved, they are remaindered intact. Such extrinsic relatedness can be represented as:

But the notion of "relatedness" defining the "eventful" Daoist world-view is intrinsic and constitutive. Perhaps "correlation" is a more felicitous term than "relation." It can be diagrammed as:

In a world of intrinsic and constitutive relations, the dissolution of relationships is surgical, diminishing both parties in the degree to which this particular relationship is important to them. Under such circumstances, people quite literally "separate," "change each other's minds," "break up," and "divorce."

The assumption that "person" is a field of constitutive relationships extends to one's physicality, one's embodied self. "Body" is a familiar way of individuating ourselves. It is significant that in the *Daode jing*, *shen* rather than *ti* is used for "body," an expression that means both "one's profile" as experienced from without and "(personally=) lived body" as experienced from within. Body does not separate one from the world, but locates one within it. As *Daode jing* 26 insists:

> How can one, being the ruler of a huge nation,
> Take one's own person (*shen*) to be less than the empire?

And *Daode jing* 13:

> Thus, only when one esteems one's own person (*shen*) more than the empire can one be entrusted with it;

> Only when one loves one's own person (*shen*) more than the empire can one be appointed to administer it.

Given the coterminous and mutually entailing relationship between person and world, to affirm and protect one's own person is to affirm and protect those same relationships that are constitutive of the empire.

The point can be generalized still further perhaps: no-thing or no-body has an *essence*, but can be defined only "correlationally" at any given time. And these relations are fluid, changing over time. We are both benefactors and beneficiaries of our friends, neighbors, lovers, colleagues, and so forth, dependent on specific circumstances.

The language of Daoism reflects this radical embeddedness, tending to describe situations rather than actions performed by discrete agents. A term such as "bright" (*ming*) does not primarily mean either perspicacity on the part of the agent or resplendence on the part of the object, but both. That is, "brightness" is first situational and then derivatively associated with subject or object. An expression such as *de* does not mean either "beneficence" or "gratitude," but both, and *xin* does not mean either "credibility (trust)" or "confidence," but again both. *Ming* characterizes an enlightened situation, *de* characterizes a generous situation, and *xin* characterizes a fiduciary situation.

For example, in *Daode jing* 23, this situational commitment of the language typically requires the same term *xin* to be translated bidirectionally, both subjectively as "credibility" and objectively as "confidence":

> Where credibility (*xin*) is insufficient
> There will be a lack of confidence (*xin*).

Shan is not "good" as some essential attribute; rather, *shan* describes being "good at" or "good in" or "good to" or "good for" or "good with," and so on. It first expresses a relationship and only then derivatively an attribute. An example: "adeptness" (*shan*) is a relational situation, not the quality of an agent. Hence, in *Daode jing* 27:

> Thus, the adept (*shan ren*) are the teachers of the incompetent
>   (*bushan ren*),
> And the incompetent (*bushan ren*) are raw material for the
>   adept (*shan ren*).

To read these terms as attributes of discrete agents rather than as descriptions and, more importantly, prescriptions for situations prompts my increasingly infamous John Dewey anecdote:

One afternoon Dewey and George Herbert Mead attended a lecture. Having taken their seats, Mead leaned over and said to Dewey, "Look at those two women on the end of the row—don't they look alike!" And Dewey, taking a long look at the two women, replied with a smile, "You know, you are right, they do look alike; especially the woman on the left."

The anecdote is humorous because Dewey, offering his most succinct and perhaps most memorable critique of substance philosophy, is purposely essentializing something that is by definition relational. From a Daoist perspective, ontological assumptions that separate out "reality" from "appearances" transform "events" into "things," and the constitutive relationships that make up these events into extrinsic bonds that obtain among discrete things.

The opposition assumed between "credibility" and "confidence," or "giving" and "getting," that requires the English language to use two terms to describe these actions reflects the primacy of agency, an assumption that is not relevant to the Daoist world.

This point about the priority of situation over agency—another way of saying process over form—is made explicit in the classical Chinese language itself. "Paronomastic" definitions appeal to semantic and phonetic associations rather than putatively literal meanings. The term "exemplary person" (*jun*), for example, is defined by its cognate and phonetically similar "gathering" (*qun*), which rests perhaps on the underlying assumption that "people gather round and defer to exemplary persons." "Mirror" (*jingzi*) is defined as "radiance" (*jing*): a mirror is a source of illumination. "Battle formation" (*zhen*) is defined as "displaying" (*chen*): a battle formation's most important service is to display strength as a means of deterring the enemy. A "ghost or spirit" (*gui*) is defined as "returning" (*gui*): presumably it has found its way back to some more primordial state. "The Way" (Dao) is defined as "treading" (*dao*): as the *Zhuangzi* says, "The Way is made in the walking."[9]

What is remarkable about this way of generating meaning is that a term is defined nonreferentially by mining relevant and yet seemingly random associations implicated in the term itself. Further, erstwhile nominal expressions default to verbal, gerundical expressions; that is, "things" default to "events"—*dao* as "the way or path" defaults to *dao*, "treading," and *jun* as "exemplary person" defaults to *qun*, "gathering"—underscoring the primacy of process over form as a grounding presupposition in this tradition.

A proposition offered by Tang Junyi that is related to the priority of situation over agency is "the inseparability of the one and the many."[10]

Throughout the Daoist texts, we have frequent reference to the "oneness" of things, or becoming "one" with things, often stated as a kind of achievement. But "one" is an ambiguous term. In meaning any "one" of the following, it can put a rather different spin on such claims:

1. "One" is the whole of (complete, all, perfect).
2. "One" is simple (uniform, homogeneous, pure).
3. "One" is identical with (one and the same).
4. "One" is individuated (a discrete unit).
5. "One" is continuous with (integrated with another).
6. "One" is unique (one and only).
7. "One" is unchanging (invariance).
8. "One" is idiosyncratic (odd).
9. "One" is superior (number one).
10. "One" is integrity (authentic, genuine).

Most translators simply abandon their responsibility to say what the text means by retaining the ambiguous "one" without further explanation.

We can appeal to *Daode jing* 39 to address the ambiguity that attends the notion of "one":

> Of those things of old which became integrated:
> The skies in becoming integrated became clear,
> The earth in becoming integrated became stable,
> Spirits in becoming integrated became spiritual,
> The river valley in becoming integrated became full,
> And the myriad things in becoming integrated flourished.

Each of these aspects of our experience—the heavens and the earth—preserve their own integrity and uniqueness (the skies are clear and the earth is stable) while at the same time effecting integration and harmony with the other operations that constitute our environment. Many of these senses of "one" listed above have relevance to the Daoist world—particularity, uniqueness, continuity. But those essentialistic connotations of "one" that would allow for a final separation between the one and the many distort the Daoist "acosmotic" cosmology.[11] That is, Daoism will not accommodate a worldview in which the "many" are determined by, and thus reducible to, a transcendent and independent "One": the *uni-* of universal, universe, unity, univocal. It will not allow for the dissolution of the particular into some higher reality. To translate the phrase *de yi* as "in getting One" for a language community in which this "One/many" sense of "One" as some underlying, unifying principle is a predominant assumption, is to fail the reader.

Tang Junyi takes this peculiar holographic, interdependent relationship between "part" and "whole"—better, "focus-field"—as the distinguishing feature of Chinese culture. He states:

> The basic spirit of Chinese culture is that "part" and "whole" are blended and mutually entailing. From the perspective of recognition, you do not divide a part out from a whole, and from the perspective of feeling, the part strives to realize the whole.[12]

When this mutuality and interdependence of foci and fields (events and their environments) is translated into the world of intellectual engagement and politics, it becomes the value of inclusive, consensual understanding rather than exclusive disputation,[13] it becomes canonically driven commentary rather than dialectically competing philosophical systems, and it becomes accommodating "right thinking" rather than an individual's "right-to-think."[14]

One and many must be seen to be in opposition before the problem of one and many will arise. The final separation between one and one, and by extention, one and many, arises from several assumptions. The transcendence of some creative principle which stands independent of its creatures establishes the familiar "One-many," "Being behind the beings," "Reality behind appearances" ontology.[15] A "real" essence posited as ground for something else less real (mind or soul for body, God for world) makes the relationships that obtain among such essences extrinsic. Discreteness also emerges in the final separation between stasis and motion, between permanence and change, between being and becoming, where the former is in each instance the originative source of the latter.

As we have seen, Daoism construes oneness and manyness as interdependent and mutually arising. In *Daode jing* 42, Dao engenders both "one" and "many," both "continuity" and "difference." Viewed as the creative source of unique particulars—what *Daode jing* 25 calls "flowing" (*shi*) and "distant" (*yuan*)—Dao entails proliferation:

Dao engenders one, one two, two three, and three the myriad things.

This might also be read:

Dao gives rise to continuity, continuity to difference,
difference to plurality, and plurality, to multiplicity.

However, because the myriad things are constitutive of Dao and continuous with each other, "distance" (*yuan*) means "returning" (*fan*). Thus, we must also understand this proliferation and difference in terms of continuity, requiring us to allow that:

The myriad things engender three, three two, two one, and one, Dao.

Or, said another way:

Muliplicity gives rise to plurality, plurality to
difference, difference to continuity, and
continuity to Dao.

That is, the multiplicity of things seen from another perspective is the continuity among things: the particular focus entails its field. A more concrete political illustration of this "inseparability of one and many" is *Daode jing* 28, in which the sage-ruler enables the many officials with their disparate responsibilities to function as one continuous operation:

If unworked wood is split up, it becomes vessels. If the Sage makes use
of these vessels, he becomes their head official. Thus, the best arranging does not divide things up.

My third Daoist proposition is: *there is a priority of historia and mythos over logos, of narrative over analysis.*

As Angus Graham observes, dispute is characteristic of people who would ask "What is the Truth?" rather than those in classical China who would ask "Where is the Way?"[16] One implication of a putative reality-appearance distinction is that it makes analysis a privileged method for seeking the truth. By contrast, the Daoist tradition seeks a way to live productively rather than the truth. That is, in attempting to correlate oneself most effectively within one's context, the Daoist project requires more of a narrative than an analytic understanding, a "know-how" rather than a "know-what." "Reason" does not function analytically as some abstract, impersonal standard available for adjudication, but rather is a reservoir of historical instances of reasonableness available for analogical comparison. Similarly, "culture" is *this* specific historical pattern of human flourishing as it is lived out in the lives of the people; "logic" is the internal coherence of this particular community's narrative; "knowledge" is a kind of "know-how" evidenced in making *one's own* way smoothly and without obstruction in *this* particular locale; "truth," or probably better its cognate, "trust," is a quality of relatedness demonstrated in *one's own* capacity to foster productive relationships that begin with the maintenance of one's integrity and extend to the enhancement of one's natural, social, and cultural contexts.

With no separation between phenomena and ontological foundations, "reality" is precisely that complex pattern of relationships which in sum constitute the myriad things of the world. Knowledge predominantly, then, tends not to be abstractive but concrete and practical; not to be representational but performative and participatory; it involves not closure but disclosure; it is not discursive but is, rather, a kind of know-how: how to effect robust and productive relationships.

The *Daode jing* does not purport to provide an adequate and compelling description of *what de* and Dao might mean as an ontological explanation for the world around us; rather, it seeks to engage us and to provide guidance in *how* we *ought* to interact with the phenomena, human and otherwise, that give us context in the world. It is not a systematic cosmology that seeks to explain the sum of all possible situations we might encounter in order to provide insight into what to do, but rather provides a vocabulary of images that enable us to think through and articulate an appropriate response to the changing conditions of our lives.

"Knowing," then, in classical Daoism is not so much a knowing *what* that provides some understanding of the environing conditions of the natural world, but is rather a knowing *how* to be adept in relationships, and *how*, in optimizing the possibilities that these relations provide, to develop trust in their viability. The cluster of terms that define knowing are thus programmatic and exhortative, encouraging as they do the quality of the roles and associations that define us.

Propositions may be true, but it is more important that teachers and friends be so.

Tang Junyi's proposition that overlaps with this characterization is captured in a phrase: "the unceasingness of life and growth," derived from the *Yijing*'s "life and growth is what is meant by 'change'."[17] The *Daode jing* 20 describes this indefatigable process:

> Placid and deep like the vast reaches of the sea;
> Whistling high above like the tireless winds.

The various terms that are used to denote the "world" in Chinese are processional: "the process of self-so-ing" (*ziran*), "succeeding generational boundaries" (*shijie*), and "worlding" (*tiandi*). Even when *Daode jing* 25 posits a temporal priority of Dao to this world—"there was something spontaneously formed and born before the world itself"—this is not to introduce an initial beginning or some metaphysical source but, on the contrary, to overrule that possibility. The language of "origins" in the *Daode jing* is resolutely genealogical: "fetal beginnings" (*shi*), "mother" (*mu*), "ancestor" (*zong*), and so on. *Daode jing* 14 observes the continuity between these origins and the present world:

> Grasp the Dao of antiquity to manage the affairs of today, for being able to understand the fetal beginning of antiquity—this is the primary cord of the Dao.

It is the experience of perpetual and appropriate transformation that is "time (and timing)" (*shi*) itself.

My fourth Daoist proposition is: *there is a contingent and negotiated harmony that attempts to get the most out of the existing ingredients, rather than a deterministic and necessary teleology.* Given that the world is without beginning or end and that it is constituted by phenomena that stretch across the full range of every continuum, the goal in any and every moment is to achieve that kind of correlationality that maximizes the creative possibilities of the ever unique situation. Since coercion, and the diminution in possibilities it entails, is anathema to such maximization, the desired character within relationships is responsive yet nonassertive activity: *wuwei*. This antipathy toward coercion is everywhere in the *Daode jing*; for example, chapter 30 advocates such an attitude even in the prosecution of war:

The adept commander looks only to attain his purposes,
And would not dare to use the situation as an opportunity to exercise
coercion.

But the achievement of an optimally productive harmony is not driven
by some predetermined mechanical process, divine design, or rational
blueprint. It is an autopoietic (*ziran*) process, where the energy for its
transformation lies within the world itself.

Tang Junyi's version of this proposition is to reject any kind of fatalism as relevant to this tradition—what he terms "anti-fatalistic" (*fei
ding ming guan*). Dao is presented in the *Daode jing* as the unfolding
of a contingent world according to the rhythm of its own internal creative processes without any fixed pattern or guiding hand (*buzhu,
buzai*). For example, in *Daode jing* 51:

Thus, among the myriad things none but revere Dao and venerate *de*.

This reverence for Dao and veneration for *de* is always so of its own
accord without anything commanding it to be that way.

It is certainly true that there is a preoccupation with "the propensity of
things" (*ming*) in this culture, but *ming* is not "fate" or "destiny." It is
the future that emerges out of the force of circumstances—out of the
self-directing and spontaneously arising propensity of things. This Dao,
always focused as one *de* or another, is the unfolding of a course of
experience that reflects the total collaboration of those participants in
the world as they dispose themselves one to another, and who, through
their conduct, influence its outcomes.

Fatalism is not an option in a cosmology where one and many are
inseparable, since each particular shapes and is shaped by its contextualizing "many." In a world where "one and many are inseparable,"
one aspect cannot have some unconditioned and determinative force
over all others.

The fifth proposition follows from the last: *there is an indeterminate aspect entailed by the uniqueness of each participant that qualifies order, making any pattern of order novel and site-specific, irreversible, reflexive, and, in degree, unpredictable.*

Given that order is an always negotiated, emergent, and provisional
harmony contingent upon those elements that constitute it, it is not
linear, or disciplined toward some given end and, hence, is not ex-

pressible in the language of completion, perfection, or closure. Rather, order is an aesthetic achievement made possible through *ars contextualis*, "the art of contextualization," and, hence, is expressible in the aesthetic language of complexity, intensity, balance, disclosure, efficacy, and so on. Since all participants in the order are correlational, the one cannot be separated from its context—focus cannot be separated from its field. Thus, any construal of order is recursive, a coming back upon itself.

There is another important sense in which order is recursive. The Tang Junyi proposition that would seem to resonate with this underdeterminacy of order is "there is no proceeding without reversion."[18] In Tang Junyi's own words, order is not linear but cyclical, so that nothing flourishes without subsequently reverting to its origins: "Returning is the movement of Dao."[19] Order is the unceasing movement of things on a continuum between extremes: empty to full, proceeding to returning, rolling out to rolling up, waxing to waning. The human experience itself is the turning of the seasons, where sixty years completes one full cycle from spring to the depths of winter, only to produce out of itself another cycle.

Daode jing 16 observes:

> The myriad things flourish together
> And in this process we observe their reversion.
> Now of these things[20] in all of their profusion,
> Each again reverts and returns to its roots.

This movement is the observed cadence and regularity of the world around us as it expresses its inherent capacity for self-transformation. This "cyclical" process, while passing through familiar phases, is not replication. It is the unfolding of an endless spiral that evidences, on the one hand, persistent and continuing patterns and, on the other, novelty, with each moment having its own particular orbit and character. It is because "the path is made in the walking" that the emerging pattern of experience is always the collaboration of the unique one and the context in which it participates.

My final Daoist proposition is: *there is a priority of a dynamic radial center over boundaries*. Given the radical embeddedness of the Daoist experience, order always begins here and goes there. There are degrees of relevance established through extending patterns of deference, what we have called elsewhere "the *wu*-forms": nonassertive action (*wuwei*), unprincipled knowing (*wuzhi*), and objectless desire (*wuyu*).[21] Hence, we can describe the person, the family, the community, and the world in terms of centripetal centers that extend outward as radial circles, in degree subjective and objective, inner and outer, local and global. Through these patterns of deference one can experience what is distant (*yuan*) and even what is at the furthermost reaches (*taiji*), but one can never extricate oneself from one's particular perspective to discover some ultimate boundary on experience. The particularity and the temporality of experience preclude the possibility of fixed boundaries. Experience is always experienced from within. It is for this reason that, repeatedly, *Daode jing* chapters conclude with a localizing expression such as "Hence they discard that and secure this."[22]

Tang Junyi has a proposition that again expresses this same idea of being born into a particular and specific context: "The natural tendencies of the human being are the Way of *tian* (Heaven)."[23] This is really only a more complex way of acknowledging the continuity between the human being and *tian* (*tian ren heyi*), a continuity that is always historicist, genealogical, and biographical. To illustrate this point, he cites a passage from the *Zuozhuan*:

> The people inheriting a world in being born into it, is what is called "the propensity of circumstances."[24]

The reality and inescapability of particular perspective is the point of *Daode jing* 47:

> Without going out their door they know the world,
> Without peeking out their window they know the Way of *tian*.
> . . . Thus the sages know without going anywhere, see clearly without looking, and accomplish without doing anything.

"Knowing," "seeing," and "accomplishing" within one's experience requires extending oneself from where one is, because one is always somewhere. This extension is achieved through responsive and efficacious participation in one's environments and through one's full contribution at home in the local and the focal relationships that, in sum, make one who one is. Said another way, Dao is not "discovered" by traveling to distant places and experiencing strange and wonderful things. In fact, this process of "leaving home" effectively puts at risk precisely those specific relationships that have been cultivated to constitute oneself as a viable perspective on the world.

These several propositions defining the Daoist sensibility make clear the understanding of relationality advocated by Daoism and together reflect its ecological commitment. Perhaps these propositions can be summarized in the claim that, in the Daoist world, "everything is local."

Notes

1. See *Anticipating China: Thinking Through the Narratives of Chinese and Western Culture* (Albany: State University of New York Press, 1995), esp. 232–37, and *Thinking from the Han: Self, Truth, and Transcendence in Chinese and Western Culture* (Albany: State University of New York Press, 1998), 23–77.

2. Tang Junyi, *Zhongxi zhexue sixiang zhi bijiao lunwenji* (Collected Essays on the Comparison of Chinese and Western Philosophical Thought) (Taipei: Xuesheng shuju, 1988). Most of the essays collected in this volume date from the 1930s, before Tang Junyi's thought and his language were so thoroughly influenced by his reading of German philosophy.

3. Yang Yu-wei and I were guilty of just such a reading in our translation of Chen Guying's popular translation, *Laozi jinzhu jinyi* (A Modern Translation and Commentary on the *Laozi*), but we are in good company with James Legge, Bernhard Karlgren, Arthur Waley, D. C. Lau, W. T. Chan, Michael LaFargue, Robert Henricks, and a score of others. Interestingly, Fukunaga Mitsuji's *kanbun* reading with the benefit of the original character is *aratamezu*, "is not reformed, revised," and his interpretative translation is the passive causative *kaerarezu*, "is not made to change."

4. This phrase does not occur in the Guodian or in either of the Mawangdui versions of the text.

5. See as another example *Analects* 1.11 and our discussion of this passage in *The Analects of Confucius: A Philosophical Translation* (New York: Ballantine, 1998); appendix 2, 279–82.

6. Tang Junyi, *Zhongxi zhexue*, 9.

7. Nathan Sivin, *Medicine, Philosophy and Religion in Ancient China* (Aldershot, England: Variorum, 1995), 1:2–3.

8. See also *Daode jing* 10 and 55.

9. *Zhuangzi* 4/2/33.

10. Tang Junyi, *Zhongxi zhexue*, 16.

11. In *Anticipating China*, David Hall and I coin the expression "acosmotic" to claim that "the Chinese tradition . . . does not depend upon the belief that the totality of things constitutes a single-ordered world" (pp. 11–12).

12. Tang Junyi, *Zhongxi zhexue*, 8.

13. Sivin, *Medicine*, 1:9.

14. Randell P. Peerenboom, "Confucian Harmony and Freedom of Thought: The Right to Think versus Right Thinking," in *Confucianism and Human Rights,* ed. Wm. T. de Bary and Tu Wei-ming (New York: Columbia University Press, 1998).

15. Tang Junyi is explicit in rejecting both "transcendence" and "absoluteness" as conditions of Chinese cosmology. See *Zhongxi zhexue*, 241.

16. A. C. Graham, *Disputers of the Tao: Philosophical Argument in Ancient China* (La Salle: Open Court, 1989), 3. See our discussion in Hall and Ames, *Thinking from the Han*, part 2.

17. *Yijing* Harvard-Yenching Concordance Series 40/*ji shang*/5.

18. Tang Junyi, *Zhongxi zhexue*, 11.

19. *Daode jing* 40.

20. The Guodian text has "Heaven's Way" (*tian dao*) for "Now things" (*fu wu*), which might be translated as:

> In the way of the world in all of its profusion,
> Each thing again reverts and returns to its roots.

21. See Hall and Ames, *Thinking from the Han*, chap. 3.
22. See *Daode jing* 12 and 72, for example.
23. Tang Junyi, *Zhongxi zhexue*, 22.
24. *Zuozhuan* Harvard-Yenching Concordance Series 234/ *cheng* 13/2 *zuo*.

"Responsible Non-Action" in a Natural World: Perspectives from the *Neiye*, *Zhuangzi*, and *Daode jing*

RUSSELL KIRKLAND

In his 1997 work, *Oriental Enlightenment: The Encounter between Asian and Western Thought*, J. J. Clarke wrote:

> [The] East-West ecological debate has undergone a palpable transformation in its brief history. In retrospect, its earlier manifestation . . . now seems somewhat naive and over-inflated, often conveying the conviction that Eastern traditions could provide a ready-made solution to Western ills. . . . In more recent years Asian traditions . . . have been brought to bear in more circumspect and critical ways. . . . Some of the earlier utopian enthusiasms for a 'new paradigm' which would sweep away and replace failed Western ways of thinking have given way to a more hermeneutic mode of discourse. . . . Thus, for example, there is an increasing awareness that Eastern attitudes to nature have in the past often been suffused in the West with an idealised glow which has tended to obscure . . . those theoretical aspects of Eastern religions which might be at variance with [modern] ecological principles. There is a recognition, too, of the need to take account of the historical and cultural distance between ancient Eastern philosophies and the contemporary environmental problematic, and not to imagine that the West can simply . . . adopt alien intellectual traditions wholesale.[1]

What follows is intended to contribute to the process of maturation that Clarke finds in current ecological thinking in relation to Daoism. In particular, I shall suggest that attitudes and values consistent with modern environmentalism should not be sought in the "classical" texts, like *Laozi* and *Zhuangzi*, which modern sinologists—ideologically guided by late imperial and modern Confucians—long taught the public to regard as the "foundations" of Daoism. I shall "take account of the historical and cultural distance" between such texts' ideals and our own by utilizing exegetical methods—that is, by critically examining those texts to find what they intended to say, irrespective of what we today might wish they said.[2] Like scholars from the late Holmes Welch to the contemporary philosophical sinologist Bryan Van Norden, I shall show that classical Daoist texts present perspectives on life that will, at times, not please the modern reader. And I shall urge today's environmentalists to pay *less* attention to the texts of "classical Daoism" and *more* attention to the later historical practitioners of Daoism, who sometimes cherished values much more compatible with those of modern environmentalism. Support for modern "ecological principles" can indeed be found in the medieval traditions of Daoism. But unless we read our own values into texts written by minds that do not share our modern values, it appears that the *Daode jing* and *Zhuangzi* present "theoretical aspects" that are "at variance with [modern] ecological principles." Yet, I propose that thoughtful analysis of those aspects have the potential to deepen and enrich modern ecological thought and to provide healthy new paradigms for a nonpaternalistic engagement with the natural world. The Daoist classics do not, as some interpreters have argued, present a cold-hearted vision of the world. In fact, read correctly, the Daoist classics usually show us a world that is far less threatening, and thus far less in need of corrective action, than is assumed by most modern thinkers. They also suggest that one can have subtle but far-ranging effects upon the whole world if one resists interventionist urges and "cultivates the Dao"—a cosmic reality that is as powerful and trustworthy as it is natural.

The Universalistic Ethic of Daoism in Critical Historical Perspective

In the entry on Daoism in the second edition of the *Encyclopedia of Bioethics*, I noted that at times Daoists expressed "a universalistic ethic that extended not only to all humanity, but to the wider domain of all living things." But I also noted that "Daoist conceptions of history, humanity and cosmos also undercut some of the paternalistic tendencies so common in other traditions. . . . Our lives are to mirror the operation of the Dao, which contrasts markedly with Western images of God as creator, father, ruler, or judge." Since "the Dao is not an external authority, nor a being assumed to possess a moral right to control or intervene in the lives of others," we ourselves cannot justify assuming such a right. And since "the *Daode jing* commends 'feminine' behaviors like yielding, as explicitly opposed to 'masculine' behaviors of assertion, intervention, or control . . . there is . . . little temptation for a Daoist to 'play God,' whether in medicine, government or law." The same applies to environmentalism: "despite all its insistence upon restoring harmony with the natural order, Daoism is not consistent with the activist tendencies of modern environmentalism. No Daoist of any persuasion ever embraced goal-directed *action* as a legitimate agency for solving problems."[3]

Because Daoism has almost always been viewed through Confucian and Western lenses, it has often been misunderstood and mischaracterized. For example, for generations Daoists were reproached for allegedly advocating selfish disregard of the needs of human society. New research has begun to demonstrate that Daoists of many periods actually practiced and taught a social ethic, though Daoist ethical values were often articulated in terms quite distinct from, and sometimes alien to, the values of Confucians or indeed of most Western interpreters.

For instance, because of the biases of modern Confucians, Christians, and secularists, few have ever noticed that premodern Confucians and Daoists shared an assumption that correct performance of proper rituals would result in a profound and positive effect upon the entire world. For example, the Lingbao liturgy called the *jinlu zhai*—which has been performed by Daoists since the fifth century—was believed to harmonize the world by forestalling natural disasters and engendering peace and harmony throughout the realm. Since few today—in Asia or in the West—believe that ritual activity can actually achieve such results, interpreters have given little attention to the fact that most people in premodern China, especially Daoists, assumed that it could.

Premodern Daoists actually expressed a wide range of altruistic ideals—some that modern minds can readily embrace and some that may seem quite alien. For instance, texts spanning several hundred years depicted the Tang dynasty Daoist Ye Fashan (631–720) as a heavenly official who had been exiled into the human world to prove his moral worth through selfless exertion on behalf of others. According to those texts—some compiled by emperors and learned government officials—Ye practiced altruistic thaumaturgy for the benefit of all members of society—curing diseases, dispelling bad weather, and even raising the dead! The texts present his lifetime of altruistic public service so as to stimulate readers to emulate his commitment to promoting the general welfare through social service.[4]

On the other hand, the texts of classical Daoism do not appear to advocate such activities. But then again, those texts have generally been read through distorting lenses, and many of their teachings have never been fully appreciated. For instance, the *Daode jing* urges the reader to emulate the idealized "sage" as follows:

> Having [involved himself] on behalf of others, he himself has more;
> Having [involved himself] in engagement with others, he himself is augmented (Wang Bi ed., chap. 81).

Passages like this one, which commends altruistic ideals, have seldom been factored in when most interpreters "explain" the teachings of classical Daoism. But when we give such passages proper attention, we see that from the *Daode jing* of classical times to thirteenth-century biographies of Ye Fashan, Daoist texts have reminded readers that the practitioner of Daoism is spiritually deficient if he neglects others' welfare in a selfish pursuit of individual goals.[5]

If the Daoists of premodern China endorsed altruistic values, would they support the altruistic agenda of modern environmentalists? Some clearly would. For example, an imperial official of the seventh to eighth century, when asking permission to retire as a Daoist priest, petitioned the throne "that a circular palace lake . . . be converted into a pond reserved for liberating living creatures." In the year 711, a Tang emperor honored the Daoist master Sima Chengzhen by decreeing that a secluded mountain area be set aside "as a blessed spot for the prolongation of the life of flora and fauna and the construction of a Daoist abbey." Another eighth-century emperor, impressed by Sima's successor, not only "prohibited hunting and fishing" at a different Daoist mountain site, but also "totally abolished the [imperial] livestock pens."[6]

But we do not find such activities commended to the reader of such classical texts as the *Daode jing, Zhuangzi,* or *Neiye.* Some interpreters, like Holmes Welch, have argued that the teachings of the *Daode jing* are actually inimicable to modern sensibilities and thus useless as a source of remedies for most modern problems.[7] H. G. Creel even went so far as to claim that such texts' image of the sage "released upon humanity what may truly be called a monster," someone callous and aloof, with no regard for others.[8] And Bryan Van Norden has recently argued that Laozi's ideal world would feature crime, economic deprivation, and unchecked infant mortality, concluding "that the *Laozi* expresses a synoptic vision which we would be ill-advised to adopt."[9]

But are such interpreters really correct? If the texts of classical Daoism do not exhibit, for instance, the environmental consciousness that seems to have surrounded eighth-century Daoism, are we really to conclude that ancient Daoists were self-centered and cold-hearted? A careful exegetical analysis of texts like the *Daode jing* actually reveals a complex array of teachings about how one should understand, and interact with, the world around us. Those teachings often commend selflessness and an altruistic concern for others—but not always on terms that may seem sensible or attractive to modern readers.[10] Nonetheless, an exploration of those teachings may suggest ways for us to expand and enrich our perspectives on the issues that confront us today.

Responsible Non-Action: A Radical Thought Experiment

Several years ago, my Stanford colleague Lee Yearley sought to impress upon students the differences between the thought of the Confucian thinker Mencius and that of the Daoist thinker Zhuangzi. His starting point was Mencius's famous insistence that human nature is such that none of us would fail to be moved if we saw an infant facing imminent death. To stimulate rumination, Yearley gave our students an assignment to write a paper beginning with the following proposition: Mencius and Zhuangzi are sitting together on a riverbank when they descry an infant floating on the river, apparently on its way to its death from drowning. The assignment was to describe what each man would do in that situation, and why.

This example came to mind when I began to ponder what the Daoists of ancient China would say about the ecological issues of the late twentieth century. If, so to speak, planet Earth is drifting in the direction of presumable disaster, what action would such Daoists as Zhuang Zhou have us take? In other words, would "Daoism," in that sense, provide happy solutions to the apparent dilemmas that face our planet today?

Careful analysis suggests that Holmes Welch was at least partly correct: classical Daoism might *not* offer happy solutions to modern problems, at least not on our own, modern terms. What if, for instance, the issue in the foregoing example was not a baby floating in a river, but rather the species of the whooping crane? If Zhuang Zhou were sitting by, watching one of Earth's species threatened with apparent extinction, what would he really do, and why? The answer, I fear, may not fill our hearts with sanguine certainty of the cranes' future or with happiness that Zhuang Zhou would share *our* desire to preserve them. In fact, the only logical answer would seem to be that Zhuang would, as it were, watch the whooping cranes float down the river on their way to apparent extinction and would take no action to interfere with the natural processes at work in the world. An honest reading of the texts reveals that the only logical answer to Yearley's thought experiment is that the Daoist sees no action required: the Daoist trusts that the world is already operating as it is supposed to be operating and that all human activity—no matter how well-intentioned—can only interfere with the course of nature as it is already unfolding. Such "radical" thoughts are uncomfortable, and Yearley has argued that most readers of such texts—Chinese and Western, past and present—have simply ignored or denied such thoughts and have instead constructed more pleasant interpretations, interpretations that make classical Daoism more thinkable for us![11]

Intellectual honesty demands that we separate what we ourselves want to believe from what we say others believe, particularly others whose beliefs took shape in ages and cultures that were quite different from our own. Yearley's analysis—uncomfortable but intellectually honest—suggests that Daoists like Zhuang Zhou did not advocate the humanistic activism that modern environmentalists embrace, or even the humanistic values of his Confucian contemporaries, like Mencius.

Like modern Westerners, China's Confucians generally assumed that the world inherently tends toward chaos and therefore requires human redemptive activity. But Daoists—of all periods—have never shared the Confucian (or Western) distrust of the natural processes of life. While medieval Daoists may at times have embraced the idea of taking steps to preserve the lives and habitats of nonhuman creatures, the texts of classical Daoism seem to reveal no fear regarding, for instance, the extinction of whooping cranes. Nor do they seem to enjoin human intervention to "save" such creatures, or even Earth as a whole, from extinction.

In fact, from the classical Daoist perspective, it is clearly morally suspect for humans to presume that they are justified in judging what might constitute "impending ecological danger," or to presume that interventional action is necessary to rectify the situation. The classical Daoist answer to ecological problems, as to other types of problems, is always found in going contrary to the Confucians, who, like Western humanists, assume that humans have a special wisdom found nowhere else among the world's living things. Whereas Confucians like Mencius might feel morally compelled to dive into the river of life's events to save a threatened species of birds, Daoists like Zhuangzi would feel morally compelled to refrain from doing so.

In what follows, I will attempt to explain the moral reasoning that would compel Zhuang Zhou to watch the cranes on their way to apparent extinction, taking no interventional action. I will propose that the ultimate principle of classical Daoism is that there is a universal reality beyond the comprehension or control of human thought or activity and that modern environmentalists should beware the assumption that we are, in humanist terms, the necessary savior of planet Earth. The classical Daoist position is that planet Earth needs no savior as such, especially not ourselves.[12]

The fundamental issue is the humanist assumption that the responsible person is morally compelled to *intervene* in events that seem to threaten "life." The position of the classical Daoists is quite the opposite: even when "life" may seem to be threatened, the responsible person is morally compelled to *refrain* from taking action. To classical Daoists, such commitment to what I shall call "responsible non-action" is the only course open to a person who truly understands and appreciates the nature of life. And to set aside such principles when one sees a baby or a flock of cranes endangered—simply because one feels stirrings of alarm or regret within one's heart/mind—must be regarded not only as contrary to sound moral reasoning, but also as a sign of what we might call spiritual immaturity.

Why It Is Wrong to Resent Unexpected Changes

In _Zhuangzi_ 18, we find two famous stories in which a man experiences a sudden change that strikes others around him as deeply troubling.[13] In one, the philosopher Hueizi goes to offer his sympathies to Zhuangzi upon the death of Zhuang's wife. In the next story, a willow (or, more likely, a tumor) suddenly sprouts from a character's elbow. In each story, a sympathetic friend is shocked and dismayed to find that the first character is not shocked or dismayed by the unexpected turn of events. In each story, the first character patiently explains the nature of life and counsels his companion to accept the course of events that life brings us, without imposing judgment as to whether those events are good or bad.

When Zhuangzi's wife dies, Zhuangzi does not argue that the world is a better place for her absence. The only issues in the passage are 1) that people are born and that they later die, and 2) that to ignore that inescapable fact would constitute culpable stupidity. Now imagine that the story of the death of Zhuangzi's wife involved, instead, the death of the species we call whooping cranes: Zhuangzi would, in that case, patiently explain to his very caring but shallow friend that he had indeed felt grief to see such beautiful birds come to their end, but had gone on to engage in appropriate reflection upon the nature of life, and had eventually learned to accept the transitory nature of all such creatures. If, according to this author, one must learn to accept with serenity the death of someone one loves (such as one's wife), wouldn't he also urge us to accept the death of some birds with similar equanimity? If life's process of change should catch up with us, even, by extension, to the extent that our planet should become devoid of all forms of life, the response of the author of these passages would logically be 1) that such is the nature of things, and 2) that crying over such a turn of events would be silly indeed, like a child crying over the death of a goldfish. The only reason that a child cries over such a death is that he or she has formed an immature sentimental bond to it. As adults, we appreciate the color and motion of fish in our aquariums, but we know that to cry over the death of such fish would be juvenile behavior. As our children grow, we teach them, likewise, not to follow their raw emotional responses but to govern those emotions and to learn to behave responsibly according to principles that are morally correct, whether or not they are emotionally satisfying.

Now, considering Zhuangzi and Mencius on the riverbank, the texts of classical Daoism warn that one never knows for sure when an event that seems fortunate is actually unfortunate, or vice versa. What if, one might wonder, the baby in the water had been the ancient Chinese equivalent of Adolf Hitler, and the saving of him—though prompted by the deepest feelings of compassion and by a deep-felt veneration for "life"—later led to the systematic extermination of millions of innocent men, women, and children? If one knew, in retrospect, that Hitler's twentieth-century atrocities could have been wholly prevented by the moral act of *refraining* from leaping to save an endangered child, then the act of letting that particular baby drown would logically represent a supremely moral course. What if, for the sake of argument, a dreadful plague soon wipes out millions of innocent people, and the pathogen is traced to an organism that once dwelt harmlessly in the system of whooping cranes? The afflicted people of the next century—bereft of their wives or husbands, parents or children—might curse the simple-minded do-gooders of the twentieth-century who had brazenly intervened in life's natural course of events and preserved the cursed cranes, thereby damning millions of innocents to suffering and death.

We today like to believe 1) that all living things are somehow inherently good to have on the planet, and 2) that saving the existence of any life-form is inherently a virtuous action. If Mencius or a modern lover of "life" were to leap into life's river to save a baby, he or she would exult in this act of moral heroism, and would obtain a sense of self-satisfaction from having done a good deed. But, Zhuangzi suggests, since human wisdom is inherently incapable of comprehending the true meaning of events as they are happening, when can we truly know that our emotional urge to save babies or pretty birds is really an urge that is morally sound? The answer of classical Daoism seems to be that we can never be sure—that the extinction of Zhuangzi's wife or the whooping crane may be natural and proper in the way of Life itself, and that to bemoan such events may be to show oneself no more insightful about life than a child who cries sentimentally over the loss of a goldfish. Zhuangzi's lesson seems to be that "things happen" and that some happenings distress us because we attach ourselves sentimentally to certain patterns of life; whenever those patterns take a drastic turn, a mature and responsible person should calm his or her emotions and take the morally responsible course of simply accepting the new state of things.

Perspectives from Other Early Daoist Texts

A critic might object that numerous passages of the *Laozi, Zhuangzi,* and *Neiye* seem to assume that death is not desirable and that there is thus at least a value in preserving one's own life. Some passages in the *Daode jing* also commend behavior calculated to benefit others. Surely, therefore, it would be inaccurate to say that Daoist principles forbid us to care about other living things.

Let us turn here to the *Daode jing*. If, for instance, *Daode jing* 49 states that the sage looks upon all people as little children, the opening of chapter 5 suggests that he might have little concern whether such children live or die:

> Heaven and Earth are not "benevolent" (*ren*):
> They take all things to be [like] straw or dogs.
> The Sage is not "benevolent":
> He takes all people to be [like] straw or dogs.[14]

Though interpretations of this passage vary, it is hard to miss the implication that one ought to live with no regard for others. Consider, for instance, that nature's rains or asteroids arrive regardless of whether any given living thing—or species—is thereby given more abundant life or whether it is drowned or vaporized.[15] The argument here is that while feelings that would seem to urge intervention may arise in the heart/minds of humans, such feelings are *not* evident in the broader world. Though "we" may be, in such an environmentalist sense, "caring" or "compassionate," Heaven-and-Earth show no compassion or shame when they send a typhoon or an asteroid toward human habitations or when they afflict a population (human or nonhuman) with an epidemic disease, or when they watch idly while humans alter the habitat of an endangered species. The lesson of *Daode jing* 5 is not that one should emulate the impartial Dao except when we feel that someone or something is threatened. Rather, the lesson is that one should emulate Heaven-and-Earth rather than those humans who cultivate such idealized emotional attachments as Mencius seems to have cherished. If one judges human activity by how well it correlates to activity seen in "Nature," then the Mencian "moral feelings"—which are absent in "Nature"—actually appear quite unnatural.

If we are honest in our reading of the texts of classical Daoism, we find little support there for the modern idealization of salvific human action and no support for the modern ideal of heroic human action. The *Zhuangzi* seems to laugh at the very notion that we can comprehend what is going on around us. And the *Daode jing* says something even more uncomfortable to the modern mind: it argues, quite persistently, that humans cannot, and ought not, intervene in life's events, because there is a greater force than us already at work in the world. The *Daode jing* asserts that the natural reality that it calls Dao is a perfect force for the fulfillment of life. Far from needing humans to complete its activity, that Dao is, despite appearances, the most powerful force that exists, and it inevitably leads all situations to a healthy fulfillment—provided human beings not interfere with it.[16] From this perspective, Confucians—and environmental activists—wrongly fear that life will end in chaos, without human redemptive activity: in truth, the *Daode jing* teaches that because of the subtle beneficent activity of the force called Dao we can rest assured that life will proceed as it ought to. Human activity thus is not—indeed, *cannot* be—redemptive at all. For that reason, our moral responsibility is to refrain from such activity, to desist from misguided interference in the inherently trustworthy tendencies of Heaven-and-Earth.

"Cultivating Life"

A recurrent theme of Daoism through the ages has been *yangsheng*, a term often translated as "fostering life." But does the Daoist ideal of "fostering life" correspond to any of our modern ideals?

It is intriguing that medieval Daoist literature abounds in stories of exemplary men and women who earned recognition—and even "transcendence"—by secretly performing compassionate acts, particularly for creatures disdained by others.[17] Is that because they recognized a biological concept of "life," that is, that any creature has life if it has been born, has not yet died, and is therefore capable of having "life experiences"?

In 1979 Norman Girardot argued otherwise: "Indeed," he stated, "the very idea of life or health, including as it does both physical and spiritual dimensions, evokes an archaic aura of religious meaning— that the fullness of life is supranormal by conventional standards."[18] Here Girardot raises a point of fundamental importance, that is, that from the Daoist perspective "life" is not merely a biological phenomenon but a meaningful process that extends beyond the visible dimension of biological activity. To argue for preserving the mere biological activity of bodies—human or nonhuman, individual or species-wide—is to deny the most vital aspect of the entire Daoist tradition—an enduring call to see our own reality as extending into the unseen.[19] What is so difficult for the modern mind to accept is that the Daoists of ancient China valued "life," but they did not value what modern minds define as "life." When classical Daoists suggest that we "foster life," they mean something utterly different from a modern person's urge to keep Aunt Emily breathing or to keep the whooping cranes breeding.

"Compassion"? A Classical Daoist Perspective

Chapter 67 of the *Daode jing* reads, in part, as follows:

> I constantly have three treasures:
> Hold onto them and treasure them.
> The first is called "solicitude *or* due consideration" (*ci*).
> The second is called "restraint."
> The third is "not daring to be at the forefront of the world."[20]

This passage is significant because it is one of the few passages in the Daoist classics that clearly commend concern for others. A key element is the term *ci*, which is usually translated as "compassion." But our modern term "compassion" has lots of ideological "baggage," and no one, to my knowledge, has meaningfully explained what this passage really says. How, one wonders, can one be expected to be "compassionate" when one is elsewhere urged to treat all humanity as "straw or dogs"? The answer involves the fact that we must beware reading this passage on the basis of Confucian assumptions, Christian assumptions, or modern humanist assumptions.

I shall be radical enough to argue that we are reading texts from people of an alien culture, ancient China, who took seriously three propositions that strike most modern interpreters as utterly preposterous: 1) that Dao exists; 2) that it operates wisely and reliably, without human assistance; and 3) that any interventional activity by humans will inevitably interfere with that operation and will lead, ineluctably, to unintended, but quite avoidable, tragedy.

One grants that such propositions are utterly at odds with fundamental assumptions of modern thought—whether secular or religious. From the modern perspective, "natural processes" are inherently untrustworthy: they either pose threats, as when one creature's expanding habitat threatens another, or they are too weak to withstand the effects of human activity. From the modern perspective: 1) there is no force involved in life's affairs that is as powerful as that of human beings; and 2) while there may be a consciousness wiser than ours, it does not systematically protect or care for living things, so there is also no wiser involvement in life's affairs than our own. Ultimately, these perspectives assume human power and wisdom to be supreme.

From any Daoist perspective, such assumptions are absurd. Many naïve misinterpreters of Daoism have happily assumed that the only human intervention that is deleterious is "their" intervention, never "my" intervention. Modern environmentalists have often simplistically interpreted the *Daode jing* as saying, for instance, that the interventional activity of a construction crew building a dam is an unwarranted imposition upon nature but that the interventional activity of a legislator or a protest group who desires to stop such construction is not interventional at all. But if we read the ancient Chinese term *wei* correctly—as denoting "human action intended to achieve results," specifically "results superior to what would result if nature were simply allowed to take its own course"—then it would follow that any action intended to stop the construction of a dam, the draining of a wetland, or the burning of a rain forest constitutes precisely such deleterious interventional activity. The only difference is that the developers and their opponents desire different results. And as everyone seems to know, the view of the ancient Daoists is that "human action intended to achieve results" is contrary to the Dao, whatever the actors' motivation.

So is the classical Daoist perspective that we ought to stop caring about the state of the world? The answer to that question is both yes and no. From the classical Daoist perspective, the only good actions are actions that are not taken, and the only good people are those people thoughtful enough, considerate enough, humble enough, and brave enough not to take any interventional action. From the Daoist perspective, it is *only* such people who can truly be regarded as enlightened and morally responsible. The basis for that contention is that, unlike all modern thinkers, the ancient Daoists took seriously an idea that all modern minds regard as preposterous: the idea that living things do not live in an uncaring world. In the classical Daoist view, "Nature" is *not* a morally insensate juggernaut that often threatens the deserved well-being of innocent living things. A flood that affects the inhabitants of a floodplain is not a catastrophe in natural terms, only in human terms. And there is no sense in which human intervention—human activity intended to control such events—could be considered wise or appropriate action. The reason for this is that the *Daode jing* contends that—contrary to the assumptions of modern interpreters, secular or religious—the natural processes of the world are themselves guided and directed by a natural force that is not only utterly benign but continuously at work in all the processes and events of the world, whether we perceive it or not. "Returning to the Dao" means learning to see that force at work in the world and learning to rely upon it—not upon our own actions—for the fulfillment of the health and harmony of all living things. We therefore "cultivate life" by practicing enlightened self-restraint and by not presuming to act. *Daode jing* 67 argues that coupling "restraint" with "due consideration" allows us to be "courageous" while we are being "expansive," rather than "daring to be at the forefront of the world."

But, one might ask, doesn't the *Daode jing* also insist that the world is currently in disarray and urge the reader to engage in new and different behaviors so that the world may thereby be redeemed from the problems that currently afflict it?

The answer to those questions is "yes." But note that those questions do not actually call for humans to take any action to intervene in worldly events. Rather, the *Daode jing* enjoins the reader to make a bold and meaningful change in the world by 1) beginning the bold and enlightened process of refraining from interventional activity, and 2) allowing the inherent beneficent forces of the world to hold sway. The only wise and beneficent behavior in which humans can engage is one of humble and enlightened self-restraint—self-restraint that is necessary to ensure that we no longer interfere with the activity of the force called "Dao."

In modern thought, as indeed, in the thought of Confucians like Mencius, humans are morally required not merely to see themselves as meaningful agents but also to act, to do what we can do to ensure that events take the desirable course. But from the classical Daoist perspective, our proper role in the unfolding of life's events is no role whatsoever. Humans are not, and can never be, agents of good in the world, and human actions can never enhance the conditions of life itself.

Because there actually *is* a benign natural force at work in the world, any extraneous action on the part of humans can logically only cause further disturbance. So, from the Daoist perspective, the "conscientious moral agent" is not, as James Rachels has argued, someone who deliberates carefully and then "is willing to act on the results of this deliberation."[21] Rather, the responsible person is someone who is willing *not* to act on the results of moral deliberation. It is not that one should act without moral deliberation, much less that one should refrain from moral deliberation. To the contrary, one should deliberate appropriately, understand the nature of life correctly, and then bring one's behavior into accord with the true nature of life by refraining from taking action. A person who proceeds in this way is not, as H. G. Creel imagined, a heartless "monster" who allows avoidable catastrophes to occur, but rather a person of enlightened restraint, by virtue of which restraint he helps prevent avoidable catastrophes.

But we should also note that by this definition, the natural order, which one should uphold through conscientious non-action, is a natural order that actually includes death. In fact, it includes death as a universal event that ends the life-process for all living things. By this definition, death is not a horrible destruction of a meaningful and desirable life-process but rather the natural and correct completion of the life-process, at least in the present dimension. Life, as the *Zhuangzi* says, is the companion of death, and vice versa. Neither can be demonstrated to be more meaningful or more desirable than the other, and both the *Zhuangzi* and *Liezi* are replete with characters who learn that personal experience after death may actually be as good as, or better than, one's experience before death. For the person who truly understands life, death is the ultimately natural event. And if this is true for individual lives, it would logically seem to hold true also for the life of a species, or even for the life of a planet.

When Zhuangzi's wife died, and even when old "Master Lao" died, the wise and enlightened characters in Zhuangzi's text put the matter into correct universal perspective, restrained their emotions, and admired the beauty of a universe wherein death is a natural and proper outcome of life. To have done otherwise would have been to demonstrate an inability to understand and appreciate the integrity and meaningfulness of Life itself. To have done otherwise would have been to demonstrate a pernicious misconception that the event that we call death somehow negates the value of what has gone before it. Such beliefs, these texts show, were common among the shallow-minded denizens of ancient China, just as they are common among both the religious and the secular minds of the modern world. To the composers of the classical Daoist texts, the death of Zhuang's wife, the death of "Master Lao," the death of the dinosaur, the death of the whooping crane—all of these are to be accepted with tranquillity and with respect for the integrity and value of the natural processes of life. If, therefore, we see a baby floating down a river—or a planet, for that matter—we should refrain from imposing our humanistic notions of value upon the wisdom of Nature itself, for to the composers of these texts, Nature is not cruel or insensate but perfectly and totally benign. Our world is *not* merely designed wisely, then left to run unattended, leading to disasters for various living things unless heroic humans intervene. Rather, our world is designed wisely and operated wisely by a force that is like a loving mother (see *Daode jing* 1, 20, 25, 52, 59). That motherly force nurtures and cares for all things; then, at the end of their natural lives, they all return to it. Treasuring tranquillity, the conscientious Daoist observes that return, with awareness and due respect, and, in due course, he or she follows that same course, returning without fuss to the immaterial state from which he or she first emerged.

The Transformative Power of the Perfected Person

The modern mind finds it easy to reject such interpretations, however sound, because they not only fail to help us solve our problems in a happy way, but they challenge us to question whether or not we really understand our own lives. Such implications can be deeply unsettling, because a fundamental thrust of Western humanism is that humans are different from and superior to banana slugs: humans can analyze problems and take action to solve them. From that perspective, refusal to engage in problem-solving activity would reduce us to the status of the slug and would thus constitute a shameful derogation from our proper moral duties. Neither secular nor religious minds in the modern West can justify the proposition that the state of affairs in the world around us is, to be blunt, none of our business. That some Daoist texts urge us to see life in those terms is unacceptable to many, so, rather than sacrifice the beauty of Daoist naturalism, they simply redefine Daoism to suit their own sensibilities, denying the very presence of teachings that offend our modern biases. How dare one suggest that the sage who follows the Dao would really take the same course as a banana slug, watching dispassionately as life's strange pageant unfolds around us? Surely, such minds opine, humans are superior to creatures like the slug, because humans are *capable* of intervening in life's events. But from the perspective of classical Daoism, modern humanism makes the mistake of assuming that the *ability* to intervene in life's events translates into a moral *duty* to do so. The unmistakable teaching of the *Daode jing* is that humans may indeed be capable of intervening in life's events, but the evidence of life, which humans constantly ignore, is that such intervention is destructive to all involved, and we thus have a moral duty to refrain from such interventional action.

Such a perspective strikes many modern minds as unthinkable. But that is because we have modern minds, and our fundamental assumptions are often at odds with those of the people who produced these ancient texts. We assume, for instance, that beneficent involvement with the world around us *must* involve interventional activity. But that is because we are not Daoists. A careful reading of the Daoist texts of ancient China, as well as of many later texts, shows that the Daoists who told us to leave the world alone were neither heartless nor defeatist: they merely advocated a form of beneficent involvement in the world that we today dismiss as impossible because, in our terms, it is wholly inconceivable that the transformation of oneself into a sagely being, in accord with the deeper realities of life, has a corresponding effect on everything and everyone around one, extending ultimately to envelop the whole world. As a matter of fact, when one transforms one's being into a state in harmony with life's true realities (that is, Dao), that state has a beneficent effect upon the world around one and facilitates the reversion of all things to a naturally healthy and harmonious condition. This teaching is vaguely suggested in the *Zhuangzi* and the *Neiye*, more clearly suggested in the *Daode jing*, and more fully expressed in texts of Han times and beyond.

Many Chinese writers, starting with the editor of the *Daode jing*, found it hard to resist the impulse to express such teachings as teachings referring to the life of the ruler. The *Daode jing*, of course, maintains quite clearly that when the ruler refrains from interventional activity and cleaves to the unseen reality called Dao, the world inevitably reverts to its natural and proper condition. Today, of course, moderns minds assure each other that such teachings are unthinkable. Apparently, there were similar responses in ancient China, for the composer of sections of the *Daode jing* laments that although his teachings are informed by a powerful ruling force, people do not believe in it and therefore do not believe in his teachings. He insists, nonetheless, that we *should* believe in them, and the respectful reader today should at least ponder such teachings, despite their variance from most modern ideas.

A passage of the *Neiye*—as usual, quite vague—states that when one has attained a proper state of well-managed tranquillity, one "sets in motion the vital breath (*yun qi*), and one's mental and physical processes become like those of Heaven."[22] The meaning of this passage remains unclear, but other passages more clearly suggest that the properly tranquilized individual can and does transform those around him or her:

> To transform without altering the *qi*,
> To change without altering the awareness,
> Only the gentleman (*junzi*) who clings to oneness is able to do
>   this!
> If one can cling to oneness and not lose it, one can master (*jun*)
>   the myriad things.
> The gentleman acts on things; he is not acted on by things.
> From the orderliness of having attained oneness
> He has a well-governed heart/mind (*xin*) within himself.
> (Consequently,) well-governed words issue from his mouth,
> And well-governed activity is extended to others.
> In this way, he governs the world.
> When a single word is obtained, the world submits;
> When a single word is fixed, the world heeds.
> This is what is called "public rightness" (*gong*).[23]

Such passages seem to suggest that a highly accomplished practitioner can, by what we might call meditational practice or prayer, achieve the ability to exert influence over the entire world. At first glance such a teaching is so alien to anything found in our modern world that our first, and indeed second, impulse is to ignore it as a silly exaggeration.

But even today followers of Christianity and other modern religions often testify to the power of prayer to alter the events that seem to be unfolding around us. And throughout Chinese history, bright and thoughtful people—untainted by Western rationalism—have found such ideas eminently sensible. In China, they were called "Daoists." Western conceptions of China, informed by hostile Confucian biases, dismiss such people as marginal or insignificant.

I propose that the idea that biospiritual practice can transform the world constitutes what moderns might be comforted to call "the positive side" of the Daoist rejection of interventional activity. The teachings of classical Daoism quite clearly tell us to keep our hands off the processes at work in the world: "the world," says *Daode jing* 29, "is a spiritual vessel, and one cannot act upon it; one who acts upon it destroys it." But what of modes of human involvement with the world that do not consist of interventional activity? The *Neiye* and *Daode jing* teach that a sufficient degree of personal self-cultivation can and will result in a beneficent transformation of other living things, a transformation that reaches ultimately to the furthest extent of the world. Such a transformation, our texts suggest, is not only benign but desirable, for it effects a therapeutic metamorphosis throughout the world, while avoiding the deleterious effects that are inherent in interventional activity. Some passages suggest that this can be so because a person who practices sufficient restraint can achieve a state of tranquillity that is qualitatively identical with that of the beneficent force called Dao, a force that achieves its ends without taking action and benefits all living things without involving itself actively with them, a force as imperceptible as the life-giving force of the natural substance we call water. In ancient China, readers of these texts were taught to have faith in the imperceptible existence and inexhaustible potency of such powers and to rely upon their cultivation of such powers to effect a positive transformation of all living things. These texts warn that any other course of action, no matter how well-intentioned, will inevitably disrupt the subtle array of natural processes that are invisibly at work in the world.

Chapter 35 of the *Daode jing* says:

> Grasp the "Great Form" and go into the world.
> In going, no harm is done:
> Peace and well-being [ensue].

Music and delicacies [can induce] wayfarers to stay.
As for talk of the Dao, how insipid and tasteless!
Looked for, it is imperceptible.
Listened for, it is inaudible.
[But] used, it is inexhaustible.

I propose that before we dismiss such messages, we should carefully reexamine our own assumptions. Before we conclude that classical Daoist teachings, when correctly understood, are impractical, we should reexamine our culturally constructed belief that there is no such thing as what the Daoists call "Dao," and that there is thus no validity to their call for us to return to it, rather than to attempt to manage the world through interventional activity. Redemptive human activity, these texts argue, is not only unnecessary but actually impossible, and the sage is someone who accepts that fact, lets the natural processes of life go forth on their own imperceptible courses, and accepts the fact that we can do nothing to improve the world's condition other than by restraining our impulse to act and our hubristic conceit that our actions can be heroically salvific.

While modern minds might refuse to follow such teachings, we should at least make sure that we understand them, and we should realize that we deny the reality of Daoist teachings if—in what Clarke calls our "utopian enthusiasms for a 'new paradigm'"—we attribute to Daoists other teachings that might make *us* somewhat happier. The purpose of studying other cultures is *not* to use them to solve our own problems. Such paternalistic attitudes rest upon the colonialist assumption that other people exist to serve our own needs—the same colonialistic assumption that environmentalists decry in non-environmentalists' attitude toward nonhuman life and the environment as a whole. If we should abandon colonialist attitudes in regard to the rain forest and allow it to live according to its own inherent integrity, should we not treat the thinkers of other ages and cultures with the same respect?

Notes

1. J. J. Clarke, *Oriental Enlightenment: The Encounter between Asian and Western Thought* (London and New York: Routledge, 1997), 177–78.

2. For an explanation of the importance of exegetical methods for understanding the Daoist classics, see my article, "Hermeneutics and Pedagogy: Methodological Issues in Teaching the Tao te ching," in *Essays in Teaching the Tao te ching*, ed. Warren Frisina and Gary DeAngelis (Oxford and New York: Oxford University Press, in press).

3. See Russell Kirkland, "Taoism," in *Encyclopedia of Bioethics*, 2d ed. (New York: Macmillan, 1995), 5:2463–69.

4. See Russell Kirkland, "Tales of Thaumaturgy: T'ang Accounts of the Wonder-Worker Yeh Fa-shan," *Monumenta Serica* 40 (1992): 47–86; "A World in Balance: Holistic Synthesis in the *T'ai-p'ing kuang-chi*," *Journal of Sung-Yüan Studies* 23 (1993): 43–70; and the entry "Ye Fashan," in *Encyclopedia of Taoism*, ed. Fabrizio Pregadio (London: Curzon Press, in press).

5. For a fuller exposition of these themes, see Russell Kirkland, "The Roots of Altruism in the Taoist Tradition," *Journal of the American Academy of Religion* 54 (1986): 59–77.

6. See ibid., 72–74.

7. See Holmes Welch, *Taoism: The Parting of the Way*, 2d ed. (Boston: Beacon Press, 1965), 164–78. Interpreting "Lao Tzu's" teachings as "too radical" to be applied to modern social or political problems (171), Welch asks, "Are any of his beliefs acceptable to us? What would be the consequences of putting them into practice? The answer to that question will scarcely brighten the Tao Te Ching in our eyes" (174). He concludes that "Lao Tzu is not the kind of thinker to whom twentieth-century Americans would turn for advice" (165).

8. H. G. Creel, *Chinese Thought* (Chicago: University of Chicago Press, 1953), 112–13.

9. Bryan W. Van Norden, "Method in the Madness of the Laozi," in *Religious and Philosophical Aspects of the Laozi*, ed. Mark Csikszentmihalyi and Philip J. Ivanhoe (Albany: State University of New York Press, 1999), 203.

10. For a full analysis of the heretofore-ignored ethical teachings of the *Daode jing*, see my article, "Self-Fulfillment through Selflessness: The Moral Teachings of the Daode jing," in *Varieties of Ethical Reflection: New Directions for Ethics in a Global Context*, ed. Michael Barnhart (New York: Rowman and Littlefield, in press).

11. See Lee Yearley, "The Perfected Person in the Radical Chuang-tzu," in *Experimental Essays on Chuang-tzu*, ed. Victor Mair (Honolulu: University Press of Hawaii, 1983), 125–39.

12. From this position, Confucian moralism appears more tolerable than that of modern environmentalists. At least Confucius did not commend sentimental concern for the horses that might be killed when a stable burns: he reserved his moral concern for the lives of beings of our own kind (*Analects* 10:17; see my discussion of this passage in "Self-Fulfillment through Selflessness.") Mencius, willing to commend a ruler's sentimental compassion for an ox being led to slaughter for a sacrifice, is closer to the position of the modern environmentalist in that he approves of the stirrings of the heart/mind raised by the imminent death of the animal. But even Mencius—willing to commend such stirrings as a moral guide—even he did not envision the modern romantic vision of the compassionate human as the moral savior of entire

nonhuman species, ecosystems, or planets. If Confucius admonished against concerning oneself with nonhuman spiritual beings until the moral needs of humans have been fully addressed, Mencius would seem prepared to exhort us to engage in appropriate moral action toward all the humans around us before we begin considering moral action toward nonhumans. The Confucian sage-king is generally one who engenders moral harmony among human beings, not one who inflames sentimental solicitude for oxen, cute puppies, or adorable dolphins. Mencius supported the king who felt compassion for an ox—not because oxen are inherently worthy of sentimental compassion, but rather because he could use that incident to teach the king how to rule his human subjects with appropriate moral concern. Mencius did not chide the king for having ordered the death of an innocent sheep!

13. See Guo Jingfan's edition in *Concordance to Zhuangzi* (Cambridge, Mass.: Harvard Yenching Institute, 1956), 46; and *Wandering on the Way: Early Taoist Tales and Parables of Zhuangzi*, trans. Victor H. Mair (New York: Bantam Books, 1994), 169.

14. Translation mine. The term "hundred clans" here is exactly the same term (*baixing*) found in *Daode jing* 49, where the sage is said to treat them as children. Wang Bi's commentary shows that the final line is to read "straw and dogs": to read the term "straw-dogs" into the text is unwarranted exegesis.

15. I would argue that it is necessary to take into account here the intellectual history of ancient China, for "benevolence" is not just a term of common discourse, but a technical term in the vocabulary of the classical Confucians, particularly that of Mencius. One can, in fact, read this passage as a direct argument against Mencius's teaching that one should cultivate a set of moral feelings (compassion, respect, shame, etc.) that he alleges to be intrinsic to human nature.

16. See further my entry, "Taoism," in *Philosophy of Education: An Encyclopedia*, ed. J. J. Chambliss (New York and London: Garland Publishing, 1996), 633–36.

17. See further my entry, "Taoism," in *Encyclopedia of Bioethics*.

18. Norman J. Girardot, "Taoism," in *Encyclopedia of Bioethics*, 1st ed. (New York: Macmillan, 1979), 1631.

19. See Kirkland, "Taoism," in *Encyclopedia of Bioethics*.

20. On the translation of the term *jian* ("restraint"), see my textual notes in "Self-Fulfillment through Selflessness."

21. James Rachels, *The Elements of Moral Philosophy* (New York: Random House, 1986), 11. I address these matters more fully in my article, "Self-Fulfillment through Selflessness."

22. Translation mine. I use the edition reproduced in *Zishu ershiba zhong* (Twenty-eight Classical Texts) (Taibei: Guangwen, 1979), 1:621–24. The passage in question (624, line 6) is found in section N of Rickett's 1965 translation (p. 167); section XIV.2 of his 1998 translation (p. 54); and section XXIV of Roth's translation (Roth translates *yunqi* as "revolving the vital breath"). See W. Allyn Rickett, *Kuanzi: A Repository of Early Chinese Thought* (Hong Kong: Hong Kong University Press, 1965); *Guanzi: Political, Economic, and Philosophical Essays from Early China*, vol. 2 (Princeton: Princeton University Press, 1998); and Harold D. Roth, *Original Tao: Inward Training (Nei-yeh) and the Foundations of Taoist Mysticism* (New York: Columbia University Press, 2000), 92.

23. Text, p. 622, translation again mine. Cf. Rickett, *Kuanzi* (1965), 161 (section D); Rickett, *Guanzi* (1998), 44 (section VII.1); Roth, *Original Tao*, 62–64.

Metic Intelligence or Responsible Non-Action? Further Reflections on the *Zhuangzi*, *Daode jing*, and *Neiye*

LISA RAPHALS

R**ussell Kirkland** has argued persuasively against reading a specifically interventionist ecological sensibility into the *Zhuangzi*, *Daode jing*, and *Neiye*, the three preeminent "Daoist" texts of Warring States and Han vintage. His discussion centers on what he calls "responsible non-action" as a central tenet of the moral reasoning of ancient or classical Daoism.[1] Classical Daoists, he argues, unlike both classical Confucians and modern Westerners, accepted *all* the processes of life and death, including the destructive ones. They thus eschewed heroic redemptive action by humans either on behalf of others or to "fix" or "improve" the world; indeed, ancient Daoist moral reasoning specifically precluded such intervention. In summary, Kirkland's arguments suggest that, however much Daoism (broadly understood) may seem resonant with an "ecological" sensibility, interventionist misreading of these texts would allow the ends to justify the means.

In response, I want to examine these problems and texts from another point of view. These texts suggest several possibilities for what I will call noninterventionist action. By this I mean modes of action that are indirect, work at a distance, and do not involve heroic, deliberative, or even necessarily deliberate intervention. In these texts, Daoists frequently did intervene in these ways. Their modes of action differed from the contemporary model of heroic intervention and included the extended effects of individual self-cultivation, transforming the people by exemplary rulership, and the use of a peculiar kind of wily and foresighted intelligence. Finally, I want to introduce a note of healthy skepticism.

Noninterventionist Action

In the "Mastering Life" chapter (19) of the *Zhuangzi,* Confucius is faced with the situation of someone apparently drowning in the flood. The falls at Lüliang are so precipitous that the water falls from thirty fathoms and races in rapids for forty *li* "so swift that no fish or other water creature can swim in it." Confucius sees a man dive into this maelstrom and, "supposing that the man was in some kind of trouble and intended to end his life, he ordered his disciples to line up on the bank and pull the man out." But, the story continues, after the man had gone a couple of hundred paces, he emerged from the water, strolled along the bank and sang a song. Confucius runs after him and asks whether he has some special way of staying afloat. The "Daoist" explains to Confucius that: "I have no way. I began with what is inborn, grew it by essential nature, and completed it by means of fate." He continues: "I was born on land so I feel at ease on land; that is inborn. I grew up on the water so I feel at ease on water; that is essential nature. I don't know why I do what I do; that is fate!"[2]

Here, the apparent victim not only does not need saving, but also is more adroit than Confucius and his disciples. Indeed, the narrative of the *Zhuangzi* uses the swimmer to show the limits and inefficacy of the heroic intervention Confucius, at least, intended. Confucius does not come off very well. He initially misunderstands the nature of the situation and mistakes the man first for a ghost (and not in need of aid) and then for a suicide. At no point does he himself intervene, beyond ordering his disciples to save the man, which they seem unable to do. (The water was so turbulent that "not even a fish could survive in it," and the swimmer only emerged after a few hundred paces.) In this apparent human crisis, Confucius lacked the means to intervene effectively.

Parenthetically, we might extend this implied critique of heroic intervention as inefficacious to Ruist persuasions on ecological matters. Our accounts of the biographies of the figures known as Confucius, Mencius, and Xunzi suggest that their official positions (when they had them) did not enable them to act, beyond attempts at persuasion. Their persuasions to rulers certainly addressed ecological concerns— the most famous is probably Mencius's Ox Mountain (6A8)—but there is little explicit record of such persuasions having specific effect. Probably more important is the inclusion of *yue ling*, or "monthly ordinance" calendrics, within the *Li ji* that gave agricultural regulation the force of ritual command, at least by Han times, when the *Li ji* was probably compiled.[3]

Here and elsewhere, the *Zhuangzi* rejects or ridicules heroic intervention. Can we, therefore, infer that ancient Daoists rejected any kind of intervention, on either their own or others' behalf? I suggest that these texts present or sanction at least two modes of action that constitute alternatives to heroic ecological intervention. The first is self-cultivation: literal self-preservation through cultivating one's vital essence (*jing*). The second is nurturing others and even "transforming the people" (*hua min*) through indirect action.

Self-cultivation

In the *Zhuangzi*, Robber Zhi upbraids Confucius for the folly of the so-called sage-kings who sought reputation, made light of death, forgot their origin, and did not "cultivate their fated span" (*shou ming*).[4] In contrast, the tree that is too gnarled for the carpenter is left on the mountains to live out its allotted fate.[5] In the *Zhuangzi*, as in other Warring States texts, everyone is born with an allotted life span, *ming* or *fen*.[6] The problems of fate and fatalism reappear in any number of Warring States philosophical debates.[7] *Zhuangzi* 12 describes the emergence of notions of decree and fate as part of the origin of the world:

> In the far beginning there was nothing and nothing had no name. Then One arose from it and there was One but it had no form. Living things acquired it in order to come alive, and it was called power (*de*). The formless had allotments (*fen*) but they were still not divided out, and they were called fates (*ming*).[8]

According to *Zhuangzi* 6:

> that life and death are decreed, that there are regularities of night and day, this is Heaven. Everything in which people cannot intervene, this is the nature of living things.[9]

Zhuangzi's Confucius advises his disciple Zigao that "nothing is as good as bringing about what has been decreed; this is what is truly difficult."[10] In these passages everything has its allotment. In Kirkland's reading of classical Daoist ethics, such decrees cannot and should not be interfered with. Classical Daoist ethics also sanctions subtle action through indirection and resonance with natural processes and the action of "fate."

The most straightforward of these modes of indirect action is active self-cultivation. It is in the *Neiye* that one first meets clear references to the personal cultivation of such forces as qi (life energy), *jing* (vital essence), and *shen* ("numen"), which "became a central theme in certain versions of modern Taoism, as well as in Chinese medicine."[11] In the *Neiye*, *de*/power is not an intrinsic force, but rather the ability to succeed (as suggested by its homonym *de*/acquire) by constant self-cultivation or "acquisitional agency."[12] The *Neiye* specifies that the cultivation of *de* must be worked on each day:

> Respectful and cautious, and avoiding excesses, he daily renews his Power (*de*).

> He comes to understand everything in the world and thoroughly examines its four extremities.[13]

These classical Daoist texts represent self-cultivation practices as active interventions upon one's own person that could, depending upon the text and period, permit one to survive through difficult times, enhance longevity, or even (in Six Dynasties Daoism) become an immortal.

Action at a Distance

Whatever their power to optimize individual survival, could the extended effects of self-cultivation practices in any sense constitute an ecological intervention, however indirect or subtle? The *Daode jing* repeatedly describes the indirect strategies by which "the sage" or sage-ruler causes the people to prosper without, apparently, doing anything at all; indeed, the extent to which *wuwei* informs the *Daode jing* (DDJ) is well known. To cite a few examples:[14]

The sage makes a dwelling in *wuwei* concerns (*wuwei zhi shi*, DDJ 2).

They use *wuwei* and nothing is undone (DDJ 48).

The words of the sage say: I use *wuwei* and the people transform of themselves (*min zi hua*, DDJ 57).

Act by means of *wuwei* (DDJ 63).

Zhuangzi 1 describes a numinous man (*shen ren*) of Guye who concentrates his *shen* and "protects creatures from sickness and plague and makes the harvest plentiful"[15] This passage suggests the view that a realized sage can have a nurturing effect on the world at large by acting at a distance. The *Zhuangzi* makes no suggestion that these salvific effects are intended; they appear to be a beneficent byproduct of self-cultivation practices.

Other passages in Warring States and Han texts suggest that self-cultivation also benefits others indirectly. The *Lü shi chunqiu* chapter "Living Out One's Lot" (*Jin shu*) describes how sages used knowledge of yin and yang to understand what benefits the myriad creatures and to live out their allotted lifespans, without augmentation or diminution of what had been allotted them:

> Sages investigate the conformity of yin and yang, discriminate what benefits the myriad creatures to live to the greatest advantage, and by making the numinous essence (*jing shen*) tranquil, they preserve their longevity and lengthen it. This lengthening is not a matter of either shortening or extending it, but rather of bringing its [allotted] number to completion (*bi qi shu*). Bringing its number to completion consists in eliminating harm. (*Lü shi chunqiu* 3.2, pp. 3b–4a)

This passage suggests a view that what fate allots is not inviolable, and it requires the activities of a sage or a ruler (discussed below) to guarantee it.

Transforming the People through Example and Efficacy

Turning to more direct forms of intervention, the rhetoric of several Warring States works on government and military strategy suggest that good rulers, like sages, brought about conditions that enabled both people and other living things to "complete their *ming*," to live out their allotted spans, in peace.

Sunzi's *Art of War* stresses that the effective general acts with speed and avoids destruction of life and matériel.[16] Xunzi's "Debate on Principles of Warfare" makes it clear that a true king does not punish the people. His military regulations

> do not seize those who offer allegiance, do not leave in place those who resist, and do not imprison those who flee for their lives (literally, who flee "for the sake of their *ming*." (*Xunzi* 55/15/60)

Similarly, in his punitive expeditions

> those who submit to the sword he allows to live, those who resist, he kills, and those who flee for their lives (for their *ming*) he treats as precious tribute. (*Xunzi* 55/15/61)

The *Huainanzi* (HNZ) contains extensive discussions of "transforming the people" (*hua min* or *min hua*), in both the "Art of Rulership" chapter (HNZ 9) and "The Great Family" (HNZ 20).[17] Correspondingly negative rhetoric attaches to vicious rulers like Zhou of Shang, whose subjects and ministers were in such terror that "none could feel certain of his fate" (*Xunzi* 57/15/85).

A Skeptical Note

Attempts to apply a paradigm of Daoist action or non-action to the contemporary ecological situation are open to both textual and philosophical objections. But we must also ask whether the present crisis is different in kind, rather than merely in degree, from previous crises, such as the agricultural straits of the Han. E. N. Anderson, Francesca Bray, and others have assessed the magnitude and effects of that crisis and considered the kind of changes it produced in Han government and social policy.[18] We may nevertheless ask: does our current situation entail a degree and kind of interference with nature/Dao that is beyond the conceptual scope of these texts? Is our technological grasp of implements of mass destruction, through atomic warfare in the 1950s and 1960s and now through environmental disaster, simply beyond what these texts envisaged? Unsustainable levels of specifically chemical pollution, new diseases, and the prospect of global warming suggest the possibility of damage beyond even the power of Dao to reverse.

If we take the current situation to be different in kind and not merely degree, we cannot apply the textual precedents of the past with any felicity to Daoist attitudes toward predictability. Skeptical instincts suggest that we simply do not *know* what Zhuang Zhou would have done. To underscore an obvious point, our sources for Warring States and Han history are such that we have mostly texts. Daoists are not texts, and texts do not act (or not act) with purpose, one way or the other. We need to be careful about inadvertently equating what we might consider to be the logical or philosophical implications of a text with the actions of any person. Texts are not persons; we cannot predict the behavior of Daoist humans from Daoist texts, especially within a tradition that prized "flexible response to circumstances" (*ying bian*) and the ability to deal resourcefully with unpredictable events. Truly unpredictable circumstances might prompt action outside of what we have in our texts.

Conclusion

Imponderables aside, the textual evidence shows that notions of self-cultivation, indirect action, and action at a distance as the basis for effective rule figured prominently in wide range of Warring States and Han thought. These modes of actions were ascribed to sages, rulers, and generals, all individuals in one or another sense charged with public welfare.

Most moderns would, of course, reject utterly the idea that self-cultivation or action at a distance could address a "real" ecological crisis. That prima facie rejection, however, would ignore the porosity of notions of selfhood in a wide range of Chinese thought: the inseparability of "inner" and "outer," the high cultural value of "selflessness," macrocosm-microcosm identifications, and constructions of individuality that differ from Western norms. Any number of practices in both these texts and Chinese folk belief reflect this belief in a porous border between inner and outer, for example, the contiguity of practices for the management of the flow of qi in both fengshui and traditional medicine.[19]

Many Warring States and Han accounts of indirect action specifically involve responses to rapidly changing or unpredictable circumstances, what the Greeks called *metis*.[20] I have argued elsewhere at length that this subtle cast of mind thoroughly informed classical Chinese moral reasoning, philosophy, and political thought, but was especially prevalent in Daoist texts such as the *Daode jing* and the *Zhuangzi*.[21] In a recent study, *Seeing Like a State: How Certain Schemes to Improve the Human Condition Have Failed*, James C. Scott argues that a single body of ideas, which he calls high modernism, lies behind a diverse range of ideas. Scott argues that the disasters of the twentieth century come from a neglect of *metis*, and only the presence of individuals of *metis* within high modern societies has prevented even worse harm being done by such grand schemes of human improvement.[22]

The cultural ecologist E. N. Anderson remarks that most of the world's traditional societies "encode in their moral teachings practical wisdom about the environment and the individual's duty to treat it with respect."[23] In the Chinese case, some of these practices may be more directly visible in the texts and practices of the Han synthesists and later Daoists than in Warring States Daoist texts.[24] We risk grave anachronism and misreading if we impute to them specifically interventionist attitudes toward "nature" or ecology that, in fact, run counter to their own ethics. Nevertheless, they may have broad lessons for our own ecological situation in the casts of mind and modes of action they represent. These include notions of porosity of self and metic intelligence. Flexible boundaries between self and other indubitably contributed to Chinese practical wisdom about the environment. Similarly, the resourceful and ingenious attitudes associated with metic intelligence in China and elsewhere may provide alternatives to heavy-handed intervention.

Notes

1. I use this term to refer to Warring States and Han texts and practices, including both Huang-Lao and immortality practitioners (*fang shi*). For discussion of this terminology, see Livia Kohn, "Two New Japanese Encyclopedias of Taoism," *Journal of Chinese Religions* 23 (1995): 261; and Russell Kirkland, "The Historical Contours of Taoism in China: Thoughts on Issues of Classification and Terminology," *Journal of Chinese Religions* 25 (1997): 59–67.

2. *Zhuangzi* 19:657–58. References are to the edition by Guo Qingfan, *Zhuangzi jishi* (Beijing: Zhonghua shuju, 1961). Translations are my own unless otherwise specified.

3. See "Li chi" entry in *Early Chinese Texts: A Bibliographic Guide*, ed. Michael A. N. Loewe (Berkeley: Society for the Study of Early China and the Institute of East Asian Studies, University of California, Berkeley, 1993).

4. *Zhuangzi* 29:998.

5. *Zhuangzi* 20:667.

6. *Ming* and *fen* have interesting differences from and similarities to Greek *moira* or *aisa*. For discussion, see Lisa Raphals, "Fatalism, Fate and Stratagem in China and Greece," a chapter in *Early China, Ancient Greece: Thinking through Comparisons*, ed. Steven Shankman and Stephen Durrant (Albany: State University of New York Press, forthcoming 2001).

7. Despite Confucius's notorious reluctance to discuss gods, spirits, and fate, the *Zhuangzi* (and the *Xunzi*) attribute discussions of fate to him. Mohists accuse Ruists of fatalism. Both the *Zhuangzi* and the *Xunzi* emphasize the importance of fate but reject fatalism, albeit in very different ways. The *Zhuangzi* stresses the importance of alignment with inevitable change in the fates of peoples and times (and, presumably, endangered species). Xunzi de-emphasizes inevitability and attempts to counter the charge of fatalism by stressing free will and the importance of individual action. For both, fate can, in different senses, be mastered and treated strategically. For further discussion see Raphals, "Fatalism, Fate and Stratagem."

8. *Zhuangzi* 12:424. I use both the terms "decree" and "fate" to translate *ming* in order to avoid introducing a kind of crypto-fatalism into the texts.

9. *Zhuangzi* 6:241.

10. *Zhuangzi* 4:160. This phrase could also be translated as "Nothing is as good as following one's destiny."

11. Russell Kirkland, "Varieties of Taoism in Ancient China: A Preliminary Comparison of Themes in the *Nei Yeh* and Other Taoist Classics," *Taoist Resources* 7, no. 2 (1997): 75. For discussion of these translations, see Harold Roth, "The Inner Cultivation Tradition of Early Daoism," in *Religions of China in Practice*, ed. Donald S. Lopez, Jr. (Princeton: Princeton University Press, 1996), 126–27.

12. Kirkland, "Varieties of Taoism," 77.

13. Guanzi XVI, 49, 4a–b; *Guanzi*, trans. W. Allyn. Rickett, vol. 2 (Princeton: Princeton University Press, 1988), 48.

14. Passages are numbered according to the Wang Bi edition. For discussion of *wuwei* in other Warring States and Han texts, see Roger T. Ames, *The Art of*

Rulership: A Study in Ancient Chinese Political Thought (Honolulu: University of Hawaii Press, 1983), 28–64.

15. *Zhuangzi* 4:160.

16. See Joseph Needham and Robin D. S. Yates, *Science and Civilization in China*, vol. 5, part 6: "Military Technology: Missiles and Sieges" (Cambridge: Cambridge University Press, 1994), 19. For Daoist influences on Sunzi's *Art of War*, see Ralph Sawyer with Mei-chün Sawyer, *The Seven Military Classics of Ancient China* (San Francisco: Westview Press, 1993), 425 n. 24.

17. For discussion, see Ames, *Art of Rulership*, 99–107.

18. For discussion, see Francesca Bray, *Science and Civilization in China*, vol. 6, part 2: "Military Technology: Agriculture" (Cambridge: Cambridge University Press, 1984); and E. N. Anderson, *Ecologies of the Heart* (New York: Oxford University Press, 1996).

19. See, for example, Anderson, *Ecologies*, 15–16.

20. For discussion, see Marcel Detienne and Jean-Pierre Vernant, *Cunning Intelligence in Greek Culture and Society*, trans. Janet Lloyd (Atlantic Highlands, N.J.: Humanities Press, 1978).

21. See Lisa Raphals, *Knowing Words: Wisdom and Cunning in the Classical Traditions of China and Greece* (Ithaca: Cornell University Press, 1992).

22. James C. Scott, *Seeing Like a State: How Certain Schemes to Improve the Human Condition Have Failed* (New Haven: Yale University Press, 1998). High modernism is the attempt to design society in accordance with what are believed to be scientific laws.

23. E. N. Anderson, *The Food of China* (New Haven: Yale University Press, 1988), 175.

24. Needless to say, Warring States texts presumably reflected, and may have in part derived from, many of these same beliefs, especially shamanism.

Non-Action and the Environment Today: A Conceptual and Applied Study of Laozi's Philosophy

LIU XIAOGAN

B ased on textual and lexicological investigations, this paper attempts to reinterpret the concept of *wuwei* (non-action) and discuss its significance in both the contexts of Laozi's philosophy and current environmental issues. In comparison with general actions, *wuwei* refers to a higher standard of human actions and their results. *Wuwei* is not term with a single meaning, but is a cluster of similar terms and phrases. In fact, *wuwei* represents a different value orientation from prevailing convention and demands the most appropriate manner of actions.

This paper differs from most other works in the discussion of environmental ethics or comparative studies—which have focused on the theoretical aspects of the relations of humans and nature[1]—by focussing on two aspects. The first is the conceptual investigation and interpretation of *wuwei*. The second is a discussion of two cases, namely, the conflagration of Indonesian rain forests and the miserable experience of the Inuit community caused by a campaign conducted by Greenpeace.[2] Superficially, the two aspects have no direct association at all; however, if we believe that Laozi's philosophy conveys profound wisdom and reflection of human societies and history, there must be something universal and applicable. This paper is an attempt at demonstrating how the significant doctrines of Laozi's philosophy may be applied to the modern ecological context. Certainly, this experiment is based on serious textual study and conceptual investigation and reinterpretation, though these could not be fully expounded in this paper because of space limitations.[3]

While this paper demonstrates that the Daoist theory of *wuwei* is relevant to current environmental discussions, it is not a miraculous cure. Actually, we should not expect any miracle in dealing with ecological predicaments. Daoism is significant for the long-term protection of the environment and for the radical resolution of ecological deterioration, not for immediately extinguishing a fire. Nevertheless, the theory of *wuwei* may help us to learn from emergency cases and to find a better way to remedy calamities and prevent tragedies from happening in the future.[4]

The analysis and application of the theory of *wuwei* in this paper is founded on the systematic reconstruction of Laozi's philosophy. According to my understanding, *wuwei* is the methodological principle to actualize *ziran*, or naturalness, the core value in Laozi's system. Dao as the ultimate source and ground of the universe provides quasi-metaphysical and axiological foundations for both *wuwei* and *ziran*, while Laozi's theory of dialectics supports *ziran* and *wuwei* mainly in the perspective of human experience.[5]

What Is *Wuwei*? A Conceptual Investigation and Its Application

The concept of *wuwei*, or non-action, is one of the most ambiguous in Chinese thought. Its denotation and connotation are very different. Let us first investigate its literal meanings and philosophical implications.

Wuwei: *Actions as without Action*

The combination of *wu* (no) and *wei* (action) superficially seems to indicate "no-action." However, the character *wu*[a] in classical Chinese is derived from three synonyms: *wang*[a], *wu*[b] (*wu*[c]), *and wu*[a].[6]

The accuracy of the theory of the three origins of *wu* remains to be further proved by etymologists and philologists. However, the three meanings of *wu* are quite clear and need no additional justification for scholars who have an essential knowledge of classical Chinese. The first is the most general denotation of *wu* (there is not), in opposition to *you* (there is). The third is an abstract concept of "not-being" which has nothing to do with the historical or practical world. Taking just these two meanings, *wuwei* indicates, literally, "no-action" or "taking no action," which is far from its connotation in the *Daode jing*, because Laozi does not promote literally doing nothing. For example, he mentions "gaining the world by doing nothing"[7] and "accomplishing the tasks and completing the work."[8] Only the second meaning, "being as non-being" is close to Laozi's thought, namely, restraining some human actions. Thus, *wuwei* could be rendered as "action as non-action."

The second character of *wuwei*, that is, "*wei*," is also among the most general words in classical Chinese. In addition to being used as the preposition "for" (when pronounced in the fourth tone in Modern Standard Chinese), it is mostly used as a verb to indicate broadly almost any kind of action and behavior. Functionally, it becomes a noun in the compound *wuwei*, while keeping its meaning as a verb. The general meaning of *wei* is doing, making, or acting, while its specific meaning derives from and varies with its context.

The concept of *wuwei* could be expressed with two forms or aspects: the negative and the positive. The word *wuwei* is apparently a negative term that suggests restraining and preventing human activities. It generally demands temperance of certain actions, such as oppression, interruption, competition, strife, confrontation, and so on. However, the concept also positively implies a special manner and style of behavior, namely, "action as non-action," or "actions that appear or are felt as almost nothing," or simply, "natural action." Accordingly, the theory of *wuwei* prefers a natural, gradual, and moderate style of conduct and opposes movements exercised intensively, coercively, dramatically, and on a large scale. Evidently, natural action is close to the spirit of *wuwei*, and intensifed action is against it.

When applying these conceptual theories to ecological problems, it is obvious that most human movements causing and intensifying environmental crises in postindustrialized societies are against *wuwei*. These movements are broad in scale and rapid in expansion. Comparatively, environments in the premodern world were not ruined, because people had acted naturally, divergently, and slowly. People are inclined to blame the burning of rain forests on the tradition of slash-and-burn cultivation. Fire, however, conventionally worked well for smallholding farmers for centuries because communities spontaneously regulated the use of fire.

One of the main reasons for the continued rain forest conflagration in Indonesia is "industrialized burning" set by plantation owners and subcontractors, which has devoured at least two million hectares of the world's second-largest region of rain forest. For every hectare of burned land, one hundred hectares is engulfed in smoke stretching from Thailand and the Philippines to New Guinea and the northern coast of Australia. Smoke has affected peoples' health right across the region. An estimated forty thousand Indonesians have suffered respiratory problems, and up to one million have suffered eye irritations. Smoke has been blamed for ship and air crashes that killed about three hundred people.[9] All of these disasters are related to the large-scale and intensified movements of economic development. The manner of these movements is in opposition to the spirit of *wuwei*.

Wuwei: *A Cluster of Related Terms*

In addition to *wuwei*, there are about fifty similar terms and phrases in more than thirty chapters of the *Laozi*: for example, no-words (*bu yan*), no-initiation (*bu wei shi*), no-possession (*fo shi*), no-knowledge (*wu zhi*), and no-competition (*bu zheng*).[10] Of them all, *wuwei* is the most often used: twelve times in ten chapters,[11] compared to *wushi* (no engagement), appearing only four times, and *wuming* (no name), which occurs five times. Thus, *wuwei* is the most representative of these negative expressions.

Wuwei is also used as a noun, which is a condition or a criterion for a term to become a philosophical concept.[12] Thus, *wuwei* is grammatically and functionally different from other similar terms. Because the meaning of *wei* is broad and the function of *wuwei* is well established, *wuwei* is generally taken as representative of many comparable expressions. To borrow from Donald Munro, *wuwei* is a clustering idea instead of a single-meaning concept.[13] Hence, it comes with ambiguity.[14]

Wuwei negates directly human behaviors and actions that cause ecological crises, such as the industrialized burning that produced rain forest conflagration in Indonesia. Nevertheless, the spirit of *wuwei* is not only applicable in analyzing the reasons for environmental problems, it is also relevant to the environmental movements themselves. A wrong way, against the principle of *wuwei*, doesn't help to maintain environmental equilibrium, but brings about tragedies to human life.

In 1983, following seven years of pressure from Greenpeace, the new European parliament outlawed baby seal pelts in Europe. This miserably affected the life of the 100,000 Inuit living in the Canadian Arctic. The seal furnished most of the Inuit diet and nearly all essentials of life, as had the buffalo for the North American Plains Indians. In the years following the seal pelt ban, an economic winter swept across the Canadian Arctic, and welfare figures soared. In Canada's tiny Clyde River, nearly half of the population was soon collecting unemployment checks. As their lives soured, their social problems escalated. Many Inuit turned to alcohol and drugs. Crime and family violence doubled. The despair led to an epidemic of suicides, mostly by young men. There were 47 suicides among Canadian Inuit in the eleven years before the ban, but 152 in the same period after it.

More distressingly, the sacrifice of the Inuit occurred not for reasons of environmental protection, but because of mistakes made by Greenpeace and the European parliament. The ringed seal lives in abundant numbers beneath the Arctic Ocean's icy lid. International scientific organizations charged with protecting fish and sea mammals had rated the seal hunt one of the best-managed in the world. Several hundred species of mammals were officially listed as endangered, but the harp seal was not one of them; its population was actually increasing. In addition, scientific groups and humane organizations commissioned by the U.S. government to investigate the way seals were killed found nothing was more humane or effective than the technique used on the ice.

What went wrong in this case? It was certainly not Greenpeace's objective of environmental protection, but the way it promoted its campaign. Unlike the mainstream conservation groups, Greenpeace, based in Vancouver, used a confrontational approach to environmental issues. Their president, Robert Hunter, an ex-newsman, was adept at using dramatic stunts in front of cameras. Greenpeace filmed the killing of thousands of "whitecoat" seal pups in the spring. Pictures of French actress Brigitte Bardot hugging a baby seal flashed around the world, and other actresses, European politicians, and U.S. congressmen followed what became a well-worn trail. The powerful images were broadcast worldwide and were called "mind bombs" by Hunter. In the last two decades, Greenpeace activists have proved themselves effective in using simple images to shape global opinion on complex environmental matters. Often, however, the impact of such campaigning is lost in media hoopla.[15]

Greenpeace succeeded in banning baby seal pelts and raising funds, but they ruined the environment and community of the Inuit, and their natural life. Greenpeace played with photographic techniques, practiced confrontational activities, and pursued dramatic effect. All of these are clearly contrasted with the deep wisdom of *wuwei*. Simplified and intensified movements may create a furor and cause a sensation, but they often mislead people and even bring disasters such as the Inuit have suffered. Environmental preservation involves serious and complicated issues affecting various groups of people and different nations and regions; thus, it demands a patient, gradual, and enduring working attitude that is in line with the Daoist wisdom of *wuwei*.

Wuwei: *A Representative of Value Orientation*

Although the concept of *wuwei* seems broadly to negate all human actions, it in fact focuses on certain actions of humankind. Generally speaking, *wuwei* is the reflection and crystallization of Laozi's methodology and value orientation or, more accurately, of Laozi's direction of methodological value. Laozi often presents his thoughts with opposite pairs of concepts. Of these oppositions, in most cases, he prefers the negative term, contrary to public opinion. For example, common belief would have it that hardness and strength conquer softness and weakness. However, Laozi repeatedly declares that "the tender and the weak conquer the hard and the strong,"[16] that "keeping to weakness is called strength."[17]

Similarly, people like glory, whiteness, and the masculine, while Laozi wants to keep to humility, darkness, and the feminine.[18] The public pursue the straight, the full, and the new, but Laozi emphasizes the significance of the crooked, the void, and the old.[19] The masses value the noble and the superior, but Laozi claims that their foundation is the ignoble and the subordinate.[20]

All of the aforementioned choices, expressions, and special terms are in accordance with the spirit of *wuwei*: to moderate and adjust human behavior and actions in general. The implications of the examples converge at the opposite points of the prevailing preference, which illustrates Laozi's peculiar, perhaps unique, value orientation. The theory of *wuwei* favors nontraditional and nonprevailing approaches and manners in governmental and social life as well as in individual activities, for a better outcome.

However, Laozi's aforementioned negative preference is by no means the highest value or the final objective. This is of methodological, and not ultimate, purport. As I mentioned at the beginning of this paper and will discuss later, the core value of Laozi's philosophy is naturalness or natural harmony. In the statement "doing nothing and nothing left undone,"[21] Laozi clearly reveals the positive objective of *wuwei*: "doing nothing" is a means, though different from regular ones, to realizing the end of "nothing left undone," namely, everything develops or is accomplished naturally. Therefore, *Wuwei* actually purports superior approaches and consummate results of human activities.

Wuwei represented Laozi's alternative methodology, which was independent of and refreshingly different from the dominating and prevailing values and trends. The theories of *wuwei* embody the spirit of naturalness and are directed toward the realization of natural harmony both among human societies and between humans and nature. *Wuwei* is superior to regular or prevalent manners of human conduct because it is based on comprehensive and humanistic perspectives and consideration, not pushed by fashions and trends for immediate benefits. Therefore, actions following the principle of *wuwei* cause less risk or fewer side effects. And, to the contrary, what induces ecological problems are just the most prevailing trends: industrialization and commercialization or global capitalism. Most nations and regions are involved in almost the same pattern of economic development: prosperity rests on high consumption that stimulates human desire for biological enjoyment and for the satisfaction of vanity.

While developed Western countries have begun to reflect the results of modernization and are beginning to promote environmental protection, Eastern developing countries are struggling to catch up with Western countries' economic standards and life-style through intensified economic competition and development plans. Indonesia is among these developing countries, and the conflagration there was actually the result of the request for rapid economic development. In 1997, Indonesia's President Suharto called for a ban on deliberate burning, but the appeal rang hollow, as many of his government's policies had led directly to large-scale forest clearance. The government is now encouraging plantation owners to double the area of land for oil palms. Large areas of the peat-bog forest of central Kalimantan in Borneo, the spot of the bonfires, have been earmarked for conversion to rice fields, and fire is the quick and cheap way to clear the land. The traditional method of slash-and-burn is too slow to meet the development plans. Since 1995 or 1996, there has been an explosion in land clearing for large-scale agricultural and forestry plantations in Sumatra and Kalimantan.

No one has the right to criticize the ambitions of the Indonesian government to develop their economy intensively and on a large scale. Many developing countries are anxious to catch up with the Western countries. Thus, in the late 1950s in China Chairman Mao launched the "Great Leap Forward" campaign that became a disaster, with at least 30 million peasants dying in the famine it caused. Neither Suharto nor Mao was wrong in pursuing prosperity for their nations. The cause of the tragedies was their belief in the prevalent and simplified formula: the faster the better, the more the better, and the larger the better. To prevent this kind of disaster from happening, the wisdom of *wuwei* may help people to be patient and self-confident, to keep their calmness, patience, and confidence both under great pressure and when challenged by worldwide trends.

Why Is *Wuwei* Important? Theory and Application

Why does *wuwei* matter in Laozi's philosophical system? How does Laozi justify its significance? What is the relationships of *wuwei* to other aspects of Laozi's philosophy? What is implied by these aspects of *wuwei* for today's environmental discussions? To answer these questions, let me illustrate how *wuwei* relates to the Dao, to dialectics, and to *ziran*, respectively.

Dao: The Quasi-metaphysical Model of Wuwei

It is a commonplace that Dao is the highest concept of Laozi's philosophy. As the final source and ground of the universe, Dao is somehow equal to a metaphysical concept in Western philosophy. The difference is that in Chinese philosophy there is no dichotomy between the metaphysical and the physical, the ontological and the axiological, the descriptive and the prescriptive, and so on. Dao runs through the whole universe and human life and is both the transcendent and the immanent. Therefore, as the model for human behavior and as the object of the ultimate concern of human beings, Dao is similar to God. The difference is that Dao has nothing to do with will, feelings, and purposes.

In the traditional version, Laozi literally says: "Dao constantly does nothing (*wuwei*) and nothing is left undone (*wu bu wei*)"[22] and, in the silk version, "Dao has constantly no name." The currently unearthed bamboo slips version reads, "Dao constantly does nothing (*wang*ᵃ*wei*)." Although the words in the versions are not the same, the spirit is identical, because *wuwei* is a cluster of related ideas, including no-name. Obviously, *wuwei* is a characteristic of Dao, and Dao is a model for human beings to practice *wuwei*. Therefore, to practice the principle of *wuwei* simultaneously means to follow the spirit of Dao.

In addition to the literal statements of the connection between Dao and *wuwei*, there is a more profound significance for people to practice *wuwei* according to the features of Dao. Dao is invisible, inaudible, subtle, formless, and infinite; in other words, Dao is beyond the capacity of human cognition and understanding. Western philosophers and theologians established many specific theories and hypotheses about metaphysical issues, such as forms, ideas, and God, but Laozi only offers a vague concept of the general basis for and origin of the universe. Laozi's attitude and theory are rational: there is an obvious coherence of the myriad things in the universe, and something must be the basis or reason for the coherence. We do not know and cannot know exactly what it is or what it looks like. The Dao is just used as a substitute to indicate the ultimate thing we do not know. Therefore, the term Dao intrinsically implies the limitation and finiteness of humankind's comprehension and knowledge. Because of the ultimate truth, the Dao is beyond human capabilities of recognition,[23] and thus human beings should generally be prudent in conduct that may seriously affect the environment.

This is a key difference between the meaning of Dao and God and the scientific spirit. Human beings, as creatures of God, know clearly God's will and are bestowed with the right to manage the universe. In addition, they have the scientific confidence and technological ability to discover and make use of all their surroundings infinitely. However, as the symbol of the origin of and the foundation for the coherence of the world, Dao constantly means the secret part of the universe, of nature, and of human societies, which people can never exhaust. Dao represents forever the unknown final reason of the world surrounding us, reminding human beings of their limitations. As average members of the ten thousand things in the universe, humans have no power to do what they wish without facing unexpected consequences. Therefore, prudent behavior and action, namely, *wuwei*, are important and beneficial.

Most ecological problems we are facing today, such as pollution, resource depletion, and environmental degradation, were unexpected and undesirable. Everything we do may produce unanticipated consequences; therefore, no human beings should be arrogant. In the Indonesian conflagration case, among many things the government did not predict were both the impact of the fires on the weather and El Niño. Indonesia is usually at the center of a zone of intense convection currents, updrafts of hot air that create storm clouds, which is helpful in controlling conflagration. However, during the El Niño years, when winds and ocean currents across the equatorial Pacific are reversed, this pattern breaks down. In 1997, El Niño was the most intense of the century. The result was virtually no rain from May to September and much less rain than normal during the monsoon season from November to March. People wanted to make use of the prolonged drought to set bonfires to clear the land, and the drought intensified the conflagration. All of these were beyond the government's expectations. No one intended the disaster; however, it was still caused by human beings, who are both responsible and blameworthy.

Greenpeace's Alan Pickaver had told Inuit hunters, in the wake of the devastation caused by the seal pelt ban: "You are right to criticize and perhaps be angry with us. The impact of our campaign on indigenous people like you was not intended." We believe he is sincere in saying this. However, too many environmental problems and human tragedies are caused by unintentional activities. Thus, we should not be satisfied with our good wishes and use them as excuses for our mistakes. While there are lessons to be learned from the accidents, we would do better to draw some wisdom from Daoism: be more considerate in general actions, including protection movements. Right directions and good intentions don't always produce good outcomes. Thus, the spirit of *wuwei* is generally a valuable guide.

Dialectics: The Experiential Foundation of Wuwei

The proposition "doing nothing and nothing left undone" expresses Laozi's dialectical theory. There is a rich tradition of dialectics in the history of Chinese philosophy. Needless to say, the dialectics tradition in Chinese philosophy has nothing to do with Hegelianism or any specific Western theories. It just indicates general doctrines dealing with the unity and transformation of the pairs of contradictions. Laozi is one of the most important dialecticians of ancient China. Chinese traditional dialectics is quite different from modern Marxist theory, because it did not claim the importance and inevitability of struggle, as do Leninism and Maoism.

Laozi's theories of dialectics may be summarized by some patterns dealing with the relations between the two parts in a contradictory pair, such as being and non-being, calamity and happiness. We will briefly discuss two patterns here. The first is the interdependence between opposite things and concepts. For example, "Calamity is that upon which happiness depends; happiness is that in which calamity is latent."[24] More examples are given in chapter 2:

> The whole world recognizes the beautiful as the beautiful,
>   yet this is only the ugly;
> The whole world recognizes the good as the good,
>   yet this is only the bad.
> Thus being and non-being produce each other;
> Difficult and easy complete each other . . .
> Front and back follow each other. . . .[25]

These illustrate the interdependence of beauty and ugliness, the good and the bad, the difficult and the easy, the front and the back, and so on. It is impossible for us to choose only one aspect of them and abandon the other aspect. In the long run, and on a large scale, we cannot expect a perfect outcome of our activities fully in accordance with our intentions and without any side effects.

The second pattern is the reversibility of opposite sides, such as "the correct could become the perverse, and the good may become evil."[26] In addition, "Bowed down then preserved; Bent then straight; Hollow then full; Worn then new; A little then benefited; A lot then perplexed."[27] Furthermore:

> He does not show himself; therefore, he is luminous.
> He does not justify himself; therefore he becomes prominent.

He does not boast of himself; therefore he is given credit.
He does not brag; therefore he can endure for long.[28]

Things are in change and in reversal. Thus, humility produces greatness, and ambitions bring about failure. The obverse side and the reverse side often exchange their positions. Happiness may become calamity; calamity in turn may induce good luck, like the story of the
old man who lost his horse.[29] Things both in human life and in the
natural world often change to become the opposite of our expectation
and will. That high speed economic development causes environmental degradation is only one, typical example.

Things may develop into opposites that we did not predict and do
not want. This is another argument that *wuwei* is reasonable. The fire
in Indonesia is a typical example that good intentions and ambition
can be unexpectedly reversed to become disaster.

In 1995, Suharto announced plans to clear one million hectares of
forested peat bog in central Kalimantan and turn it into rice field. Peat
is highly organic soil containing more than 50 percent combustible
decayed vegetable matter. It can be cut and then dried for use as fuel.
The peat in these bogs is ten thousand years old—dating from the end
of the last Ice Age—and is twenty meters deep in many places. These
massive stores of organic material can burn for years, producing much
more smoke than a conventional forest fire. Thus, in 1997 fire became
an international accident, more serious than the better-known smoke
pollution resulting from the Kuwait oil fires of 1991. Following the
fire, most of the central Kalimantan peat forest was burned and black,
but, what is more worrying, the fires still smolder beneath the surface,
impossible to extinguish. No one predicted or warned of this result,
the reverse of what was intended. However, this offers no excuse for
or resolution of the disaster.

In the Inuit case, the intended protection of animals and the environmental equilibrium becomes instead the destruction of not only the
Inuit people's way of life but also of their whole community. The lesson here is that environmental balance as a whole includes mutually
influencing aspects. If we focus on only one aspect, without a comprehensive vision, unexpected reversibility and undesirable consequence
will surely happen. Therefore, human beings should be more considerate and take into account the unexpected reversibility of intentions
and consequences.

Ziran: *The Objective of the Principle of* Wuwei

The concept of *wuwei* is also important because it is the essential means to actualize the core value of Laozi's philosophy, namely, naturalness, or *ziran*. Laozi contends in chapter 25:

> Man models himself on the Earth;
> The Earth models itself on Heaven;
> Heaven models itself on the Tao;
> And the Tao models itself on naturalness (*ziran*).[30]

What is noteworthy here is the position of humans in the world. Both traditional Daoists and Confucians believe that humans are generally inferior to Earth and Heaven, which are the symbols of the natural world and the universe. The theory that humankind should conquer nature was never mainstream thinking in classical China. In fact, Heaven, Earth, and their phenomena are often used analogously as the philosophical foundation of theories about human life, moral and political. This is diametrically opposed to the Western Christian tradition criticized by Lynn White, Jr.[31]

The key point of the above quoted chapter is the last sentence: Dao models itself on naturalness (*ziran*). This sentence explicitly states the importance of naturalness as the core value of Laozi's philosophy. Humankind, Earth, Heaven, and the Dao—all of them follow the principle of naturalness. However, that Earth, Heaven, and the Dao follow naturalness should be understood only anthropomorphically, because Laozi does not think of them with will and purpose. The true meaning and message of the sentence is that humans should practice the principle of naturalness. Here, naturalness is just a token of *ziran* that indicates the principle and the state of natural harmony of human societies and of the whole universe.

Natural harmony and natural order are especially valuable in comparison with forced order or chaos. Human nature prefers natural order to forced order. Needless to say, few people like the chaotic state. However, naturalness as the highest ideal could be actualized only through the practice of *wuwei*. *Wuwei* finds its significance in the value of naturalness. Laozi argues in chapter 64:

> He who takes an action fails.
> He who grasps things loses them.

> For this reason the sage takes no action (*wuwei*) and therefore
> does not fail.
> He grasps nothing and therefore he does not lose anything. . . .
> The sage desires to have no desire.
> He does not value rare treasures,
> He learns to be unlearned, and returns to what the multitude has
> missed (Dao).
> Thus he supports the naturalness of ten thousand things
> and dares not take any action.[32]

The first half of the quotation states that the advantage of *wuwei* is to prevent negative consequences from happening. The second half introduces the model of *wuwei*, namely, the sage's actions as non-action, such as desiring no desire, valuing no treasures, and daring not to take any action to interrupt the naturalness of the ten thousand things. The last sentence reveals the positive purpose and advantage of *wuwei*: to support and help all things in their natural development. The natural development of the ten thousand things indicates the harmonious relationship of all creatures in the universe. This is the essence of naturalness and the ideal of Laozi's philosophy. Obviously, this is also an ideal ecological environment for modern people.

While *wuwei* may lead us into natural harmony, its opposite, taking reckless or aggressive actions, often damages the natural order and atmosphere. In the Indonesian case, although indigenous people practiced slash-and-burn cultivation for centuries, they lived in harmony with the rain forest. According to local forest scientists, burning is a legitimate land management practice. Fire eliminates field debris, slows regrowth of weeds, reduces pest and disease problems, adds fertilizer in the form of ash, and loosens the soil to make planting easier. Burning didn't cause environmental problems, because the operations were divergent and often shifted from area to area. When people later returned to the same place, the forest in that area had already recovered. However, from the 1950s to the 1980s, the government moved 730,000 families into the rain forest according to the transmigration plan. New immigrants neither knew nor practiced the conventional slash-and-burn cultivation, and they set bonfires to clear the land with no misgivings. Moreover, whereas traditional shifting cultivators burned one hectare, modern plantation owners burn 1,000 hectares at a time. Since the transmigration plan was implemented, the area of forest has been reduced by 250,000 hectares each year.[33]

In the Inuit case, seal hunters kill seals solely to meet essential needs to maintain their normal life. They don't kill seals for profit accumulation, and neither do they brag about or waste their capture. Thus, indigenous folk enjoyed a harmonious environment. It is the dramatized propaganda of Greenpeace that has ruined the harmony between the Inuit community and the natural environment. One of the lessons derived from the Inuit story is that preservation movements themselves may destroy the natural harmony and cause a new imbalance between human beings and their natural surroundings, if the movements operate in an unnatural way. In this sense, the spirit of *wuwei* is also significant as a curative as well as preventative strategy.

How Is *Wuwei* to Be Applied? Approaches and Implications

For Daoist sages, *wuwei* is a natural practice and the performance of life. For true Daoists, there is no self-restraint or self-control. However, the situation is different if we try to promote the spirit of *wuwei* in modern societies concerned about environmental protection and improvement. For average people, trying to follow the principle of *wuwei* and to realize natural harmony, self-restraint is necessary. Therefore, for them, *wuwei* may be interpreted as the self-restraint of behaviors and actions. The first aspect is external restraint to prevent interruptions and to protect individual independence and subjectivity. The second aspect is internal restraint to moderate and control selfish desires and emotions that are often the root of unnatural behavior and conflicts. Self-restraint, especially internal restraint, is essential for average people to practice *wuwei*. Nevertheless, *wuwei* does not actually mean purely passive or negative attitudes, but a high standard or criterion, which means the most appropriate behavior.

Wuwei: *Restraints of External Actions*

Most discussions of *wuwei* concern its external aspect, namely, the restraint on external activities, such as competition and wars. In the *Laozi*, external *wuwei* is often mentioned as no-competition (*bu zheng*) and no-advance (*bu xian*). Chapter 68 reads:

> One who excels as a warrior does not appear formidable;
> One who excels in fighting is never roused in anger;
> One who excels in defeating his enemy does not join issue;
> One who excels in employing others humbles himself before
>   them.
> This is known as the virtue of non-contention;
> This is known as making use of the efforts of others;
> This is known as matching the sublimity of heaven.[34]

All of these—not formidable, not angry, not joining, being humble—
are external aspects of *wuwei*. This chapter of the *Laozi* suggests that
wuwei by no means indicates doing nothing, but indicates external
restraint; thus, it could be applied to everything, even in wars. The
spirit of these performances is summarized in the expression "the virtue of non-contention" (*bu zheng zhi de*), which matches the natural
perfection of Heaven. Laozi's "three treasures" also embody the spirit
of *wuwei*. We read in chapter 67:

> I have three treasures. Guard and keep them:
> The first is deep love (*ci*),
> The second is frugality (*jian*),
> And the third is not to dare to be ahead of the world.[35]

"Deep love" is a translation of the Chinese word *ci*, which has also
been translated as compassion, gentleness, motherly love, commis-
eration, pity, or love.[36] All of these words are, in a sense, right for *ci*.
However, while the basic meaning is love, *ci* is deeper, gentler, and
broader than love. The meaning of *ci* is, interestingly, quite close to
A. H. Maslow's concept of "love knowledge" or the "Daoistic objectivity" that permits people to unfold, to open up, to drop their de-
fenses, to let themselves be naked not only physically but psychologically and spiritually.[37] In short, it leaves people who are loved
uninterrupted. Thus, *ci* means doing little toward people because of
deep love. This is in accordance with the principle of *wuwei*. The second treasure, frugality, suggests that people do as little as possible for
their personal lives, also matching the meaning of *wuwei*. The third
one, "not daring to be ahead of the world," is the most important point
in the spirit of *wuwei,* implying "not to compete for the first," "not to
claim to be the first even if you are indeed the first." In short, it is "no
competing," "no contending," nonaggressiveness.

Indeed, if people, following the model of the Daoist sage, had not competed so much and so intensely, they would have suffered fewer disasters, and the environment would have been better preserved. In the Indonesian case, the government wanted the country's economic development to catch up with that of developed countries as soon as possible. There is nothing wrong with this goal. However, when more and more non-Western countries are involved in the globalized economic competition for so-called modernization, developing countries have no chance or no patience to develop gradually, comprehensively, and naturally. This is the main reason that environments in Asia are degrading rapidly and natural resources are depleting speedily.

In the Inuit case, Greenpeace wanted to do better than others and wanted to be noticed by the world. They competed to protect animals effectively. They forgot the higher purpose of animal protection: the equilibrium between humans and nature. It is easier for people to become blind to the higher objective and the comprehensive picture in a competitive situation than in a natural environment.

Wuwei: *Restraints of Internal Actions*

In addition to restraining overt actions, the concept of *wuwei* also indicates a restriction of covert activities, such as desire, intention, vanity, and ambition. Laozi repeatedly mentions that sages claim nothing for their function. For example, Laozi describes the style of the sage:

> The sage manages affairs without action (*wuwei*),
> And spreads doctrines without words.
> He lets all things arise, but claims no authority.
> He creates the myriad, but claims no possession.
> He accomplishes his task, but claims no credit.
> It is precisely because he does not claim credit
>   that his accomplishment remains with him forever.[38]

The sage practicing *wuwei* claims no authority, no possession, and no credit although he creates and protects the natural atmosphere for the myriad things. He claims no credit, thus the credit he has earned or his accomplishment will never be challenged. In reality, when projects are accomplished, many conflicts happen because someone wants to claim more merit than others think he or she deserves. Those who practice *wuwei* claim no credit or merit because they have no desire for it. In the final analysis, many disasters, including environment crises, relate to human desires. Thus, Laozi declares:

> There is no crime greater than lavish desires,
> There is no disaster greater than discontentment,
> There is no misfortune greater than greed.
> Hence in being content, one will be always contented.[39]

Desire, in a normal situation, is natural or reasonable and would not cause any trouble. However, if one's desire becomes too much, and even infinite, it will certainly bring about dissatisfaction. If he or she cannot control it, it may induce miserable consequences.

In Indonesia, plantation owners continue to set bonfires without any misgivings, even after the president's ban, because they want to gain more profits through their business. The selfish desire for wealth and success is the main impetus for setting the fires. In premodern societies, people did not need too much food, meat, and cotton. The accumulation of wealth was limited. In the modern financial system, wealth, as numbers in bank accounts, accumulates rapidly and infinitely, stimulating people's infinite desires. In the Inuit case, it seems that success and popularity that may introduce more donations are more important to Greenpeace than the equilibrium between humans and nature. For Greenpeace, protecting the environment becomes a means to actualize their desires.

Wuwei: *The Most Appropriate Actions*

To sum up, as above demonstrated, *wuwei* as external and internal restraints means better results, not pure negating of all actions. Although non-action, or *wuwei*, comes in a negative form, its meaning is highly positive: it aims at a higher standard and outcome of human actions. Its true spirit is to pursue appropriate actions and consummate ends. Needless to say, the criteria of appropriateness and consummation depend on specific circumstances and situations. There is no fixed and miraculous equation or formula of *wuwei*.

Regularly, an action is an action without special suggestions. However, non-action, or actions without action, is different from general actions. It is qualified action toward a higher criterion. In the famous statement "doing nothing and nothing left undone," *wuwei*, or doing nothing, is only a means for a consummate outcome. Laozi contends in chapter 48:

> Doing nothing, nothing will remain undone.
> Taking over the world: only by not working.
> A person who sets to working,
> Doesn't have what leads him to gain the world.[40]

Laozi confidently believes that, to realize a purpose, non-action is better than regular actions. It seems impossible for laymen to understand that one can win a state without struggles or wars. However, we find instances in the history of *Spring and Autumn:* some states gained other states or territories without conflicts because these states voluntarily wanted to join the better states.[41] Obviously, to win the world without struggles is the best manner, though it is not easy. For Daoists, it is not worthy to conquer a country by wars and killing. If one wins a war, one should "treat the occasion like a funeral ceremony," because too many people are killed.[42] The concept of *wuwei* pursues a high standard of human behavior and actions through unconventional approaches. Thus, Laozi further argues:

> The large state is like the lower part of a river,
> It is the feminine of the world;
> It is the convergent point of the world.
> The feminine constantly overcomes the masculine with tran-
>   quillity.
> Because she is tranquil, therefore she is fittingly underneath.
> The large state could take over the small state
>   if it is below the small state.[43]

The below, the underneath, the feminine, the tranquil, all of these are the opposite or the negative side of the common preference, which is in line with "doing nothing," or *wuwei*. However, Laozi believes that these seemingly negative features can conquer the obverse side and win the benefit, which means "nothing left undone," or *wu bu wei*. *Wuwei* means better approaches and results. A large state takes over a small state in a humble way: is this not much better than military victory? A female overcomes a male through tranquillity: is this not better than fighting and struggling?

Commonly, people superficially compare the opposites and take into account only the immediate impression: thus, a stone is harder and stronger than water. However, after a long time, water drills and crushes stones. Air or wind is soft, yet it penetrates everything and ruins iron and concrete. Thus, we have the saying from Laozi: the soft overcomes the hard, and the weak conquers the strong. The soft and the weak seem to do nothing, but they can actually conquer the hard and the strong. *Wuwei*, as a special way to deal with difficulties, is more effective and comes with not as high a price; hence, it is better and higher than general actions.

Therefore, *wuwei* is actually a positive doctrine pursuing the optimum or consummate outcome of human activities. To render it into modern words, *wuwei* means the balance between minimal effort and best result. In an ideal society, according to Laozi, people enjoy their free life and hardly feel the influence of sages and need not even praise them.[44] The duty of Daoist sages is only to help the ten thousand things to develop their naturalness.[45] This is not a utopian but an ideal principle of doing less and gaining more of what is better. To prepare for the worst, the spirit of *wuwei* means actions with less risk or fewer side effects. Therefore, *wuwei* is by no means passivity, pessimism, or escapism. In fact, to believe in natural harmony and to practice *wuwei* requires great courage, confidence, and wisdom.

Even if we could not reach the best result, *wuwei* is still helpful. The bottom line of the significance of *wuwei* is to avoid serious mistakes and disasters: no single human being makes no mistakes. With higher position and more power, one may make more serious mistakes and induce greater tragedies. Therefore, the wisdom of *wuwei* is more important for leaders than for average people. This is the reason Laozi always uses sages as agents of *wuwei*. From this, an easy inference is that a higher and more powerful position demands more comprehensive consideration and more prudent decisions. Unfortunately, the reality is often the opposite: high position often leads to rash decisions and then to disasters on a large scale. For example, Mao once decided to take agriculture as the first task, and the Party encouraged the whole country to reclaim farmland from pastures and lakes. However, the movement resulted in one million hectares of deserts and bogs. Suharto wanted to double the area of land for oil palms, to 5.5 million hectares by the year 2000, which would involve clearing an area of forest equal in size to Wales. In the Inuit case, only powerful persons could dispatch congressmen, politicians, and actresses and place pressure on the European parliament.

Generally, human beings make two kinds of mistakes. One is when they don't make enough effort, and the other is caused by overdoing. The former mistake is easy to remedy because it does not waste too many resources. The second one is much more difficult to correct, as in the examples of the Cultural Revolution in China, the conflagration in Indonesia, and the tragedy in the Inuit community. *Wuwei* is a good principle to follow to avoid irreparable mistakes and to pursue the consummate outcome.

To conclude, if we understand its essential meaning and can apply its wisdom to modern issues, *wuwei* could be positive and significant in helping with environmental protection and improvement.

Notes

1. In addition to other pioneering works, I have to mention *Nature in Asian Traditions of Thought*, ed. J. Baird Callicott and Roger T. Ames (Albany: State University of New York Press, 1989), and the special issue of *Philosophy East and West* 37, no. 2 (April 1987). This paper is greatly inspired by those works, though I don't quote them.

2. I wish to thank Professor Jessie Poon of the department of geography, the State University of New York at Buffalo. She has provided me with the main part of the materials on the Indonesia conflagration that I quote in this paper.

3. On the comprehensive examination and interpretation of Laozi's philosophy, see Liu Xiaogan, *Laozi: Niandai xinkao yu sixiang xinquan* (A New Investigation of the Date and Thought of Laozi) (Taipei: Great East Book Co, 1997).

4. In his response to the draft of this paper, James Miller raised the question whether *wuwei* is helpful as a curative as well as preventative strategy. There is not space here to elaborate upon my thinking. Briefly, my case of the Indonesian rain forest fire argues for *wuwei* as a preventative strategy; while the case of the Inuit community suggests that the spirit of *wuwei* is also significant in the process of correcting the ecological tragedy. I hope to discuss these questions in the near future. I appreciate all comments and questions from participants of the conference on Daoism and ecology, as well as those of the readers.

5. On the reconstruction of Laozi's philosophy, please see Liu, *Laozi*. On studies in English on the concept of *wuwei*, I would like to mention Roger Ames, "Putting the Te Back into Taoism," in *Nature in Asian Traditions,* ed. Callicott and Ames, 113–44; and "Taoism and the Nature of Nature," *Environmental Ethics* 8, no. 4 (1986). See also Roger T. Ames, *The Art of Rulership* (Albany: State University of New York Press, 1994): 28–64. On the development of the concept of *wuwei*, see Liu Xiaogan, "*Wuwei* (Non-action): From Laozi to Huainanzi," *Taoist Resources* 3, no. 1 (1991): 41–56.

I use the prefix "quasi" before "metaphysical" because Dao runs across the realms of physics and metaphysics. Chinese philosophy differs in this way from Western traditions. Similarly, I use the term "axiological," or "normative" in this paper in the pure sense of what one ought to do, not necessarily indicating moral behavior.

6. According to Professor Pang, the first is *wang*[a], which suggests that the state of something has disappeared or someone has escaped. Sometimes it also suggests the state before the emergence of something or someone. This was the earliest form of *wu* and was found in oracle bone inscriptions of the Shang dynasty (ca. sixteenth–eleventh century B.C.E.) The second is *wu*[c] (dance, borrowed as *wu*[b] before its coinage), which indicates a medium who is dancing as a means to communicate with spirits or gods. Thus, the word suggests something that exists but cannot be seen, touched, or heard by human beings. From this, the second meaning of *wu*[b] is explained as "there-essentially-is" looking like "there-is-not," or "being as non-being" *(you er si wu)*. The character *wu*[b] was also found in oracle bone inscriptions and in bronze inscriptions of the Zhou dynasty (eleventh century–221 B.C.E.). The third is another *wu*[a], which implies absolute "non-being" or "there-is-not." This meaning of *wu* is philosophical and appeared in *Mohist Classics (Mo jing)* in the later Warring States period.

The author of the *Mohist Classics* defined its meaning as "non-being-after-non-being" (*wu zhi er wu*), compared with "non-being-after-being" (*you er hou wu*). See Pang Pu, "Shuo wu" (On Nothingness), in *Zhongguo wenhua yu zhongguo zhexue* (Chinese Culture and Chinese Philosophy), ed. Institute of Sinology of Shenzhen University (Beijing: Dongfang chubanshe, 1986), 62–74. See also his new book, *Yi fen wei san* (One Divides into Three) (Shenzhen: Haitian chubanshe), 270–83.

7. *Daode jing* 48, 57. See Liu Xiaogan, "On the Special Meaning of Ch'ü in the Tao-te-ching," *Hanxue Yanjiu* (Chinese Studies) 18, no. 1 (2000): 23–32.

8. *Daode jing* 17, as well as 2, 9, 77.

9. Data and information about the Indonesian rain forest fire are based on "Indonesia Ablaze: The Forests Are Burning Again," *Straits Times*, 29 March 1998, 40; Dominic Nathan, "The Ball Is in Jakarta's Court," *Straits Times*, 4 April 1998, 56; Janet Nichol, "Bioclimatic Impacts of the 1994 Smoke Haze Event in Southeast Asia," *Atmospheric Environment* 31, no. 8 (1997): 1209–19; and Yanchun Guo, "Ranshao de diping xian" (A Burning Horizon), *Lianhe Zaobao*, 13 April 1998, 12.

10. Other examples: no-claim (*fo ju*), no-desire (*wu yu*), no-selfishness (*wu si*), no-body (*wu shen*), no-establishment (*bu li*), no-bragging (*wu jing*), no-boasting (*wu fa*), not-conceit (*wu jiao*), no-engagement (*wu shi*), no-mind (*wu xin*), no-taste (*wu wei*), no-learning (*bu xue*), no-violence (*bu wu*), not-angry (*bu nu*), and no-daring (*bu gan*).

11. *Wuwei* occurs nine times in seven chapters in the Mawangdui silk version and seven times in six chapters in versions A and B of the recently published Guodian bamboo slips. However, the number is not very significant in bamboo versions because they consist of a total of only 2,000 characters, far less than the 5,000 in the silk and current versions. Among the three versions of the bamboo slips, *wuwei* is always written as *wang[a]wei* in versions A and B. In version C, in the repeated chapter 64, *wuwei* is written as *wu[b]wei*.

12. For example, "By acting without action (*wuwei*)"(*Daode jing* 3, 63); "I know the advantage of taking no action (*wuwei*) . . . few people in the world know the advantage of taking no action (*wuwei*)" (43). In these contexts, *wuwei* is used as a subject or an object. See Liu, *Laozi*, 113–14.

13. Donald J. Munro, *The Concept of Man in Contemporary China* (Ann Arbor: University of Michigan Press, 1977), 26–56.

14. Laozi, like most classical Chinese thinkers, doesn't define his concepts specifically and clearly. The ambiguity of Chinese philosophical terms has both advantages and disadvantages. It is disadvantageous because its meaning is uncertain and difficult for modern people to learn and apply. It is advantageous because it is not a pre-fixed formula or a prescription that leads people to use it without considering the varied conditions. Differing from scientific and technological formularization, *wuwei* demands subjectivity and wisdom instead of obedience and skill, thus preventing people from becoming slaves of mechanization and computerization.

15. John Dyson, "The Seal Hunter's Tale," *Reader's Digest*, January 1997, 68–74. I analyze this case not to criticize Greenpeace, but to use the case as an example to discuss the relevance and significance of Daoist wisdom in modern ecological issues. I respect the general efforts of Greenpeace in environmental protection.

16. *Daode jing* 36.

17. *Daode jing* 52. In addition, "the weak overcomes the strong and the soft overcomes the hard" (78), and "the most submissive thing can ride roughshod over the hardest" (43). While most people believe that the male always overcomes the female because of the male's dynamism, Laozi believes that "the female often overcomes the male with tranquillity"(61). While average people favor courage (*yong*), magnanimity (*guang*), and advance (*xian*), Laozi embraces benevolence (*ci*), frugality (*jian*), and retreat (*hou*) as "three treasures" (67).

18. *Daode jing* 28.

19. *Daode jing* 22.

20. *Daode jing* 39.

21. "Doing nothing and nothing left undone" is my paraphrase of "*wuwei er wu buwei*," for better understanding its essential meaning. The sentence appears twice in current versions (*Daode jing* 37, 48), and once in the bamboo versions (48). Literally, there is no such sentence in the silk versions; however, it may occur in chapter 48, according to the damaged space and corresponding part in the silk version. One of the readers of an earlier version of my paper suggested the sentence should be literally translated as "The Dao constantly does not act (do), but there is nothing it does not do," in order to keep both "*wei*" in the active mode. Though this is acceptable, its philosophical meaning is not clear. According to my study, "*wuwei*" is the principle of action, "*wu buwei*" is the desirable result. The two Chinese words are not necessarily in the same mode.

22. *Daode jing* 37.

23. Human beings may reach Dao through transcendent experience, which needs special self-cultivation.

24. *Daode jing* 58; *A Source Book in Chinese Philosophy*, trans. Wing-tsit Chan (Princeton, N.J.: Princeton University Press, 1963), 167. In this paper, my translations of the *Laozi* are chosen and adapted from: Chan, *Source Book*; D. C. Lau, *Lao Tzu Tao Te Ching* (Harmondsworth, England: Penguin Books, 1987); Robert G. Henricks, *Lao-Tzu Te-Tao Ching* (London: Rider, 1991); or Michael LaFargue, *Tao and Method: A Reasoned Approach to the Tao Te Ching* (Albany: State University of New York Press, 1994), according to my own understanding and the context.

25. Lau, *Tao Te Ching*, 58.

26. *Daode jing* 58; trans. Chan, *Source Book*, 167.

27. *Daode jing* 22; trans. Lau, *Tao Te Ching*, 79.

28. *Daode jing* 22; trans. Chan, *Source Book*, 151.

29. *Huainanzi* 18.

30. Henricks, *Lao-Tzu*, 77.

31. Lynn White, Jr., "The Historical Roots of Our Ecologic Crisis," *Science* 155 (March 1967): 1203–07.

32. Chan, *Source Book*, 170.

33. Guo Yanchun, "Ranshao de diping xian."

34. Lau, *Tao Te Ching*, 130.

35. Chan, *Source Book*, 171.

36. Translations are from Lau, *Tao Te Ching*; Henricks, *Lao-Tzu*; LaFargue, *Tao and Method*; Ellen Chen, *The Tao Te Ching: A New Translation with Commentary* (New York: Paragon House, 1989); Chan, *Source Book*; Arthur Waley, *The Way and*

Its Power (New York: Grove Press, 1958); and Lin Yutang, *The Wisdom of Laotse* (Taiwan: Confucius Publishing Co., 1994).

37. Abraham H. Maslow, *The Farther Reaches of Human Nature* (New York: Penguin Arkana, 1993), 17.

38. *Daode jing* 2; trans. Chan, *Source Book*, 140.

39. *Daode jing* 46; trans. Lau, *Tao Te Ching*, 107.

40. LaFargue, *Tao and Method*, 398.

41. Liu, *Laozi*, 133–34; and Liu, "On the Special Meaning of Ch'ü."

42. *Daode jing* 31.

43. *Daode jing* 61; trans. Henricks, *Lao-Tzu*, 144.

44. *Daode jing* 17.

45. *Daode jing* 64.

Sectional Discussion: What Are the Speculative Implications of Early Daoist Texts for an Environmental Ethics?

RUSSELL B. GOODMAN with JAMES MILLER

The focus of sections two and three was ecological theory and practice in Daoist religious texts and cultural contexts. In section four we have turned to the area in which Daoism and ecology have in the past been most commonly related, that is, in discovering an environmental ethic in classical Daoist philosophy. There is a serious hermeneutical question at stake here: How can we derive an environmental ethic from ancient Daoist texts? All of the papers in this section have taken this hermeneutical problem extremely seriously, but the ways in which they have chosen to deal with it fall into two basic categories. David Hall and Roger Ames have followed the route of philosophical reconstruction as a means to explaining the meaning of these ancient texts. They claim that our disastrous modern attitude toward the environment is itself symptomatic of a broader crisis in Western philosophy that prevents us from understanding these texts properly. David Hall argues that the study of Daoist philosophy involves learning to value "deference" over "reference." The implications for environmental ethics of this philosophical shift are readily apparent.

On the other hand, the papers by Liu Xiaogan, Russell Kirkland, and Lisa Raphals have taken an alternative mode of speculation by considering what sort of ethical principles relevant to the contemporary situation we can reasonably extrapolate from these ancient texts. Lisa Raphals is right to sound a skeptical note with regard to this speculative task: we cannot know for certain what Zhuangzi's attitude would be to the destruction of the whooping crane. But the task of this brief response is less demanding: to refine some ideas derivable from these early Daoist texts that will actually help us to formulate an environmental ethics.

Aesthetics and Pluralism in Classical Daoism and American Philosophy

David Hall[1] characterizes the Daoists as having an aesthetic understanding of the natural world, and he is correct, though it is necessary to note the obvious at the outset, that "the aesthetic" is a Western, not a Chinese, let alone a Daoist, category. The European notion of the aesthetic has its origins in Kant's *Critique of Judgment*, where Kant writes:

> by an aesthetical idea I understand that representation of the imagination which occasions much thought, without however any definite thought, i.e., any *concept*, being capable of being adequate to it; it consequently cannot be completely compassed and made intelligible by language.[2]

This passage is an origin point both for Hall's idea that "from the aesthetic perspective," the world is in some sense "noncoherent," and for his explorations of "allusive metaphors." It is helpful at this point to enter into the discussion certain passages from Emerson's "The Poet," both because of their role in the development of aesthetic theory and because of their relevance to Hall's central theme of deference. As regards the former, Emerson falls into the Romantic tradition heralded by Kant, emphasizing imagination, reconstruction, and process. For Emerson, the fundamental value is the active soul, the stimulus and sense of possibility provided by a metaphor more valuable than its settled interpretation. Emerson's poet allows for and participates in the imaginative transformation of interpretations, while what he calls the "mystic" fixates on one meaning only. "[T]he quality of the imagination," Emerson writes,

> is to flow, and not to freeze. . . . Here is the difference between the poet and the mystic, that the last nails a symbol to one sense, which is a true

sense for a moment, but soon becomes old and false. For all symbols are fluxional; all language is vehicular and transitive, and is good, as ferries and horses are, for conveyance, not as farms and houses are, for homestead.[3]

How typical of an American, but not untypical of certain Daoists, to think of transportation and movement as the basic metaphor for life, including our life with language. Zhuangzi would have understood Emerson's idea of "a true sense for a moment" which then "becomes old and false."

Now as to deference, consider Emerson's statement in "The Poet" that a poem is "a thought so passionate and alive, that, like the spirit of a plant or an animal, it has an architecture of its own, and adorns nature with a new thing."[4] The "architecture of its own" that Emerson finds in the poem is akin to the objects of deference Hall finds in the Daoist view of nature. One may defer by recognizing the "thisness," "suchness," or "individuality" of something; but also more actively, by entrusting oneself to it, as a boat to a current. Emerson's word for this more active form of deference is "abandonment," by which he means something akin to Daoist ideas of letting go and giving up attempts to control things. Emerson writes:

It is a secret which every intellectual man quickly learns, that, beyond the energy of his possessed and conscious intellect, he is capable of a new energy . . . by abandonment to the nature of things; that, beside his privacy of power as an individual man, there is a great public power, on which he can draw. . . . The poet knows that he speaks adequately, then only when he speaks somewhat wildly, or, "with the flower of the mind;" not with the intellect, used as an organ, but with the intellect released from all service, and suffered to take its direction from its celestial life; . . . As the traveler who has lost his way, throws his reins on his horse's neck, and trusts to the instinct of the animal to find his road, so must we do with the divine animal who carries us through this world.[5]

One cannot but think of Zhuangzi here, of his story of Liezi riding the wind, for example; nor can one help considering an imaginary history of philosophy, or history of imaginary philosophy, in which Emerson had read the *Daode jing* and the *Zhuangzi*, just as he read the *Mengzi*, the *Bhagavad Gītā* and the *Viṣṇu Purāṇa*.[6]

Another imaginary intersection between the American philosophical tradition and classical Daoist philosophy arises when considering the opposition between the coherent and the incoherent. In *Experience and Nature*, John Dewey writes: "We live in a world which is an impressive and irresistible mixture of sufficiencies, tight completenesses, order, recurrences which make possible prediction and control, and singularities, ambiguities, uncertain possibilities, processes going on to consequences as yet indeterminate."[7] In Hall's terms, the world is a mixture of coherencies and incoherencies; it is a "pluriverse," as William James says, with no one perspective or "take" on the world adequate to its totality. In a real sense it does not form a totality. I think Zhuangzi has something to say about this: "Everything has its 'that,' everything has its 'this.' From the point of view of 'that' you cannot see it, but through understanding you can know it. So I say, 'that' comes out of 'this' and 'this' depends on 'that' which is to say that 'this' and 'that' give birth to each other."[8] Yet for Zhuangzi, the sage reaches a "state in which 'this' and 'that' no longer find their opposites... called the hinge of the Way."

What would such a state be like? And what is the proper deference to the thises and thats of this world? As a contribution to the discussion I submit my final entry from the American tradition, an anecdote from William James's essay "On a Certain Blindness in Human Beings:"

> Coming into a man-made clearing in the beautiful and refreshing woods, he is horrified at the destruction of forest life. Then he speaks to the homesteader who has ravaged the pristine forest: I instantly felt that I had been losing the whole inward significance of the situation. Because to me the clearing spoke of naught but denudation, I thought that to those whose sturdy arms and obedient axes had made them they could tell no other story. But, when they looked on the hideous stumps, what they thought of was personal victory. The chips, the girdled trees, and the vile split rails spoke of honest sweat, persistent toil, and final reward. . . . I had been as blind to the peculiar ideality of their conditions as they certainly would also have been to the ideality of mine, had they had a peep at my strange indoor academic ways of life in Cambridge.[9]

This amusing and profound story points to what one might call the conundrum of tolerance: to what degree can we tolerate those whose pursuits of happiness restrict our own possibilities? Can we defer to all? Daoist pluralism is compatible with the most acute responses to the natural world. Think of Zhuangzi's Cook Ding, able to find the spaces between the joints of every unique animal with which he is presented, and of the natural observations scattered throughout his wild and "reckless" verbal wanderings: "When the tailorbird builds her nest in the deep wood, she uses no more than one branch. When the mole drinks at the river, he takes no more than a bellyful."[10] There is no totality of coherence here, but a plurality of particular, precise coherencies.

Roger Ames presents an interpretation of this plurality of particular coherencies (*dao*s) based on the *Daode jing*. The world is not to be considered as a transcendent and independent universe, in the sense of implying a unity and univocity of meaning for all its constituent parts: there is no universe of objects all tending toward a coherent teleological goal. But the multiple orders of the world return to the Dao, just as the Dao gives birth to the myriad things. Daoist philosophy thus holds the promise for an environmental ethics based on attention, deference, and the local, but at the same time on the plurality of interacting ecological systems (*dao*s) that together constitute our evolving "pluriverse."

The Ethical Implications of Deference

Liu Xiaogan, Russel Kirkland, and Lisa Raphals attempt to describe what the ethics of deference might actually look like. Liu Xiaogan notes that human beings make two kinds of mistakes: "not enough" and "too much." "Not enough" is difficult to prevent and easy to rectify. "Too much" is difficult to rectify and easy to prevent. It is quite clear that the bias of modern science and technology is toward the problem of rectification and not prevention. Although AIDS is simple to prevent and terrifying to manage, billions of dollars are spent on the search for a cure rather than on strategies for prevention. And although the global ecological crisis is fundamentally a problem of too many people in too little space, billions of dollars are spent on genetically engineering crops to maximize yields rather than on strategies for reducing the population. The direct result of both these policies is the deaths of millions of children each year. The "deferential" answer to all of this might be the condom!

Russell Kirkland argues in his interpretation of "deference" in the *Zhuangzi* that there can be no conscientious moral act toward another or the world that is not harmful. Any deliberate act interferes with the benign natural transformation of things. This transformation includes death, but there is a difference between the natural death of an old woman and the unnatural death of a child from AIDS or starvation. Shall we speak of the natural extinction of a species as nature evolves and not of the unnatural extinction of flora and fauna as we carpet the earth with freeways and shanty towns?

The Daoists caution us to stop interfering, to defer to the well-worn paths taken by nature, even as we live our human lives. They also suggest the intriguing idea of nonaggressive forms of power: what Lisa Raphals called the "action at a distance" of the sage who practices self-cultivation.

Conclusions

Is any of this practical? Is it possible to defer, to refuse action and practice self-cultivation? It is clear that for any environmental ethics to be adequate to the task, it must in some sense be radically different than anything anyone is expecting. As Albert Einstein famously observed, "We cannot solve the problems we have created with the same thinking that created them." It is not some small error compounded over time that has caused this global crisis, but a fundamental flaw in the way we have come to understand—and refer to—the world. No one expects Daoist philosophical texts to provide concrete answers to problems that their authors could not have envisaged, just as academics are not in the business of consulting the *Yijing* to determine the date on which they should sell their stock in eBay. These essays have contributed to the task of transforming environmental ethics by deferring to the dissonant tones of early Daoist philosophy, rather than imposing a scheme of easy coherence and simple harmony. In doing so they have exemplified a genuinely Daoist ecological ethic that we would do well to take seriously.

Notes

1. This section was originally written by Russell B. Goodman as the response to David Hall's paper given at the conference on Daoism and ecology.

2. Immanuel Kant, *Critique of Judgment*, trans. J. H. Bernard (New York: Hafner, 1968), 157.

3. *The Collected Works of Ralph Waldo Emerson*, ed. Robert Spiller et. al. (Cambridge, Mass: Harvard University Press, 1971–), vol. 3:20; hereafter CW.

4. CW 3:6.

5. CW 3:15.

6. For an account of Emerson's interest in Indian thought, see Russell B. Goodman, "East-West Philosophy in Nineteenth Century America: Emerson and Hinduism," *Journal of the History of Ideas* 51, no. 4 (1990): 625–45.

7. Quoted in John J. McDermott, *The Philosophy of John Dewey* (Chicago: University of Chicago Press, 1981), 282.

8. *Chuang Tzu: Basic Writings*, trans. Burton Watson (New York: Columbia University Press, 1964), 34–35.

9. William James, *Some of Life's Ideals* (New York: Henry Holt, 1900), 8–9.

10. *Chuang Tzu*, 26.

V. Practical Ecological Concerns in Contemporary Daoism

Respecting the Environment, or Visualizing Highest Clarity

JAMES MILLER

Introduction

One of the chief goals of the conference series of which this volume is a part has been to establish "world religions and ecology" as a new field of academic inquiry. This requires scholars to cross traditional disciplinary boundaries and, largely, to invent from scratch ways in which investigation into this area may proceed. It also invites them to enter the unfamiliar territory of policy and practical application. A common line of inquiry has been to view religious traditions as repositories of moral power that can legitimately be applied to modify human behavior so as to save the environment. The environmental ramifications of an ethic of "actionless action" (*wuwei*), for example, have been treated in a highly practical way by Liu Xiaogan in this volume. In this essay, however, I want to propose the anachronistic but absolutely serious possibility that the spiritual technology of the fourth-century C.E. Daoist movement known as Highest Clarity (Shangqing) has an important contribution to make to the question of what it means to respect the environment. In doing so I hope to retrieve and reauthenticate the original insights of this ancient religious tradition in a way that is relevant for the contemporary global situation.

The spiritual power of Shangqing Daoism rested on the notion that visualization was the key to ultimate self-transformation. By this I mean that the process of spiritual transformation in Shangqing Daoism consisted largely of learning to see one's body as it truly, most perfectly is: an intricate network of biocosmic energy that has the potential for becoming fully transparent to its cosmic field or environment. The problem was how this could be visualized. In this essay I would like first of all to examine what actually took place in this process of "learning to see." I then turn to a contemporary theory of the transformative power of religious "theory." In so doing I aim to argue that "respect for the environment" may profitably be understood as a category of spiritual aesthetics as much as morality or politics. Confucian learning traditionally deals well with the ethical dimension of these questions, but I believe that Shangqing Daoism has some profound insights to offer into the aesthetic dimension of "respect." In this essay, therefore, I aim to find a small but definite way of bridging the intellectual gap between "Daoism" and "ecology."

Daoism

Shangqing Daoism developed a sophisticated, esoteric array of spiritual technologies the purpose of which was to realize the transformation of the adept into a perfected, spiritual being. This objective was achieved not by transcending or denying the world but rather by pursuing ever more deeply the latent cosmic connectivity that is implicit in having life at all. Simply stated, the method for doing this involved visualizing gods in the body. In abstract terms this meant visually coordinating and harmonizing the body's major energy networks (sometimes mistranslated as "organs") with one's spatiotemporal cosmic field and with powerful spiritual forces that are manifestations of the primordial Dao, the cosmic generativity which recursively multiplies to produce the universe of myriad processes (*Daode jing* 25). By learning to perceive the correlation of physiology, cosmology, and spiritual power, adepts were concretely able to actualize the transformation of themselves in their cosmic context. In fact, the Shangqing techniques of meditation are predicated precisely on the understanding of visualization as actualization. The root meaning of the Chinese character for "visualize" that is used in these texts, *cun*, means "to be actual," "to be present."

For Shangqing Daoists, there was, however, no immediate way of realizing the harmony with their cosmic environment, symbolized most potently in the constellations of stars in the heavens: the only way to achieve this transparency of communication was through religious texts and talismans. In fact, the scriptures themselves were seen as revelations of the very fabric of the universe.

> It is by leaning upon the Tao that the *ching* [scriptures] have been constituted; it is by leaning upon the *ching* that the Tao manifests itself. Tao is substance and the *ching* are function.[1]

Thus the *Shangqing* texts revealed to the religious visionary, Yang Xi, were considered to be "only the material aspect of texts that were formed from the primordial breath that existed before the origin of the world."[2] This material manifestation, or "function" (*yong*), of the Dao as substance (*ti*), to use the Chinese philosophical terms in the quotation above, indicates that the religious texts were in fact regarded as the patterns for the cosmic textuality or fabric into which the universe is being continuously woven. Thus, the religious text functioned as the medium through which the adept was able to weave himself into a transparent unity with the field of cosmic power that constituted his environment. Person, text, and cosmos were understood as being united in a single complex whole that could be grasped and imaged. To understand this better, I would like to present a translation and study of one such text, the *Jiuzhen zhongjing* (Central Scripture of the Nine Perfected).

The Central Scripture of the Nine Perfected

The textual history of the *Jiuzhen zhongjing* is complex.[3] Two texts in the Daoist canon bear its name, and neither of them can be said to be more authentic than the other.[4] Between the two of them, the text contains a biography of the Central Yellow Lord (*Zhongyang huanglao jun*), a "Method of the Nine Perfected" (*Jiuzhen fa*) accompanied by instructions for "dragon-script" (*longwen*) talismans, and alchemical recipes. Each text contains passages in common and also variations.[5] Although the text is not included in the *Zhen'gao* (Declarations of the Perfected), the compilation of Shangqing revelations made by Tao Hongjing toward the end of the fifth century C.E., this does not at all mean that it is apocryphal. Isabelle Robinet's conclusion is that much of the text is an authentic part of the Shangqing revelation, though she doubts the authenticity of the alchemical recipes in the second text, attributed to Zhang Daoling (second century C.E.), the first Celestial

Master.[6] For the purposes of the present study the most important section is the visualization technique described in the *Jiuzhen fa* (Method of the Nine Perfected) contained in the first text.

The methods prescribed in the *Jiuzhen fa* consist of forging systematic correlations between the body and its cosmic environment through a series of mental visualizations. By learning to perceive the fundamental connections between one's body and the universe that gave it life, these connections will be reinforced, and the cosmic energy with which life is imbued will strengthen and be refined. The network of connections is natural rather than mythic: it joins together general principles of medical theory, the calendar, and the particular cosmology that was the content of the Shangqing revelations. It does not make reference to the mythic architecture of a cosmic drama.[7]

The method consists of visualizing the Imperial Lord (*dijun*), the Great Unity (*taiyi*), and the five spirits (*wushen*) merging into one of nine great spirits on each of nine separate occasions over the period of a year. These nine spirits produce an energy of a particular color in nine layers around a particular energy system of the body. This physiological system is thus spiritually vitalized, and the energy then ascends to the *niwan*, the upper "cinnabar field" in the head. This process of visualization is thus an act of inner alchemy that is typical of the Shangqing revelations. The ordinary body is being transformed into an immortal body not by means of the ingestion of natural substances endowed with magical properties,[8] but by means of a spiritual technique that is intellectual (taking place in the mind), corporeal (transforming the body), and textual (based on texts and talismans). All three dimensions are brought together in an act of interior visualization.

By way of example, I offer a translation of the first of the sequence of nine visualizations that were to take place over a year:

> In the first month, on your birthday, the *jiazi* day, or the *jiaxu* day, at dawn, the five spirits, the Imperial Lord, and the Great Unity merge together into one great spirit which rests in your heart-system. The spirit is called the Lord of Celestial Essence, his style Highest Hero of Soaring Birth, his appearance is like an infant immediately after birth. On whichever day at dawn, enter your oratory, place your hands on your knees, control your breathing, close your eyes. Look inside and visualize the Lord of Celestial Essence sitting in your heart. His name is called Great Spirit. Make him spew forth purple energy to coil thickly around your heart in nine layers. Let the energy rush up into the *niwan*.

Inner and outer [dimensions] are as one. When this is done, chatter your teeth nine times, swallow saliva nine times, then recite this prayer:

> Great Lord of Celestial Essence,
> Highest Hero of Soaring Birth,
> Imperial Lord, transform inside me,
> come into vision in my heart.
>
> Your body is wrapped in vermilion garb,
> your head is covered with a crimson cap.
> On your left you wear the dragon script;
> on your right you carry the tiger writing.
>
> Harmonize my essence with the threefold path,
> unite my spirit with the Upper Prime.
> To the Five Numinous Powers I present a talisman,
> with the Imperial [Lord] may I be wholly identical.[9]
>
> Your mouth spits out purple florescence,
> to nourish my heart and concentrate my spirit.
> As my crimson organ spontaneously becomes alive
> may I become a soaring immortal.

When this is done, repeat the practice at midnight following the same method as above. These are the times of correspondence when the Five Spirits of the Imperial Lord spontaneously combine and form one great spirit. [They correspond to] the so-called periods of stimulation that match one's birth. Do not wait over and over for auspicious signs to practice successful meditation.[10]

The overall purpose of the visualization is to actualize the flow of energy that has been made possible through the synchronization with the cosmic order. Having made appropriate physical and mental preparations, and being appropriately synchronized with the calendar, the adept is ready for an encounter with a god, that is, a symbol of pure cosmic power. The visualization of the god is the mechanism whereby this spiritual "transfiguration" takes place. By means of "figuring," that is, visualizing, the form of the god and the concomitant regeneration of one's physical body, the actual transformation takes place. The condition for the possibility of this is that the human imagination is cosmically profound—able to engage the depths of cosmic power in the way that our ordinary sensibility is unable.

The effect of visualization is actively to harmonize the individual with the field of cosmic power in which he or she is situated. Through this harmonization a specific religious transaction takes place in which the superficial order of human existence is reconfigured or transfigured according to the deep potentiality of cosmic power inherent in the root of the Dao. Clearly, there is at work here what we could anachronistically label a powerful ecological sensibility. Spiritual transformation depends upon nothing less than seeing that one is intimately connected to the cosmic rhythms of time and space. But this Daoist conception of "ecological" religious practice is in one sense far from unusual: In fact, it is simply a tradition of perception or iconography. But it is particularly interesting from the perspective of "religions and ecology" because of its conception of the perfected person or spiritual accomplished being as being fully transparent to his or her environment. That is, the person, in his or her body becomes the icon through which the perfect reality of the cosmos is disclosed.

Theory

I want now to move away from the particular language of Shangqing Daoism to consider a contemporary "theory of theories" presented by Robert Neville in his book *Normative Cultures*.[11] In this book Neville's main task is to articulate a way of comparing "normative cultures" or traditions of religious theories. To do so he must first consider what a good "theory" is. Neville's theory of theory is relevant to the present study because it argues fundamentally that a theory is a *theoria*, or a way of seeing things together. Specifically, a theory is a method of synopsis (seeing together) in which a diversity of phenomena are brought together in some particular unifying vision.[12]

A significant problem of comparative religions with which I am not concerned here is whether or not it is possible to have a synoptic view of religious traditions that respects their unique importance, unity, and diversity. But the parallel problem, with which I am concerned, is whether or not a particular religious "theory" is capable of respecting the unique importance, unity, and diversity of the reality of the natural environment. The fact that we have a disastrous environmental problem suggests that our current "theories" or ways of perceiving the environment do not, in fact, properly "respect" (its) nature or reality.

The criteria of a good theory, according to Neville, are that it achieves a synoptic unity while respecting the singular importance of the diversity of things being theorized about. Theory, therefore, is not just an act of synoptic perception: it is also a moral act of valuation. This is the overall burden of Neville's project of the "axiology of thinking." Theories "provide us with new signs for interpreting the world,"[13] and by means of these signs we come to orient ourselves in the world. Our orientation to the world determines the way in which we pursue our responsibility toward it. Thus, imagination, orientation, and responsibility are all integrated in Neville's theory of theory. If Neville is correct, which I believe he is, then it is vital for those who are concerned about "respecting the environment" to develop new "theories" or ways of perceiving nature that properly respect its singularities and diversity.

Reductive theories (whether scientific formulae or the grand narratives of human culture) do not need to respect the singularity of value of their objects; they just need to explain what is important from the point of view of the theory itself, and do not need to tell the whole story about the things involved. It is this lack of *respect*, rather than the imposition of unity, that formally engenders a predatory or colonialist attitude to the environment. In fact, unity is vital if we are to have understanding.

It will be immediately apparent that avoiding this problem of "disrespect" is in fact very difficult. The proper respecting of unity, diversity, and importance cannot likely be derived from some Archimedean vantage point. More likely is that a theory will be developed, tested, reformulated, and retested in a continuous process of experimentation or theorizing. The difficulty is that this requires that theories expose themselves to the possibility of correction, and religious theories do not have a good track record in this department! Perhaps one of the most important aspects of Neville's theory of theory is that it underscores the fact that theory ideally is a concrete *process*. It therefore breaks open the hermetic seals that tend to protect theories from the threat of improvement and places the process of theory in the context of concrete engagement with reality. If a religious tradition can make any claim to truth, it must, in fact, be constantly transforming itself.

A good theory-process is therefore wholly organic and properly ecological. It is organic in that it evolves in a life that can be narrated. It is ecological in that the "subject," "object," and iconic theory are related in a whole in which each part is constantly adapting itself to the others. The ideal theory is one which is imperceptible, that is, one which distorts the reality of its objects as little as possible. It is in light of this fact that Shangqing theory or visualization technology has something useful to contribute to the debate about religions and ecology.

Clarity

If we try to understand the architecture of the religious dynamics of Shangqing Daoism, we quickly discover that a distinguishing feature of human beings is that we have an inherent communicative possibility (*de*) with the cosmic field (*dao*) in which we find ourselves. Despite the esoteric nature of the texts, the goal of Shangqing spiritual transformation was to exploit this possibility of communication so that body and mind were completely integrated with their environment. To become spiritually transformed means, in light of Neville's theory of theory, to become the embodiment of the perfect theory or vision of the cosmos. Shangqing Daoists did not aim to respect the environment in a moral sense but, in fact, aimed to become the embodiment of perfect "respect." This meant achieving a sort of perfect cosmic clarity. Like Neville's theory of theory, this process begins in the imagination, extends to one's orientation to the world, and culminates in a form of practical responsibility toward it.

If Shangqing Daoism can be persuaded to contribute something to the modern problem of ecology without disrespecting its historical particularity, it is that it holds up this unique vision of a spiritual person. Fully in accord with the visions of many Daoist traditions, such cosmic beings travel lightly through the world because they have learned to mentally visualize and physically embody the broad extent of their cosmic environment within their own being. They are transparent individuals who have achieved the highest clarity.

Respect

Shangqing Daoism understood as a way of embodying "theory" or vision demonstrates that respecting the environment should begin first and foremost in the imagination but should encompass the whole of one's being. This vision of what it means to "respect the environment"—

resolutely religious *and* physiological—is perhaps too strange or too alien to be adopted as a practical solution to a global political problem. But it is not too strange to suppose that it might spark the imagination of a few individuals to wonder what it might be like for every fiber of their being to be perfectly harmonized and fully transparent to all that is around them. If this essay achieves only that, then it will have legitimately begun to frame the question of what Daoism can contribute to the problems of ecology. The answer to this question, however, lies only in the concrete actuality of individual lives.

Notes

1. *Duren jing* (Scripture of the Salvation of Humankind), DZ 1, preface, 4a; trans. Isabelle Robinet, *Taoist Meditation* (Albany: State University of New York Press 1990), 23.

2. Ibid.

3. See Isabelle Robinet, "Introduction au *Kieou-tchen tchong-king*," *Society for the Study of Chinese Religions Bulletin* 7 (fall 1979): 24–45; Isabelle Robinet, *La Révélation du Shangqing*, 2 vols. (Paris: École Française d'Extrême-Orient, 1984); Masayoshi Kobayashi, *Rikucho Dōkyo shi kenkyū* (Research into the History of Daoism in the Six Dynasties) (Tokyo: Sōbunsha, 1990).

4. The two texts are the *Shangqing taishang dijun jiuzhen zhongjing* (The Highest Clarity Supreme Imperial Lord's Central Scripture of the Nine Perfected; DZ 1042) and the *Shangqing taishang jiuzhen zhongjing jiangsheng shendan jue* (Spirit Elixir Formula for Ascending to Life [revealed in] the Highest Clarity Supreme [Imperial Lord's] Central Scripture of the Nine Perfected; DZ 1043). Neither of these texts can be said to be more or less authentic than the other (Robinet, *Révélation*, 2:24).

5. Robinet has reconstructed the probable order of the original sections, along with a table of their inclusion in other Daozang texts and anthologies ("Introduction," 43; *Révélation*, 2:73-74).

6. Robinet, *Révélation*, 2:82–83.

7. This does not mean that there was no such mythic consciousness in the Shangqing revelations. In fact, a clearly apocalyptic vision did enter the Shangqing School and thoroughly penetrated Lingbao Daoism. Kobayashi attributes this to the advent of the Buddhist mythic structure of the kalpa. But there is no evidence for this in this particular text. See the Lingbao text *Zuigen pin* for an excellent example of this melding of Buddhist and Daoist cosmology.

8. See Robert Ford Campany's essay in this volume, "Ingesting the Marvelous."

9. The character *tong* (identical) depicts one mouth in a single enclosure, thereby communicating all the warmth and comfort of sharing a home with someone within a single family unit. The character *quan* (complete) depicts a king under a roof and gives a sense of the cohesive stability of a single nation united under a single leader. Together the two characters convey far more than the English "identical and whole" and point toward the beauty and strength of the intimate translucence shared by the adept and the divinity.

10. DZ 1042: 1.3a–4a. For a complete translation of the *Jiuzhen fa*, see James Miller, "The Economy of Cosmic Power: A Theory of Religious Transaction and a Comparative Study of Shangqing Daoism and the Christianity of St. Augustine" (Ph.D. diss., Boston University).

11. Robert Cummings Neville, *Normative Cultures* (Albany: State University of New York Press, 1995).

12. Ibid., 3.

13. Ibid., 33.

A Declaration of the Chinese Daoist Association on Global Ecology

ZHANG JIYU
Translated by David Yu

Introduction

Daoism is an indigenous religion of China and a dynamic system of thought that constitutes one of the pillars of Chinese traditional culture. Although blended with many general aspects of Chinese culture and religion, Daoism has nevertheless preserved its own unique characteristics. The relationship between Chinese traditional culture and Daoism is like that of the ocean and smaller seas—where the ocean refers to traditional Chinese culture and one of the seas to Daoism. Historically, Daoism has made a wide and deep imprint upon the culture of the Chinese people. Many of its ideas and precepts, which have undergone almost two thousand years of successive development, have greatly shaped the thought, life-style, and behavioral patterns of the Chinese people.

Daoism greatly contributed to the making of Chinese culture and society and has continued to evolve for the past two thousand years. Today Daoism is one of the five living religions of China. It also thrives in overseas Chinese communities. In fact, the significance of Daoism has reached greatly beyond the confines of the Chinese language, thus enabling it to play a global role. These challenges have encouraged Chinese Daoist professionals[1] to dedicate their lives to the solemn task of promoting the spirit of Daoism and to foster contact with Daoist advocates throughout the world. They desire to spread the good teachings of Daoism by assisting people to purify their hearts and to transform their lives.

Daoism, like the other great religions of the world, has a religious ideology that reflects its worldview, moral precepts, and ultimate concerns. Due to its close association with Chinese culture, Daoism has characteristics different from other religious traditions. Two of these important characteristics will be explained below:

Respecting Dao and Greatly Valuing *De*

The *Daode jing* says: "Dao produces things. *De* fosters things. Matter gives things their physical forms. And the circumstances of the moment complete things. Therefore, the ten thousand things worship Dao and give high value to *de*."[2] Dao is the highest object of pursuit for the Daoists. A Daoist believes in Dao, relies upon Dao, cultivates Dao, and practices Dao. *De* refers to the particular conduct of the believer as she practices Dao. One may say that *de* is the practice of Dao in the believer's life. Hence, to the extent that one follows the practice of *de*, one is in fact paying homage to Dao. Thus, the core of Daoism consists of respecting Dao and giving high value to *de*. Although Dao permeates everything and is present at all times, it cannot be identified as either Being or Non-Being: Dao is forever "betwixt and between" all things in their continuous process of transformation. When Dao takes a physical form, it is called the "Lord of the Most High" (Taishang Laojun)—the progenitor of the Daoist tradition. At various times, he descends upon the world to teach and to save human beings. He says to the people: "When you are a leader, you should not contrive to dominate your followers. When you rely upon your own ability to do things, you should not value that ability as of your own making." These words represent the spirit of a true Daoist. She lets the myriad things grow according to their own natures (*ziran*). She lets things and events accomplish their results according to their natural endowments (*benxing*). She does not impose her selfish mind or evil ideas upon things. The foundation of Daoist teaching is naturalness (*ziran*) or non-action—the principle of Dao. This principle implies purity, tranquillity, and simplicity, as well as softness and noncombativeness—the spirit of humility or vacuity as expressed by the image of the valley. The bosom of Dao is like a great ocean capable of containing myriad things and of connecting hundreds of rivers. Although the ocean stays low, it benefits the multitude of things. The spirit of Dao is found in mercy and frugality; it does not dare to be ahead of the world. These are beautiful and noble virtues that all Daoists ideally respect, admire, and ceaselessly pursue.

The Way of Immortality (*Xiandao*) Gives High Value to Life

The Daoist book *Scripture of Salvation* (*Duren jing*) says: "The way of immortality gives high value to life; the recitation of this scripture can bring salvation to a limitless number of people." Daoists advocate immortality; they pursue longevity and disdain death. They admire a life of careless wandering and a flight-like freedom. Their ultimate goal is to let their lives and spirits become one with the Dao—the way of immortality (*xiandao*). This is the reason why Daoists treasure life and value it as the most worthy thing on Earth. They want to live long and practice gazing as a form of meditation. Zhang Daoling, the Daoist patriarch, once said: "Because life is another manifestation of Dao in the world, cultivation of Dao means cultivation of the Dao of life." This is the reason why the way of immortality gives high value to life and why Daoists take it as their basic tenet. Following the doctrine of giving value to life, Daoist sages in history have ceaselessly experimented and practiced the cultivation of life. Their practice has proven the correctness of the following dictum: "My life depends on myself, it rests entirely upon my own cultivation." Through the regimen of self-cultivation, human beings can obtain the fullness of bodily and mental health; they can reach the goal of immortality in the Daoist sense.

This kind of self-cultivation includes both the cultivation of the character/mind and the cultivation of the body. Cultivation of the character/mind demands the exercise of willpower through the elimination of selfish ideas and the desire for self-profit and through doing good deeds and accumulating many merits. Ultimately, the practitioner likes to see that her character/mind is in harmony with the spirit and sentiments of Dao. She expects to see that her character becomes continuously sublimated or refined and that she will become a person of high moral sentiments. Daoism says that the cultivation of Dao depends on the cultivation of *de*. Thus, the cultivation of *de* is the foundation of

Daoist moral engineering. This means that the cultivation of the body depends on the cultivation of the character/mind. Daoists are free to choose a philosophy of life and a methodology for the cultivation of the Dao as long as they are agreeable with the essential teachings of Daoism. Generally, this involves the cultivation of a Daoist's essence (*jing*), breath (*qi*), and spirit (*shen*) in her body, making sure that these three bodily treasures flourish continuously and do not decay. Through the cultivation of the body, the practitioner attempts to return these three "post-celestial" elements back to their "pre-celestial" state. In Daoist terminology, this process of reversion is called the return of the body to the state of an infant. Numerous experiments have shown that the Daoist cultivation of the body not only offers one longevity but also energetic youthfulness. However, this cultivation of the body requires a wholesome ecological environment as its external condition.

Keeping the Spontaneous (*Ziran*) and Natural (*Benxing*) Character in the World Process

Daode jing 25 says: "Humankind models itself after Earth. Earth models itself after Heaven. Heaven models itself after Dao. And Dao models itself after the natural." *Daode jing* 55 says: "To know harmony means to be in accord with the eternal [Dao]. To be in accord with the eternal [Dao] means to be enlightened." Chapter 44 says: "He who knows when to stop is free from danger. He who is contented suffers no disgrace."

As humans face an ecological crisis throughout the world, they realize increasingly that problems concerning environmental protection are not derived from industrial pollution or technological expansion alone. Rather, these problems are also derived from people's worldviews, ideas of value, or theories of knowledge. This is because the powerful industrial and technological expansion across the world has also generated systems of value and theories of knowledge affecting people's ideas about the world. These ideas have conditioned people's thought-patterns and have caused a split in their minds regarding the unity between humans and nature. These thought-patterns of contemporary people exaggerate the subjective will, causing people to think that nature can be treated as subservient to the willful desires of humans and that nature would never respond negatively to these conditions. Although these contemporary thought-patterns have achieved a high degree of social productivity throughout the world, they have also given humankind a greatly inflated image of itself. Daoists believe that this inflated image of the self is an important cause of the serious ecological crisis confronting the modern world. Today, as we face the real possibility of imminent world destruction, we should soberly examine these dominant thought-patterns associated with the contemporary worldview.

We believe that the ancient doctrines of Daoism are eminently able to remedy the deficiencies caused by contemporary ethical theories. According to Daoism, Heaven, Earth, humankind, and all things in the universe—including birds, animals, insects, trees, grass, and other existing entities—come into being through the transformation of the breath of Dao (*daoqi*). By endowing various degrees of pure or turbid breath in the myriad things, the breath of Dao gives rise to individual things. As a result, a diversity of existing things is formed in the world. Human beings, composed of the blending of the breath of Heaven with the breath of Earth, are the most spiritual and intelligent creatures among the myriad things. They are one of the four fundamentals of the universe (Dao, Heaven, Earth, humankind). How should humankind deal with the myriad things of the universe? Let us return to the ancient *Daode jing* for the answer. Since this text asserts that Dao models itself after the natural, it follows that men or women should model themselves after the earth which they inhabit. By the same token, "Earth modeling itself after Heaven" means it must be in tune with the changes of the universe. And "Heaven modeling after the natural" means that it must follow the operations of Nature—operating spontaneously in accordance with its self-so (*ziran*) character. This also means that human beings must not employ contrivances to coerce the self-so character of Nature to conform to human desires. Humans must nurture the spontaneous character of non-action (*wuwei*)—"Dao modeling after the natural"—in the innermost part of their hearts and practice it unceasingly. Only by so doing can humans solve the global ecological crisis. And if these problems of the natural environment are solved, then the well-being of humankind is assured.

The Lord of the Most High leaves it to us to emulate the doctrine of self-so (*ziran*), that is, non-action (*wuwei*) revealed in nature. Another teaching of Daoism, concerning the relationship between humans and nature is: "To be in accord with the eternal [Dao] means to be enlightened. To know the eternal is called enlightenment."[3] Closely related to this Daoist sayings is: "Not to know the eternal [Dao] is to act blindly which results in disaster."[4] Now we shall attempt to apply this group of Daoist sayings to ecological considerations. Thus, "To be in accord with the eternal [Dao] means to be enlightened" denotes the fact that the myriad things of the world are mutually connected and interdependent, and that people must maintain a harmonious relationship among things in order to enhance their longevity. Daoism affirms that each of the myriad things contains the polarity of yin and yang and that life is engendered only when each thing's yin-yang poles are allowed to commune with each other. In other words, the viability of life depends upon the mutual harmony of the yin-yang polarity of all things. This also means that the continuous development of world civilization depends on the harmonious relationship among all things. People who understand this principle can be called wise persons. Conversely, those who do not understand this principle or violate it—contradicting this teaching through human contrivances, breaking the laws of nature, destroying a species by artificial means, or promoting the overproduction of a certain species of things—are the violators of the principle of Dao.

The Daoist classic, *Baopuzi* (The Master Who Embraces Simplicity), a work of the fourth century c.e., makes a distinction between two kinds of people concerning their relation with nature: one type is called "those who enslave the myriad things" and the other is called "those who emulate nature."[5] It says that the Dao, which creates the myriad things through the principle of "self-so," makes humans the most intelligent beings among all the creatures of the earth. Those who have only a superficial understanding of the relationship between people and nature are the ones who "enslave the myriad things." They subjugate nature completely to themselves. Contrariwise, people who have a profound insight into the mystery of the relationship between humans and nature are friends of nature and derive their understanding of longevity from nature by meditative observation and gazing. Because people of this kind are intimately acquainted with nature, they understand that turtles and cranes live long lives. Therefore, they try to emulate the calisthenic methods of the turtles and cranes in order to strengthen their bodies.

From a long range perspective, the abuse of Nature will result in a disastrous destruction of our natural environment. Chapter 6 of the *Baopuzi* provides the following passages regarding this issue:

> "Flying birds are shot with bullets, the pregnant are disemboweled and their eggs are broken. When there is hunting by firearm in spring and summer. . . . Each of these destructive actions constitutes one sin, and according to its severity the Controller of Destiny will deduct a certain amount from one's life (one's natural destiny or longevity). A person dies when there are no more units of time left in one's life." Contrarily, "if one treats things with compassion, forgives others as one wishes others to forgive oneself, is benevolent even to the creeping insects . . . harms no living things. . . . In this way, one becomes a person of high virtue and will be blessed by Heaven. One's undertakings are sure to be successful, and the desire to be an immortal will be obtained." Prosperity and longevity fall upon one who extends her love to the myriad things in the natural world.[6]

The passage, "He who knows when to stop is free from danger. He who is contented suffers no disgrace,"[7] is important for ecological ethics. "He who knows when to stop is free from danger" refers to the fact that there are limits in the natural environment as to how much abuse Nature can take from human beings. This consideration necessarily calls for human restraint against any act which might interfere with the ecological balance—even if that action might generate a huge immediate profit. In this way, we can avoid Nature's retaliation, prevent a long-term disaster for the sake of some immediate gain, or escape the predicament of losing what we have already gained. "He who is contented suffers no disgrace" means that people should have a correct understanding of what success means. One should realize that in order to avoid the overproduction of natural resources, one must not pursue material benefits in an unsatiable way. The *Taiping jing* (Scripture of Great Peace) contains a passage conveying the importance of conserving natural resources for the production of wealth.[8] It proposes that the criterion of wealth depends upon whether or not a society can protect the ability of all nature's species of things to grow and prosper. The passage says: the sage-king teaches "that people should assist Heaven to produce living things and assist Earth by giving nourishment to things to form proper shapes." Wealth requires that the living things under Heaven and of the world are all allowed to grow and flourish:

[In antiquity], in the Higher August Period, because 12,000 species of living things grow and flourish, it is called the era of Wealth. In the Middle August Period, because the living things are less than 12,000 species, it is called the era of Small Poverty. In the Lower August Period, because the number of living species was less than that in the Middle August Period, it is called the era of Great Poverty. If the august breath keeps producing fewer species of things and if no good omens are seen to assure the growth of good things, the next era will be called Extreme Poverty.

According to the above passage, the reason why all things can grow in the Higher August Period is because the earth is properly nourished and is not injured by people. On the other hand, because in the Lower August Period the earth is not given proper nourishment and is greatly injured by its inhabitants, it produces fewer living things. From the perspective of the sage-ruler, if he can make the myriad living things grow and flourish, he can insure the rise of a wealthy country. On the other hand, if half of the myriad things in Nature in the sage-king's country are injured, it portends that his country is on the road to downfall. The *Taiping jing*'s insight—that the conservation of living species in the natural world is the criterion of wealth—is a real contribution to the ethics of ecological thought.

We believe that the Dao—the principle of self-so/non-action (*ziran/ wuwei*)—is the fundamental Daoist teaching. This and many other Daoist concepts have positive implications for the contemporary world. We hope that the various ideas of the different world religions, which in the best sense promise to bring benefits to humankind, will attract the attention of both religious believers and nonbelievers. Given this kind of common ground, a harmonious relationship between human beings and their natural environment, between humans and society, and among diverse groups of people can be established.

Tasks and Plans

Both Daoist doctrines and disciplinary rules provide principles aimed at protecting Nature in its overall ecological context. These considerations have prompted hermitage Daoists in China both to continue and to augment their ecological concerns. For many years, China's

Daoists, in the confines of their hermitages and within the limits of their capacities, have devoted themselves to the tasks of protecting trees and fruits, preserving and rebuilding forests, safeguarding the hermitage culture and ancient relics, and other ecological activities. They have also gone beyond the confines of their hermitages to establish closer relations with their lay Daoist followers. The hermitage Daoists have, therefore, taken the time to talk to their congregations of followers on the importance of protecting trees and forests in the mountains and on the relationship between Daoist doctrines and nature (including all manner of fowl, birds, and animals). They are helping their followers appreciate the oneness between the creativity of Dao and the survival of humankind. The hermitage Daoists believe that they have planted the seeds of ecological responsibility and service in the hearts of their congregations. For example, a certain Daoist master of Mount Thundergod (Leishen shan) in Baoji municipality, Shaanxi province, has spent several decades working with his followers in the voluntary maintenance of over eight hundred mu^9 of trees and forests. He is well recognized for his work by people in the region. There is also a Daoist congregation associated with the hermitage neighborhood of Wudang, in Yuedu county, Qinghai province, who for more than a decade has made the barren hills into green fields by planting trees and growing grass. This congregation has greatly improved the physical surroundings of the neighborhood. Then there is an elderly Daoist master in Yunnan province who devoted more than sixty years of his life to the cultivation and nursing of what some international botanists call a "living fossil"—that is, the rare plant called the "Bald Pine Tree" (*tushan*). This Daoist master's work has been recognized by the staff of the Bureau of Arboriculture in Yunnan province. Deeds of this kind are too numerous to be mentioned. Environmental concern is a tradition of Daoism and deserves further promotion. For this reason, the Chinese Daoist Association in 1993 sponsored a national ecological conference that encouraged all Daoist followers to plant trees and build forests, to beautify the natural environment, and to engage in works benefiting social welfare.

The Daoist ecological agenda in the immediate future includes the promotion of the teaching of self-so or non-action for the transformation of people's lives. This is a strategy conducive to the development of environmental protection. It also involves the sacred mission of coop-

erating with the world community in maintaining harmony with Nature. The following are three specific tasks that all Daoists are called
to perform:

- We shall spread the ecological teachings of Daoism, lead all
  Daoist followers to abide in the teachings of self-so or non-
  action, observe the injunction against killing for amusement
  purposes, preserve and protect the harmonious relationship of
  all things with Nature, establish paradises of immortals on
  Earth, and pursue the practice of our beliefs. In our evangelis-
  tic efforts, we shall nurture the people by teaching the importance in Daoism of maintaining harmonious relationships
  among things in the natural world. This means we shall promote the mutuality between valuing life and ecological concerns. We will raise the awareness regarding ecology among
  various social groups, resist the human exploitation of Nature
  and the abuse of natural environments, protect the earth upon
  which human survival depends, and generally make the world
  a better place for humans to inhabit. At the same time, we
  shall promote the nonviolent and pacifist ideals of Daoism—
  thus helping to preserve world peace and to free the world
  from environmental degradations due to war.
- We shall continue the Daoist ecological tradition by planting
  trees and cultivating forests. Using traditional hermitages as
  an organizational base, Daoists will conscientiously plant
  trees and build forests, thereby making the natural environment beautiful and transforming our hermitages into the paradise worlds of the immortals.
- We shall select some famous Daoist mountains as exemplars
  of the systematic task of environmental engineering. We expect to reach this goal by the early years of the new century.

Since the publication of this declaration in 1995, we have received
responses from groups abroad indicating their interest in exchanging
ideas with us. For example, the Religions of the World Alliance for
Environmental Protection in 1996 cooperated with the China Daoist
Association, in conjunction with China's State Bureau of Religious
Affairs and its branch offices in Sichuan and Shaanxi, to conduct a

field survey of two famous Daoist mountains: Qingcheng shan in Sichuan province; and Huashan in Shaanxi province. This survey has produced a survey report on these two sacred mountains. It deals with a critical study of these two mountains, their natural resources, famous Daoist architecture, and ancient cultural relics. The report also discusses the host hermitages' efforts to open these two mountains to the public and their work of environmental maintenance. The report concludes with some suggestions regarding the ongoing maintenance of, and public access to, these two mountains.

In sum, present-day Daoists in China have diligently worked toward disseminating Daoist teachings and in maintaining the famous Daoist mountains and hermitages, planting trees and cultivating forests, and protecting the natural environment. We believe that as the Chinese state and society today are paying greater attention to ecological problems, educational programs concerning public health issues will be further fostered and developed. We pray that tomorrow's world will be better than today's, and that, by following the principle of mutuality among all things in nature, a new harmonious world will emerge. It is this kind of harmonious natural and social environment that will be conducive to the balanced growth of the world's population.

Notes

1. In the present translation, "Daoists" generally refers to the priests of Chinese mountain hermitages; "Daoist professionals" denotes primarily the staff members of the Chinese Bureau of Religious Affairs who are assigned to work with the hermitage priests; and "Daoist followers" or "congregations" refers to laypersons, particularly those who reside in the neighborhoods of the famous Daoist mountains, who through their deeds and words express their faith in Daoism.

2. *Daode jing* 51. For the translation of the passages of the *Daode jing,*the translator has used Wing-tsit Chan's book with some minor changes: *The Way of Lao Tzu* (Indianapolis: Bobbs-Merrill Company, 1963).

3. *Daode jing* 55.

4. *Daode jing* 16.

5. For the translation of passages of *Baopuzi*, the translator has used the book by James R. Ware with some emendation: *Alchemy, Medicine, and Religion in the China of* A.D. *320* (Cambridge, Mass.: MIT Press, 1966).

6. Ibid.

7. *Daode jing* 44.

8. For the translation of the *Taiping jing*, the translator has consulted Wang Ming's Chinese text: *Taiping jing hejiao* (Beijing: Zhonghua shuju, 1960), 30. On ecological significance of the *Taiping jing*, see also the paper by Lai Chi-tim in this volume.

9. A *mu* is about one-sixth of an English acre.

Change Starts Small: Daoist Practice and the Ecology of Individual Lives

A Roundtable Discussion with Liu Ming, René Navarro,
Linda Varone, Vincent Chu, Daniel Seitz, and Weidong Lu

Compiled by
LIVIA KOHN

The Roundtable Participants

On the evening of the second day of the conference, a roundtable was convened that included six practitioners of Daoism or Daoist-inspired practice in the United States. The participants can be described as representing three different styles of Daoist practice: first, traditional Daoist ritual and inner cultivation as handed down from masters of the established schools; second, arts and techniques of life improvement that are not in themselves Daoist, but share the same worldview as Daoism and play a role in its religious practice; and third, methods of Chinese traditional medicine that also share the religion's cultural and experiential background and function as fundamental techniques in the religious quest.

The first representative of religious Daoism was Liu Ming, formerly Charles Belyea, who described himself as follows:

"I am a Euro-American Daoist priest who teaches and practices in the United States. My instruction in Daoism came through the Daoist family tradition of the Liu clan, into which I was adopted and chosen as a generation successor on Taiwan in 1979. When I was adopted, I was given over 130 manuals concerning observances and practices, few of which, I was told, appear in the standard Daoist canon, printed in the Ming dynasty (1445 C.E.). Though not always parallel to the historical mainstream of Daoism in China, this family tradition traces itself back to Liu An, the author or editor of the *Huainanzi*, a classic of ancient Daoism that dates from the mid-second century B.C.E. As Liu An was a relative of the ruling family of the Han dynasty, so the Liu tradition claims ancient imperial connections. In addition, they consider themselves a part of the Orthodox Unity (Zhengyi) tradition of Daoism, which goes back to the second century C.E. and is also known as the Celestial Masters (Tianshi) school. I follow their practices and have founded an American branch, the Orthodox Daoism of America (ODA), which is currently in the process of building an orthodox *Daotan*, or clan-style temple, in Santa Cruz, California, and composing a curriculum for the training of Daoist priests in America."

Liu Ming's Daoist practice consists of three different types of activities: study and observance of moral precepts; hygienic practices, diet, and Qigong exercises; and specific religious practices. The latter include: nonconceptual meditation (*zuowang*), alchemical meditation (*jindan*), ritual and the summoning of spiritual intermediaries through trance, meditative gymnastics (*daoyin*), not unlike Indian yoga, and dream meditation (*yun*). His ritual activity allows him to interface with the One, while his conduct and daily life link his personal with the communal life. He calls his religious life "full-time." In addition, Liu Ming performs services for a growing parish community, teaches his religion in private and public settings, publishes a Daoist calendar and a newsletter, and is currently editing the first volume in a series entitled *Willow Record*.

René Navarro participated in the discussion as the representative of Mantak Chia, the founder and master of the International Healing Tao Center who practices a form of internal alchemy (*neidan*). This is the dominant form of self-cultivation in the other great school of religious Daoism today, the Quanzhen, or Complete Perfection, which goes back to the twelfth century and is practiced predominantly in mainland China. Mantak Chia, the well-known teacher of this practice in the West, is orginally from Thailand and has his main center in a rural estate in Chiangmai. His organization has over four hundred instructors worldwide and offers year-round activities and workshops in many different countries. René Navarro is one of these accredited instructors.

He describes himself: "I come from the Philippines, a country that was colonized by Spain and the United States. A wag describes Philippine history in a nutshell as 'three hundred years in a Roman Catholic convent and fifty years in Hollywood.' Philippine culture is a mishmash of Malayan, Indonesian, Chinese, Hispanic, and U.S. influences. Whatever Asian culture was encountered by the West was almost decimated during the Spanish Conquista and U.S. Manifest Destiny. Some scholars call the culture a hybrid. I myself was a child of this culture. My name René—after Descartes—is evidence of it. As I wrote somewhere else, 'Instead of Buddhism and its rituals of meditation and chanting, the Filipino followed Christianity and its novenas and processions; instead of Chinese, he spoke English; instead of Monkey, he told stories of Ulysses and King Arthur.'

"In the early 1960s when I was just twenty years old, I began to study Chinese Shaolin boxing in Manila. Then I studied Taiji quan, of the Yang family style, and Bagua zhang, another internal martial art of Daoism. Through these disciplines I gained a glimpse of the depth and variety of Chinese culture, but my Westernized friends considered me an anomaly. Later I moved to the United States, where I began to study with Mantak Chia in 1983. I had just returned from an intensive course in martial arts in Sichuan when I came upon two of Mantak's books. I was very impressed and began taking his weekend courses in New York's Chinatown, gradually rising through the curriculum."

René Navarro still sees himself as a mere "beginner in Daoism and its related arts," and feels that all he has learned in the past fifteen years "is just a drop in the proverbial ocean of knowledge." Nevertheless, he gave a good description of the Healing Tao path, which consists of a combination of practices, including life-style techniques, such as fengshui, nutrition, and sexology; martial arts, such as Taiji quan, Bagua zhang, and Xingyi; a number of different Qigong gymnastic exercises and massages; as well as meditations that focus on the internal forces and move the energies in the body. As an integrated system, Healing Tao emphasizes the cultivation of the inner garden and considers the universe as an integral unit. Through their own efforts, practitioners realize that the inner and outer worlds are a continuum of matter and energy that they as stewards of the earth have to love and respect.

Among the second group of practitioners, representatives of Daoist-inspired cultivation, there were Linda Varone and Vincent Chu. Linda

Varone, of Cambridge, Massachusetts, is a practitioner of fengshui and the director of Dragon and Phoenix Fengshui. In addition to receiving an education in interior design at the Boston Architectural Center and an M.A. in psychology, she has studied with Liu Yun, the foremost teacher of fengshui in the West, as well as with major practitioners, such as Steven Post, William Spear, and Roger Green. She advises private citizens and businesses on the best way to make use of their space and also conducts workshops at a variety of institutions, including the Boston Design Center, the New England School of Acupuncture, and the University of Vermont.

Fengshui is the art of creating a physical environment that is best suited to the qi, or vital energy, of the person, and blends human activities harmoniously with the energy flow of the earth. It is based on the idea that the earth is not an accumulation of dead matter but a growing and living organism whose energies influence everything people do with it and on it. As a result, the siting of houses—especially their front doors—the design of gardens, and the placement of furniture inside the home have a direct impact on the well-being and success of their inhabitants. Good fortune and health in this system are defined not as a matter of hard work or genetic makeup but as dependent on the interaction with the environment and on one's ability to create harmony all around. The practice is widespread in East Asia, where special fengshui masters have advised governments, businesses, and individuals for many centuries. Often practitioners are also Daoists, and Daoist practice in turn begins with establishing a harmonious setting for oneself and one's family or community.

Vincent Chu, the next panel speaker, is the son of Gin-Soon Chu, one of the first to introduce the practice of Taiji quan to the Boston area. He was the disciple of Grandmaster Yang Sau-Chung, the fourth lineage holder of Yang-family Taiji quan and heir of the legendary Yang Cheng-Fu. He first arrived in Boston in the 1960s, where he founded the Gin Soon Tai Chi Club in 1969 and was named best Taiji quan instructor in 1995 by *Boston Magazine*. In the 1960s, nobody had ever heard of Taiji quan, which is now one of the most popular Daoist practices in the area.

Vincent Chu follows in his father's footsteps and works today as the chief instructor in the Gin Soon Club and as writer for various martial-arts magazines. He started practicing at age seven, also studying with Yang Sau-Chung as well as with Master Ip Tai-Tak of Hong

Kong. He is a master of Yang-style Taiji quan (both slow and fast forms), Push Hands, various weapons styles (sword, broadsword, and staff), and forms of Qigong. He describes Taiji quan as "a relaxed exercise in slow motion" that is good for people's health as well as their psychological and spiritual wholeness. The idea of Taiji goes back to the ancient Chinese diagram of yin and yang, which shows the two fundamental energy forces at the root of the universe in interlocking harmony. The practice of the exercise, which moves in circles of various kinds, aims to create the kind of harmony that was present at the beginning of the universe and is at the root of all existence. Recovery of this harmony, also known as oneness with the Dao, plays a key role in religious Daoism, and various senior Daoist masters and patriarchs have traditionally been associated with the practice in its different styles.

Those in the third group of participants in the roundtable, Daniel Seitz and Weidong Lu, are active in the practice of traditional Chinese medicine. Daniel Seitz is president of the New England School of Acupuncture. A former Peace Corps volunteer in Malaysia, he was an undergraduate at the University of Chicago and received a law degree from Boston University. In addition to running the School of Acupuncture, he also serves on a number of national committees, including the Accreditation Commission for Acupuncture and Oriental Medicine. Weidong Lu is the head of the Department of Chinese Herbal Medicine at the New England School of Acupuncture. He was trained at the Chejiang College of Traditional Chinese Medicine in Hangchow, where he graduated in 1983, and has worked in the United States for over a decade. He is also the consultant of research projects at the Center for Alternative Medical Research at Beth Israel Deaconness Medical Center in Boston and serves as the vice chairman of the Committee on Acupuncture under the Massachusetts Board of Registration in Medicine.

At the roundtable, both agreed that Chinese medicine incorporates many ideas and practices of the Daoist religion and that it represents a more holistic and ecologically conscious form of health care than Western biomedicine. Weidong Lu in particular emphasized that medical techniques were traditionally used in the attainment of Daoist immortality and that many famous Daoists were also medical practitioners. He mentioned three examples. First, the well-known Daoist and champion of immortality, Ge Hong (283–343), also wrote the medical work

Zouhou fang (Prescriptions for Emergencies), where he not only listed interesting medicinal formulas but also gave a detailed description of the clinical characteristics of smallpox—the first in the world. Next, Tao Hongjing (456–536) was most famous as the first patriarch of Highest Clarity Daoism and the compiler of its main scriptures in the *Zhen'gao* (Declarations of the Perfected). Also a practicing alchemist, he in addition compiled the *Bencao jing jizhu* (Collected Notes to the Materia Medica), in which he doubled the number of known Chinese medical herbs to 730. Third, Wang Bing of the Tang dynasty (ca. 762) was both a practicing Daoist and a physician. He became most famous as the editor of the medical classic *Huangdi neijing* (Yellow Emperor's Classic of Internal Medicine), and his Daoist background, as reflected in the book, helped to intensify the Daoist dimension of Chinese medicine. Relying on the works left behind by these great men, the two medical practitioners at the roundtable strive to create a Daoist-inspired health vision in Western society.

Ecology from a Religious Daoist Perspective

Among the participants, the one to comment most extensively on problems of universal ecology was Liu Ming, the founder of Orthodox Daoism of America. He said:

"The fundamental teaching of Daoism is the truth of Oneness (Dao), which means that the unknowable Dao, the source of the myriad things, is not separate from all beings. The uniqueness of each phenomenon, from a speck of dust to the entire universe, is based on a temporary magnetic charge which we describe as qi. This spins tightly into a pattern, from there grows into substance and form, only to unravel and return to the undifferentiated source of the Dao. This natural and universal pattern of qi is the basis of our religious observances and practices. These observances and practices naturally express the Oneness already present; they do not help us 'achieve' it as an outside goal. As a result Daoism, unlike other religions, neither mends or heals any real or imagined schisms between people and their origin nor acknowledges an internal division between the human mind and body. This 'no fix it' view makes Daoism unique.

"When asked to speak about 'Daoism and Ecology,' I am first and foremost concerned with the aggression that is expressed in the premise that all the world is experiencing or about to experience an ecological or environmental 'crisis,' and which also is at the basis of this conference. It reflects the idea that 'fixing' or 'saving' the planet and its people from an inevitable ecological disaster with 'ecological commandments' is imperative.

"This premise is unacceptable to me and, as a Daoist, I cannot condone such a view. It seems to me that environmentalists have a strong, indeed religious, viewpoint and agenda which consciously or unconsciously parallels popular missionary and millenarian Christianity. Am I, as a Daoist and guest, being asked to 'convert' to this kind of assessment of the world and join the mission? If so, I decline to do so. A few presenters have mentioned this Scientism/Christianity parallel, but none have actually addressed it. Perhaps we are being too polite to our hosts, or is it that we find this kind of aggressive 'fix it' or 'save it' viewpoint so common that we think it is 'harmless'?

"As a modern American Daoist, I do not see the aggression in the viewpoint of 'crisis and cure' or 'sin and salvation' as harmless. This view of life promotes a pattern of unconscious behavior (sin), that leads to crisis (damnation), in order to justify the pressing need for salvation and a savior. It seems to me that environmentalism simply 'reworks' this popular Christian sense of life by equating sin with the greedy consumption of resources, damnation with an ecological crisis which, in turn, demands some ingenious and passionate effort of reversal—environmentalist salvation. Who is to provide the salvation this time? Is the environmentalist version of the Apocalypse inevitable? Must we all 'convert' to the environmentalist interpretation of science and support the new missionaries?

"For me, this 'crisis' thinking and the lifestyle it practices—while schizophrenically criticizing it—is a kind of madness, a madness that attempts to aggressively disempower us. If we reject this 'crisis' view of life, are we primitive or naïve or even backward? I have experienced this kind of 'soft' prejudice today, right in this room. Modern environmentalist Scientism has created the same 'blind faith' and the same imperative that characterized eighteenth- and nineteenth-century missionary Christianity. Must we have this rerun? Is there room for a different view? Can the hosts rethink the premise?

"As a Daoist, I do not share the notion that life is a series of problems to be solved. I do not see human beings at odds with nature. Not in the past, not now, and not in the future. I see no crisis. What I do see are human beings who, having the option to be at odds with themselves, choose to practice greed, wastefulness, and fear. I hear them crying 'crisis' and 'save me.' It seems they have always been around. I do not wish to curtail their rights, confiscate their lands, or condemn their beliefs. If they ask me about Daoism, I say that it has provided me with the qi-experience, and not just the belief or hope, that all our small, apparently insignificant actions of every day are in fact part of the Dao and its unknowable, enormous power. I know this because of the constancy and ease I have found using the qi guidelines in my three forms of practice. So I say, calm down. I say, small is not powerless. I say, the future is never certain.

"I agree with many of you that Daoist cosmology does not revolve around human beings. The world we live in as humans is a very small part of the Daoist sense of nature and has never been assessed by Daoists as particularly significant. Earth without humans? Well, true, not a dog would bark. According to this view, the Dao is a great Recycling Center and we are all certainly going to be recycled. But that kind of an idea can never have the feeling of 'news' or be anywhere as urgent as the supposed crisis. It is obvious that we are all to be recycled—and the crisis would be over quickly if people adopted this view.

"I am asked: 'What does Daoism have to offer a world in the ecological crisis?' Perhaps nothing. My experience has shown me that my teachers were right when they said the life of a Daoist is a full-time job. Observances, regulations, and practices cover every aspect of my life. Every truly religious person I have met has agreed that their religious life is full-time—and this may well be the one thing that all religious people have in common. But offering everyone this full-time occupation is not an answer. So, is there some part of Daoism (or any religion) that can be used to heal the environmentalists' wound? My answer is: The parts are fantasies, and so is the wound. There can be no piecemeal approach to life.

"As Dr. Schipper pointed out in his keynote address, orthodox Daoist communities in Chinese history lived by regulations that expressed a deep sense of appropriateness in relationship to their natural resources. Models for the sane uses of resources and relationships abound in the lives of Daoist hermits, married clan priests, their well-managed villages and monastic communities, but these are not models for a fragmentary or patchwork remedy—they are not symptomatic cures. These models cut away at the ignorance and aggression that produces the greed behind ecological inappropriateness, but they also take away the ground of the crisis mentality.

"Having stated that there is no crisis, how then does a Daoist evaluate the statistics? The answer is: Things change. That is what science tells us. Scientific data do not create crisis, and statistics do not create the future. According to orthodox Daoism, every single human action evolves from a view. So, reflect on your view first, realize that you do have one. Then check yourself. You will find that when you go looking for a crisis you will find one.

"The greedy consumption of resources is, like all aggression, inspired by a view that assumes that we are surrounded by shortages. We find deficiencies in the world that parallel deficiencies in ourselves. It is a superficial view that traps us with its so-called facts. To understand properly, we must look deeper, try to understand that below the surface we are all One (or Dao). Then what deficiencies could there be? This is where we truly live together; from this view, we can join all those of past, present, and future who in a similarly sane way share this planet. From there we can live an inspired life which does not misuse our resources."

Ecology and the Health of the Individual

A similar distinction as that made by Liu Ming in the above section, between a sense of crisis in the West and a more holistic and integrative approach, is also found among the medical practitioners. Daniel Seitz says:

"From the Western perspective, medical problems have a material cause, such as a virus. In general, the Western approach is to treat symptoms directly, often with powerful drugs or highly invasive surgical procedures that may cause debilitating side effects. Western medicine also often tries to stave off death at any cost, which is part of the reason that health care expenditures have increased so greatly in the United States. This is not to say that Western medicine has not had many major successes—certainly it has developed many life-saving procedures—but it also has its limitations and, from the Chinese perspective, lacks subtlety and an awareness of the natural laws that govern health and illness.

"Chinese medicine, on the other hand, seeks to promote the body's natural healing ability by stimulating and balancing the body's energy, or qi. The methods of Chinese medicine are much gentler than Western medicine, relying on such things as acupuncture and herbal remedies that rarely cause side effects. Chinese medicine also recognizes that illness can result from spiritual and psychological disharmony in a person as well as from external causes. It recognizes that no two persons are alike and that each person must be approached individually and holistically. It also recognizes that death is an integral part of life. Finally, Chinese medicine has always emphasized prevention over cure of disease, a perspective that is only recently becoming accepted within Western medicine.

"While Western medicine declares itself to be 'scientific,' in a sense there is a no more scientifically based medicine than traditional Chinese medicine. It grew out of centuries of careful observation of natural phenomena, and trial and error. Based on this observation and an understanding of philosophical principles, Chinese physicians developed a thoughtful and consistent body of theory that guides the practice of the medicine and that allows for innovation to meet new challenges. Chinese medicine is, in short, the ultimate empirical science. Yet the Chinese recognize that medicine is an art as well as a science. Just as each human body is unique in certain ways, so is each healer. To be a great healer, the physician cannot rely only on mastery of technique and theory; he or she must have a highly refined intuition and sensitivity and live in accordance with the universal laws, the Dao. Most physicians in the West, unfortunately, do not yet understand that the mastery of medical arts requires self-cultivation and awareness that transcends mere technical expertise and practical experience.

"I consider traditional Chinese medicine to be an 'ecologically conscious' form of medicine. The relationship that traditional Chinese medicine bears to Western biomedicine is similar to the relationship between organic and chemical-based farming methods. Organic farming methods recognize that the earth is governed by certain natural laws; if we are to promote its fertility over a long time period, we must discover and work gently and respectfully within those laws. Chemical-based farming methods, on the other hand, seek short-term productivity at the expense of the long-term fertility of the earth; these methods are often aggressive and coercive and lead to adverse ecological side effects, such as soil depletion and erosion, groundwater contamination, and disease of livestock. Whether we are dealing with the earth or the human body, if we choose to act contrary to such Daoist principles as balance, harmony, and living in accordance with natural law, we do so at our own peril.

"So the gift that traditional Chinese medicine brings to us in the West—besides alleviating symptoms or curing illnesses untreatable by Western methods—is philosophical awareness and wisdom that will contribute to our personal wellness and to the well-being of our planet as well."

Following the comments by Daniel Seitz, Weidong Lu in his remarks concurred with his overall view and added some concrete details on the practice of Chinese medicine, which he described as the longest continuously practiced traditional form of medicine in the world, and its second largest medical system. According to Lu, Chinese medicine is based upon yin-yang theory and follows a basic three-step format. He says:

"The first step involves a general medical diagnosis plus the identification of the current status of the host, i.e., the present constitution of the patient. Three basic types of body constitutions are commonly seen: the regular type, the yin type, and the yang type. The yin-type body means that the patient tends to be cold, clammy, weak, and depressed; a person with a yang-type body is typically warm, active, excessive, and agitated; a regular type is located between the extremes of yin and yang. These three body types, moreover, can be linked to ancient Daoist thought, in that Laozi said: 'The Dao gave birth to One, the One to the two, the two to the three, and the three to the myriad beings.' Nature generated the human being, i.e., the One; the human being is either male or female, i.e., the two; and each human body comes in one of the three basic constitutions, i.e., the three.

"After clarifying the patient's overall body constitution and current state, the second step in the practice of Chinese medicine involves the identification of the properties of the herbs, foods, or drugs to be administered. They again come in three basic types: warming, cooling, and neutral. Cooling herbs or foods have the ability to reduce fever, calm the spirit, and clear toxins. Warming herbs are used to stimulate blood circulation, speed up the metabolism, and increase overall energy. Neutral foods are placed in the middle between cooling and warming ones. Do you know what kinds of food you eat most? Is it cooling or warming? Are you aware that antibiotics are cooling while

Prozac is warming? Studying this aspect of our being in the world helps us increase our awareness as natural and interactive beings who constantly exchange energies with our environment and live only by being in harmonious support with nature around us.

"The third step of Chinese medical practice, finally, involves the matching of the body type with the properties of the herbs or foods in order to achieve therapeutic effects. A warm or hot constitution needs to be treated with cooling or cold herbs to achieve health, which is defined as the harmony of yin and yang. A cooling or cold constitution similarly needs warming or hot herbs to establish proper balance.

"Once a balanced state of health is achieved, one should 'remain relaxed and void, in peace and tranquillity, and free from stray thoughts, so that true qi can rise' (*Huangdi neijing*). This will extend the basic health of the body into a balanced lifestyle in the world and a spiritually rewarding sense of oneness with the universe, or Dao—a state extolled in Daoism and reflected strongly in the medical classics that link the human being through his very personal health with the larger environment of nature and the cosmos."

Creating a Harmonious Environment

A similar sense of harmony and health is also achieved through the practices of fengshui and Taiji quan. Of the former, Linda Varone says that "it is ecology with a small 'e.'" She continues:

"Starting small and based on the key concept of qi, I focus on the individual, his or her family and their immediate environment, such as their home, office, and garden. I do this in the belief that change can start small and spread. I talk with my clients and students about the connection of personal qi, environmental qi, and earth qi as part of raising awareness about nature, and I feel that this may be called convert Daoism. I emphasize the importance of connection to nature to nourish ourselves. For interior spaces the need to bring elements of nature indoors and how this can be done with plants, light, water fountains, aquariums, and pets.

"I also talk about the need for 'decluttering,' which is the first true step of fengshui adjustments. This is what I call 'Cosmic Recycling.' I say: 'Have nothing in your home that is not genuinely useful or truly loved!'

"In addition to houses and their energy patterns, I also focus on gardens, showing people where their wealth area is and where best to locate their compost pile. In my experience, a healthy environment brings forth a healthy person, so that when someone is tending a garden and helping it grow, the garden is also growing the gardener. To give an example, one of my clients had to move from a much-loved home and garden to a new house. As she was struggling with a life-threatening illness, she wanted her new garden to be healing for her. Despite years of gardening experience, she was unsure how to do this. She then shared an insight with me: 'I will just let the garden tell me what it needs, and then I will do it.'

"This is the ultimate stage of applied fengshui—when the rules and systems are left behind and a spontaneous sense of oneness and mutual growth with the environment takes over. It has to be practiced by the individual to work in his or her home. It can be applied to larger environmental concerns only in a limited way, as proposed, for example, by visionaries, such as Steven Post and William Spear, both eminent practitioners and teachers of fengshui. Steven is teaching the "Fengshui of Cities," and William is launching the "Silent Oceans Project," starting this year with nine minutes of silence, a cessation of all manmade noise, in all the oceans of the world. This applies the rules of fengshui to entire complexes of human habitation or the natural world and constitutes a true effort at helping the planet to maintain its balance. But equally as important is the effort undertaken in every person's house and yard and office, our environment being what we have around us, what we live with day in and day out."

The next speaker at the roundtable was Vincent Chu. His main area of expertise is Taiji quan, which he describes as follows:

"Taiji quan is a gentle and slow movement of the body that is continuous and concentrated and serves to create concentration, relaxation, and balance. It is the physical interpretation of Daoist philosophy and, as the Dao cannot be told, cannot be appreciated by description but only by demonstration or, even better, by practice."

To show this, Vincent Chu then proceeded to give a brief performance, after which he asked the audience to stand up and do some of the exercises themselves. In addition, he reported on two of his students, both senior citizens. One noted that after undertaking the practice of Taiji quan for some time she could travel for an hour or more in her car without getting tired, whereas before she had been very exhausted after making such a trip. The other student found that her hands, which had been shaking severely due to Parkinson's Disease, were almost completely steady after about one year of practice.

"These effects," Vincent Chu pointed out, "are due to the workings of and applications of qi, the vital force that resides in all of us and our environment. Through the practice of Taiji quan, we can learn how to move this qi, how to control this qi, how to emit this qi into our environment. We can create harmony from within ourselves and project this harmony into our surroundings, feeling healthier and more whole within and living in better alignment with the world and the people around us."

The Inner Smile

Another, again more religiously centered way of creating harmony both within and without is through the practice of the Healing Tao as taught by Mantak Chia. René Navarro describes this teaching as follows:

"To Master Chia, practice is the heart of Daoism. He does not just put emphasis on theory and philosophy, which to him are just the finger pointing at the moon. Rather, the actual experience of daily meditation and Qigong, internal alchemy and the transformation of qi, ecstatic flight, and dream practice is what constitutes the core of Daoism. 'The Dao that can be told is not the true Dao.' The literature is replete with this view, insisting that we move from words to images to stillness in a process of 'reversal' and 'return,' finding along the way an oscillation between instruction and practice that guides us further.

"Master Chia divides his curriculum into two areas: health and longevity exercises, and internal alchemy and immortality techniques. The first covers Qigong, sexology, massage, nutrition, and the martial arts; the second covers the fusion of the Five Phases, of the trigrams Kan and Li, and various other advances spiritual practices, leading to the Congess of Heaven and Earth and the Reunion with the Dao. In concrete terms, the Healing Tao curriculum begins with the Inner Smile, a simple but profound meditation that requires the practitioner to smile into and contact the different parts of the body—organs, digestive system, skeleton—and to build a coccoon of light around it for self-protection. When I first came in contact with this, it was a radical practice for me, with the practitioner visualizing his or her body and bathing it with the warm glow of an inner smile. Then I studied the Six Healing Sounds, a technique that guides adepts to use a particular sound to stimulate and harmonize each of the five major organs in the body. After that, I went on to practice the Microcosmic Orbit. This establishes a circle of energy through the body which opens and bridges the two primary qi-meridians, or energy channels of the body—running along the spine and straight through the center of the front torso.

"To me, doing all these was an increasingly profound experience which brought me inward into the energetics of the body—the concept of yin-yang and the Five Phases as well as the energy centers and the meridians of qi. I realized that there is a universe within that embraces all the different correspondences—phases, colors, flavors, plants, animals, officials, emotions, and many more.

"Then I went on to study what is called 'Fusion of the Five Phases'— a formula of balancing the negative emotions in the Four Trigrams, i.e., the navel, Gate of Life (in the abdomen), and left and right sides of the body. Through it I learned to connect the senses to their respective organs, to nurture and grow the virtues, and to open more qi-meridians, including the Thrusting Channels, the Belt Routes, and other psychic channels.

"This work served to accomplish several goals: 1) to develop respect and love for my body; 2) to develop a relationship with and sensitivity to my inner organs, including their related functions, emotions, and healthy activities; 3) to learn the different types of breathing, including relaxed breathing to quiet the spirit (*shen*), bellowing breathing to activate and regulate the qi, reversed breathing to strengthen the essence (*jing*), and embryonic breathing which means to breathe like a fetus and be like a child again; 4) to 'wire the etheric body' so that the qi from within and without can move freely; 5) to develop stillness and quiet; 6) to be able to connect to the environment—trees, oceans, mountains, lakes, and stars; and 7) to stop the drainage of energy and attain full healing of the self.

"In the mid-1980s, I went on to practice internal alchemy, which involved following formulas of inversion or reversal that had been transmitted to Master Chia from the White Cloud Hermit in Hong Kong. It strives for the Enlightenment of the trigrams Kan and Li, standing for cosmic Water and Fire, and begins by using the energies of the kidneys/sexual organs and the heart/chest to initiate a quiet steaming. This arises first in the Lower Cauldron or Crucible, then ascends to the Upper Cauldron in a gradual regimen of meditation that involves the addition of more alchemical ingredients. The latter include things like the sun and moon energy, the power of the Big Dipper and North Star, and other astral forces. As one progresses along this path, one finds that the polarities of the body are gradually dissolved and original patterns begin to invert. Fire, an originally rising energy, starts to go down; water, a naturally downward moving force, begins to go up. They couple through the Central Thrusting Channel, following the model set out in Hexagrams 64 and 63 of the *Yijing* (Book of Changes), and create a new level of body and cosmic energy.

"Since 1997, Mantak Chia has also taught the 'Sealing of the Five Senses,' a formula which brings the alchemical work higher into the Crystal Room and requires the closing of the senses to prevent drainage and distraction. The complete sensory absorption achieved in this practice is similar to the original state of chaos described in the Hundun story of the *Zhuangzi* (chapter 7) and can be linked with traditional forms of mystical union described by religious Daoists of other schools. In addition, this 'return to the source' is also attained with other formulas of internal alchemy, such as the Congress of Heaven and Earth and the Reunion of Heaven and Humanity. They all serve to bring the individual back to a state of chaotic oneness at the root of all existence, before yin and yang separated and the world was created. Resting in this ultimate peace and oneness, practitioners find a wholeness in themselves that makes them more complete and fully enlightened human beings, able to withdraw into the quietude of universal beginning and to interact fruitfully with society and the environment.

"Advanced followers of the Healing Tao, then, also become healers, so that active Daoists often are also serious medical practitioners. I myself am a licensed acupuncturist and Chinese herbalist. Spreading the Healing Tao to others, I have taught massage, meditation, Qigong, and Taiji quan for many years. I find it essential at the present time that we lay a solid cultural foundation for Chinese medicine and Daoist-inspired holistic healing in the West. Many acupuncturists only know acupuncture; therefore, their treatment is focused on acupuncture. A holistic, Daoist-oriented medicine should have not only a holistic diagnosis but also a holistic therapy. The treatment should include not just one but several approaches. This requires teachers and practitioners of healing to expand their interest and expertise so that a truly holistic medicine can be practiced. If a cultural foundation is not established, people will continue to look for a quick fix, a silver bullet, to take away all their ill health and dis-ease.

"Building such a cultural foundation for Chinese medicine and Daoist healing requires, first of all, learning different healing modalities, but it also involves making these modalities a part of our lives as healers. That is, the healer must have a life that revolves entirely around these healing modalities, while the general population who use them should have at least a working knowledge of them. As Daoist practitioners, we undertake different practices at different times of the day in our everyday lives. Even when asleep, we do forms of dream practice or a technique of Qigong. It never ends—healing is our life. I agree fully with Liu Ming when he says that Daoism is full-time work."

Conclusion

Although the six speakers at the roundtable were from widely different backgrounds and undertaking different kinds of Daoist or Daoist-inspired practices, there was an amazing degree of consensus among them. They all agreed that what they did was related to Daoism and owed a great deal of its worldview and/or technical application to the religion. They also agreed that the goal of their practice was the establishment of harmony on different levels: within the individual and his or her body; between the individual and his or her surrounding environment and society; and between the individual and the cosmos at large. In addition, they all emphasized that the ultimate, ideal level of attainment was one of oneness, of merging of self and environment in a sense of cosmic union that alone could lead to the complete fulfillment of the self in and beyond this world.

Beyond these striking commonalities, the most amazing agreement among the practitioners is their unquestioned emphasis on the individual. None spoke about larger social units, and the only one to mention the planet as a whole, Liu Ming, did so only to reject the present sense of crisis and universal urgency as an artificial construct of the Christian West. He, as did all participants in the roundtable discussion, stuck to personal, individual practice as the key for creating a saner, healthier, and more environmentally conscious world. Every single individual, the practitioners agreed, has to work on his or her personal life by learning how to be healthy, how to live in harmony with the environment, and how to attain a sense of cosmic oneness. The environment, although not taking center stage, is always present. But it is not "The Environment," some abstract, far-off entity that we see on television or discuss in academic conferences. Environment for contemporary Daoist practitioners is the environment of our own bodies, our own homes, our own backyards, our own immediate social relations. And that is where change begins—must begin and can only begin—with each individual living a healthy life, following morality, and decluttering his or her home and personal environment.

This emphasis on the individual, which can be described as a nuclear form of environmentalism, is contrary not only to the Western-inspired sense of crisis and planetary perspective. It also stands in contrast to our commonly accepted vision of the Daoist tradition. Daoism is most commonly understood either as an ancient philosophical system, as represented in the works of Laozi and Zhuangzi (discussed at length in this volume), or as a communal religion of the kind Kristofer Schipper describes in his contribution. Daoism as the healing of the individual, as a way in which a single human being can create harmony for himself or herself, is not commonly discussed or acknowledged. And yet this form of the religion is what all practitioners find themselves representing. Personalized practice for them is the key to achieving not only well-being for oneself, but also healing for the community, the environment, and eventually the planet. "Change starts small," they say while pursuing the practice, as Linda Varone says, of ecology with a small "e."

Daoist Environmentalism in the West: Ursula K. Le Guin's Reception and Transmission of Daoism

Introduction

A generation ago, few Westerners knew much about Daoism, and sinologists were scrambling to reserve just a small place for the subject on university syllabi and bookstore shelves. But now, as scholars begin to speculate exactly how we can make Daoist ideas relevant to modern problems, there is a certain temptation to overlook the fact that there is currently an unprecedented cross-fertilization of Daoism with Western popular culture. Recent years have witnessed an extensive proliferation of self-proclaimed Daoist journals, meditation centers, and even websites, as well as new literature on Daoism and renderings of Daoist texts by nonspecialists. For example, *The Empty Vessel: A Journal of Contemporary Taoism*, published by an organization called the Abode of the Eternal Tao, is a periodical "dedicated to the exploration and dissemination of nonreligious Taoist philosophy and practice."[1] Along these same lines, the Fung Loy Kok Daoist Temples, which actually trace their lineage back to the tenth-century figure Chen Tuan, are sprouting throughout the United States and Canada. The Western fascination with Daoism is perhaps nowhere more evident than on the internet, a vast but uncoordinated library of webpages such as "Tao Secrets of a Peaceful Life," "The Western Reform Taoist Congregation," "Gay Taoism," "Center for Taoist Thought and Fellowship," "13 Chapters on Celtic Taoism," "Taoism and Atheism," and many other sometimes transitory cyberspace gathering places. In short, Daoist influences now occur in a range of unexpected Western contexts. My own wedding included a recitation from the *Daode jing*, an application of the text quite distinct from anything Laozi or even Zhang Daoling could have imagined.

Not surprisingly, the sinological community (myself included) has generally greeted this overall phenomenon—what I will from here on label "popular Western Daoism"—with varying degrees of indifference, amusement, and derision. Russell Kirkland has characterized Stephen Mitchell and others as "self-indulgent dilettantes who deceive the public by publishing pseudo-translations of the *Tao Te Ching*, without having actually read the text in its original language."[2] In a more generous vein, Kirkland labels Benjamin Hoff's coffee-table books "mindless fluff" and urges colleagues not to "Pooh-pooh Taoism." However, the spectrum of Western appropriations of Daoism is not a homogeneous field, and I believe it is premature to dismiss it all with one calligraphic brushstroke. In this essay, I will first argue briefly that scholars of Chinese religion would be well-served to take some of this material seriously, and then discuss in some detail one unusually interesting and scholastically responsible example of these popular Daoist forays and relate it directly to the topic which is the centerpiece of this volume.

The principal complaint against popular Western Daoism is that it is ahistorical and inauthentic, and much of it does indeed combine a tenuous link to the historical tradition with a promiscuous blend of Western individualism and New Age universalism. For example, The Western Reform Taoist Congregation, a typical "new religion" founded by a disaffected Methodist named Michael Torley, lists on its website a "creed" which professes belief "in the formless and eternal Tao" and a set of doctrinal positions on subjects ranging from abortion and euthanasia to sexual orientation and extraterrestrial life. But while a determination of questionable scholarship would be sufficient for rejecting a scholarly work, popular Western Taoism is not exclusively or even primarily a scholarly phenomenon. Rather, it is an *aesthetic*, *cultural*, and *religious* phenomenon. Any historian of religion would agree—or at least should agree—that religious resources continually undergo reinterpretation and recontextualization, and it is simply unrealistic to expect religious history—even modern religious history— to conform to scholarly expectations. To a great extent, the history of religions is a history of (sometimes unintentional) hermeneutic decisions, and it is most appropriate not to judge and dismiss a modern phenomenon because of its supposed lack of authenticity but to identify its hermeneutic and situate it properly in its own historical and thematic contexts. To deem popular Western Daoism unworthy of study because it is not sufficiently continuous with the Chinese tradition is to repeat the kind of parochialism that once relegated *Daojiao*—China's oldest indigenous institutional religion—to the status of a corrupt degradation of original Daoism.

There are, of course, other dynamics in play that explain why sinologists are generally reluctant to take on this material. First, there is a pragmatic concern, the fear that dignifying popular Western Daoism will tacitly encourage a contemporary audience to view it as an accurate scholarly representation of Chinese Daoism. In short, the legitimation of apocryphal versions of Daoism simply makes it that much harder for us to teach real Daoist history. Kirkland playfully notes that many students "are happy deluding themselves that they *themselves* are Taoists, because generations of irresponsible Westerners have made good money by telling the public that being Taoist just means being Pooh-brained."[3] Norman Girardot echoes this sentiment when he notes the challenge of contending with "student wayfarers much too certain of the method and destination of their Taoist journeys," who easily relate Mitchell, Hoff, and "the excremental vision of the *Chuang Tzu*" to a kind of Generation X spontaneity.[4] This is indeed a problem, but I would point out that it is one that pales in comparison to the problems faced by our colleagues in Western religion who must constantly bring their scholarship into competition with both the media and religious institutions unaware of their own interpretive lenses. Part of our task as sinologists has been to bring a Western audience into dialogue with things Chinese, but it is not entirely within our control—nor do I believe it should be—exactly what direction that dialogue will take.

A corollary to this is the idea that popular Western Daoism is more a part of American cultural studies than of sinology, and thus it is simply not our scholarly responsibility. While this is ostensibly correct—most existing scholarship on these Daoistic cultural crossovers is indeed being produced by specialists in English literature, fine arts, and popular culture—it is only we who have the expertise to explain how the Western appropriations relate back to their Chinese counterparts, to identify continuities and discontinuities, and to articulate exactly how the Daoist ideas or practices are being transformed. Furthermore, and I do not make this point lightly, we might even learn something from these transformations. It is now considered crucial in the studies of Chinese and Japanese Buddhism that they be viewed not as misunderstandings of Indian Buddhism but as the bringing to the surface of characteristics latent in the original tradition.[5] Similarly, I have elsewhere argued that Martin Buber's encounter with the *Zhuangzi* provides some useful intellectual tools through which to view the original Chinese text.[6] And now, as this volume sometimes blurs the line between historical scholarship and the constructive mining of Daoist resources, it is entirely possible that some of the questions that scholars and environmentalists are bringing to the tradition may find their most creative answers within its existing modern Western transformations.

A "Novel" Approach to Daoism

Fantasist and science fiction author Ursula K. Le Guin is an especially interesting case in point, as her involvement with Daoism reflects not a passing fancy but a lifelong passion. In fact, it would not be presumptuous to note that Le Guin, who was born in 1929, has been taking this subject seriously since before many or most of the contributors to this volume were born. Le Guin was first exposed to the Laozhuang tradition by her father, noted anthropologist Alfred Kroeber, who read frequently from the Paul Carus version of the *Daode jing* and even requested that passages from it be recited at his funeral. That marked the beginning of Le Guin's engagement with the texts of Laozi and Zhuangzi, and though she is cautious about representing herself as a "Daoist" or her work as intentional illustrations of Daoist themes, she readily acknowledges the significance of Daoism in the formation of her worldview and the evidence of this in her fiction. In *The Left Hand of Darkness*, for example, she challenges conventional dualistic thinking by offering a range of yin-yang complement dualisms,[7] most noticeably in her portrayal of a race of genderless beings who temporarily take on either male or female sex characteristics at different times in their reproductive cycles. This novel also introduces a mysterious cult that takes unlearning as a goal and prizes darkness and the creative potential of chaos. In both *The Word for World Is Forest* and *The Lathe of Heaven*, Le Guin explores the relationship between dream and reality and, as if to footnote her source of inspiration, quotes Laozi and Zhuangzi throughout the latter. In fact, the title—*The Lathe of Heaven*—is actually drawn from James Legge's anachronistic translation of *tianjun*, a phrase from the *Zhuangzi*. And in *The Dispossessed*, Le Guin imagines a quasi-anarchistic Daoist utopia, where ethical norms are recognizable as variants of *wuwei*. These same motifs also recur, though less overtly, in Le Guin's many published works on literary technique, travel, feminism, and various social issues.[8]

Le Guin scholars, while making discrete but limited use of sinological resources, have correctly identified Daoist themes in these and several of her other works, notably *City of Illusions*, the three novels making up the Earthsea trilogy, and the short story "Vaster than Empires and More Slow." Interestingly, the scholarly discussions of Daoism in Le Guin's novels make up something of a cottage industry in science-fiction and English literature circles. For instance, Douglas Barbour states that *The Lathe of Heaven* "is practically a primer in Taoist thought," and that its hero "is an unconscious Taoist sage."[9] Elizabeth Cummins Cogell, one of Le Guin's many bibliographers, sees a fluid progression of psychological, social, and political applications of Daoist ideas, with these motifs developed most maturely in *The Dispossessed*. Cogell detects three major themes—"following the model of Nature, the Theory of Letting Alone, and the eternality of change"[10]—woven into both the structure and content of the novel. She writes, "Living in a period of rapid change and search for meaning, [Le Guin] has rediscovered, updated, and brought to our attention the philosophical responses of ancient China to similar challenges."[11] Dena C. Bain views the *Daode jing* in particular as the foundation of much of Le Guin's fiction, arguing that there is "a basic mythos underlying [three of] the novels based on the Quietist philosophy of Lao Tzu's Taoism: the concepts of wholeness, of presence, of reconciling forces which appear totally opposed, but which, in the moment of complete reduction and return to the Uncarved Block, are invariably revealed to be necessary complements."[12] Taking a different approach, Robert Galbreath suggests that the maturation of the hero from *A Wizard of Earthsea* parallels a movement from religious Daoism to philosophical Daoism, from "the need to overcome death by controlling the dead and gaining immortality" to "the integration, not subjugation, of his powers," from a magic based on "manipulation and coercion" to one based on "acceptance and recognition."[13] John H. Crow and Richard D. Ehrlich attribute much of Le Guin's Daoist explorations to the influence of Holmes Welch's *Taoism: The Parting of the Way*, noting that "Welch's book seems to have had a profound effect on Le Guin, from matters of philosophy to suggesting the title of and theme for a short story ['Field of Vision'] and the scene in *The Lathe of Heaven* in which [two characters] discuss giving or withholding snakebite serum."[14] Thomas J. Remington, David L. Porter, James Bittner, and several others have also found examining Daoist parallels helpful tools in understanding Le Guin's work.

But while much of this scholarship is interesting and insightful, it is only intermittently relevant to a specialized discourse on Daoist philosophy and religion. At the very least, the material reflects a selective and incomplete understanding of Daoism in its historical contexts. For example, Elizabeth Cummins Cogell is obviously indebted to Joseph Needham's interpretations when she states that "out of the concept of Tao comes the belief in the unity and spontaneity of nature which results in a pantheistic view."[15] And in some cases, the authors approach the subject with an overly broad, even romantic, view of the wider history of religions and the comparative study of mysticism. Dena Bain frames her article by claiming, "Unlike Western religious thought, which sees the universe as real and God as a person, Eastern tradition sees the universe as illusory and God as an impersonal force. In both Eastern and Western mysticism, however, the mystical experience is nothing less than direct intuition of Ultimate Reality—a supreme being in the West, a supreme state in the East."[16] But perhaps most importantly, scholars of Le Guin's work are interested in Daoism only to the extent that it helps them better to understand her novels and short stories, and their general methodology—that of combing the writings for traces of Daoist influence in order to elucidate particular literary themes—often resembles a treasure hunt for Daoist bread crumbs. This task may be a valuable one in and of itself, but it represents a very particular hermeneutic agenda. Daoism does indeed play a major role in Le Guin's work, but the details of her reception and transformation of it, as well as her place in the creative development of the tradition, are not really transparent from discussions of the fiction alone.

The Left Hand of Laozi

The culmination—at least for the time being—of Ursula K. Le Guin's involvement with Daoism is her recent version (which she insists is not a "translation") of the *Daode jing*. This rendition of the text reflects a deeply personal project which Le Guin had been undertaking for many years, and it was only at the urging of J. P. Seaton, with whom she periodically consulted, that she considered publishing it.[17] Although Le Guin has not studied literary Chinese and characterizes her own text commentaries as "idiosyncratic and unscholarly,"[18] the volume does reflect painstaking research and considerable methodological self-consciousness, and her overall approach in many ways resembles that of liberal biblical exegesis. On the one hand, Le Guin makes every effort to reconstruct the authorial intention behind the text. Paul Carus's translation provided her with a character-by-character transliteration and translation of the original Chinese, which allowed her to assess the interpretive choices made by several important translators from past and present generations. And despite her lack of official sinological credentials, she astutely determines that the translations by Arthur Waley, Robert Henricks, and D. C. Lau are the most trustworthy, that the hermeneutic experiments of Michael LaFargue are "quirky" but "useful," and that the poetic appropriations of Stephen Mitchell and Chang Chung-yuan are vague and imprecise.[19] But on the other hand, Le Guin is perhaps less interested in recovering the motivations *behind* the text than she is in imagining what awaits *beyond* it or, in Paul Ricoeur's words, "the horizon of a world toward which a work directs itself."[20] Thus, she is unapologetic about omitting, rearranging, reinterpreting, or interpolating passages that jeopardize her ability "to make aesthetic, intellectual, and spiritual sense" of the text. She presents the text in its traditional order—while noting judiciously the disparity between the "original" and Mawangdui versions—and appends evocative titles (for example, "The Small Dark Light," "Children of the Way," "Being Obscure") to each chapter, frequently including concise commentaries and chapter notes to explain the thinking behind her translations. Le Guin seems to be engaged in a constructive conversation with the text in a way that positions her between the extremes of positivism and postmodernism as she seeks neither an objective reconstruction of authorial intent nor a simple projection of her own propensities onto the text.

The most striking example of this is that Le Guin's Laozi—and, by extension, Le Guin's Daoism—is unambiguously egalitarian. The authoritarian, hierarchical, and masculine language that recurs in many modern editions is excised here in order that the text may speak to "a present-day, unwise, unpowerful, and perhaps unmale reader, not seeking secrets, but listening for a voice that speaks to the soul."[21] This is evident in Le Guin's translation choices for several staples of "Hundred Schools" jargon, where she does not alter the meanings so much as redirect them. She renders *shengren* not as "the sage" but as "wise souls" or "the wise"; *tianxia* not as "the empire" or "under heaven" but as "the commonwealth," "the public good," or "the body politic"; *xiao* not as "filial piety" but as "family feeling"; and *li* (as a problematic quality) not as "ritual" or "propriety" but as "obedience." In other places, Le Guin is more bold with her glosses, as when she renders a key passage: "The Way is great, / heaven is great, / earth is great, / and humankind is great; / four greatnesses in the world / and humanity is one of them."[22] Her failure to translate *wang* as "king" reflects neither a misreading nor a misrepresentation, as she earnestly explains that "the king garbles the sense of the poem and goes against the spirit of the book," adding, with much the same gusto with which she wrote "the king is pregnant" in *The Left Hand of Darkness*, "I dethroned him."[23] Along these lines, Le Guin maintains that the "spirit of the book"—described as "funny, keen, kind, modest, indestructibly outrageous, and inexhaustibly refreshing"[24]— does not jibe with descriptions of autocratic rulers or military strategy. Thus, she simply excludes a famous and troublesome passage about the ruler concealing "the state's sharp weapons," dismissing this "Machiavellian truism" as an "anticlimax" and "intrusion."[25] Likewise, she expunges references in an otherwise pacifistic chapter to ceremonial wartime courtesies and jettisons obscure claims about the needs of large and small states.[26] When Le Guin does preserve the problematic language, she nevertheless offers imaginative commentary. The chapter beginning with the oft-quoted "Rule a big country / the way you cook a small fish" is considered not as a political maxim but as "counsel to the individual."[27] Again, Le Guin is never less than forthcoming about her own revisioning of the material, as she acknowledges that a virtually rewritten passage—"In looking after your life and following the way, / gather spirit"—would be more conventionally translated as "In controlling people and serving heaven / it's best to go easy."[28]

Interestingly, while Le Guin does insist on an egalitarian reading of Laozi, she does not advocate an explicitly feminist interpretation of the text or exaggerate the significance of the recurring feminine and mother imagery, as many less careful readers are inclined to do. One might argue that Le Guin's feminist writings elsewhere bear the stamp of Daoist influence, but with the *Daode jing* itself she offers minimal commentary on the passages that are usually cited as proto-feminist, and when she does address the prescription for "knowing man and staying woman," she situates it in a broader context.

> The reversals and paradoxes in this great poem are the oppositions of the yin and yang—male/female, light/dark, glory/modesty—but the "knowing and being" of them, the balancing act, results in neither stasis nor synthesis. The riverbed in which power runs leads back, the patterns of power lead back, the valley where power is contained leads back—to the forever new, endless, straightforward way.[29]

Likewise, the well-known verse about the "valley spirit"—"The mystery, / the Door of the Woman, / is the root / of earth and heaven"[30]—avoids the almost Jungian convention of evoking some archetypal "mysterious female" and reverses the "heaven and earth" dyad, thus literally taking on a more palpable and earthy flavor. Le Guin seems implicitly satisfied that the inclusiveness of Laozi's vision deems unnecessary any impulse to romanticize feminine imagery that speaks sufficiently and profoundly for itself.

But this is not to suggest that Le Guin is not ultimately drawn to the more mystical elements of the text, for she repeatedly portrays Laozi as a mystic even without directly saying what that means, beyond observing that "mysticism rises from and returns to the irreducible, unsayable reality" of the way. She offers easy comparisons with Emily Brontë, Emily Dickinson, Henry David Thoreau, William Blake, William Wordsworth, and Gerard Manley Hopkins, which gives the cumulative impression that she may see Laozi as one representative of a perennial mysticism, and the language she chooses often takes on an almost mythic quality (much like some of her own fiction). "Once upon a time / people who knew the Way / were subtle, spiritual, mysterious, penetrating, / unfathomable."[31] This mythic mood recurs especially in passages concerned with returning, reversion, or unlearning. "Mysterious power / goes deep. / It reaches far. / It follows things back, / clear back to the great oneness."[32] Le Guin is fascinated with the experience of Dao, "a pure apprehension of the mystery of which we are a part."[33] She circumscribes the unspeakable way as the wellspring of bounteous potentiality and creativity, glimpses of which may offer one "a sense of serene, inexhaustible fullness of being."[34] This is embodied through the paradoxical practice of *wuwei*, which she defines as "uncompetitive, unworried, trustful accomplishment, power that is not force"[35] and characterizes as "a concept that transforms thought radically, that changes minds."[36] As Le Guin imagines the practical meaning of *wuwei*, her Daoism seems to tread carefully over three very subtle tensions. First, it is naturalistic without being primitivistic. She views Laozi's nostalgia for less complex times as a rejection not of the material world but of both "material dualism" and "religious dualism,"[37] and she simply and undramatically translates *pu* as "uncut wood" and *ziran* as "what is" or "as they are." Second, it is utopian without being romantic. Le Guin envisions a community or communities of people in spontaneous harmony, yet she also acknowledges the role of rightful political power and implicit social structures. Third, it is deconstructionist without being nihilistic. She identifies positivism—the certainty "that our beliefs are permanent truths which encompass reality"—as "a sad arrogance" and observes that "conviction, theory, dogmatic belief, opinion" all fall under the rubric of the wrong kind of knowing.[38] But she does not see this as resulting in an intellectual, moral, or social paralysis, and she maintains an implicit confidence throughout in a liberating form of perception and experience. As Le Guin's translation reads, "To know without knowing is best," her commentary calls such a sentiment "obscure clarity, well-concealed jade."[39]

Daoism, Imagination, and the Environment

Thus far, I have not related Le Guin's reception and transmission of Daoism to the subject of ecology, and she is quick to admit that any discussion of environmental consciousness in the Laozhuang tradition must be left to extrapolation. Neither Laozi nor Zhuangzi was writing at the time of an ecological crisis—indeed, they were sufficiently concerned with other contemporaneous crises—and one may only assay an intelligent speculation as to how either one of them might have addressed such an issue. Still, the qualities Le Guin explicitly or implicitly ascribes to Daoism—egalitarianism, earthiness, naturalism, utopianism, mystical and mythic ambiance—are almost certainly consonant with a responsible environmentalism, and to press with such an argument here would no doubt rehearse ground already covered elsewhere in this volume. In any event, Le Guin does view Daoism as unambiguously placing humankind as one of the ten thousand things, with a mandate to harmonize with the planet rather than master or utilize it. This recalls the writing of feminist theologian Beverly Harrison, who notes that ecologists quickly make the connection "that we are part of a web of life so intricate as to be beyond our comprehension."[40] And while Le Guin does not directly attribute her own environmental ethic to Daoist influence, it is clear in her work that there is a strong resonance between them. In a self-conscious reflection on what is arguably her best-known work, *The Left Hand of Darkness*, Le Guin considers the implications of her reconstructed gender identifications and, in so doing, makes evident some significant connections:

> [It] seems likely that our central problem would not be the one it is now: the problem of exploitation—exploitation of the woman, of the weak, of the earth. Our curse is alienation, the separation of yang from yin (*and the moralization of yang as good, of yin as bad*). Instead of a search for balance for integration, there is a struggle for dominance. Divisions are insisted upon, interdependence is denied. The dualism of value that destroys us, the dualism of superior/inferior, ruler/ruled, owner/owned, user/used, might give way to what seems to me, from here, a much healthier, sounder, more promising modality of integration and integrity.[41]

Through her invocation of one foundational Daoist theme, the integration of yang and yin, Le Guin relates Daoist philosophy to both ethical and ecological issues, thereby blurring any artificial distinction between social justice and environmental responsibility. The impetus to integrate man and woman, to break down hierarchical social structures and utilitarian attitudes, is the same impetus to integrate humankind with the planet and the cosmos. Along these same lines, though without specifically invoking Daoism, Le Guin elsewhere discusses the environmental implications of alienation, separation, and false dualism. "The whole science of ecology . . . describes exactly what we're doing wrong and what the global effects are. The odd twist is that we become so enamored of our language and its ability to describe the world that we create a false and irresponsible separation. We use language as a device for distancing. Somebody who is genuinely living in their ecosystem wouldn't have a word for it. They'd just call it the world."[42] By extension, even if Le Guin's Daoist sages are not specifically concerned with ecological matters per se, they would, as integrated participants in the world, "mingle their life with the world."[43] In other instances, Le Guin's connection between Daoism and ecology is more overt, as when she characterizes Laozi as a "kind, wise guide" into "real mystical feeling and real moral responsibility" and immediately follows this with a discussion of the habitual misuse of the environment:

> All through my lifetime we've been wasting our world as fast as we could, using up, throwing away, polluting, burning, cutting down, killing off. I knew a beautiful place called California, when I was kid. These days it's not there. It's housing developments and WalMarts and Golden Arches and eight-lane highways. The automobile will strangle America if overpopulation doesn't do it first. In America and all over the difference between the few rich and the ever-increasing billions of poor grows greater and greater. Efficient capitalism is just as destructive as blundering communism, maybe more so.[44]

Le Guin's advice, to the extent that she offers any during this particular conversation, is "try to survive, and maybe try to make things a little better in our local corner." In order for us to do this, she adds, "we need our imagination—well exercised and in good working order." And the landscapes of Le Guin's imagination are one place to find Daoistically flavored, environmentally responsible worlds. To bring this discussion full circle, I would suggest that this is best developed in one of Le Guin's relatively recent novels, *Always Coming Home*, a story that on the one hand is easily recognizable as a Daoist reverie and on the other hand regularly appears on university syllabi and in bibliographic collections as an "ecotext." In this "archaeology of the future," Le Guin envisions a people who live harmoniously without an underlying myth of progress, who value a kind of primordial wisdom without seeking and applying knowledge. They are aware of the people who live across the border and even have official relations with them, but for the most part they are satisfied not to wander beyond their own horizons. And while they have a basic technology and a vast archive, they only infrequently avail themselves of these resources. What is most striking about Le Guin's speculative phenomenology of Daoism is that it reveals a world that is surprisingly comfortable with some modicum of technology, though she longs for an appropriately *ziran* manifestation of it.

> I hope I've never been, and am never, perceived as being in any way "anti-science" in my work. Confusion often arises concerning what science and technology are. For example, I thought *Always Coming Home* was a rather interesting work in the technological mode; I had tried to think out carefully and consistently a highly refined, thoroughly useful, aesthetically gratifying technology for my invented society of the Valley. Being an anthropologist's daughter, I think of technology as encompassing everything a society makes and uses in the material sphere. However, a lot of people now use 'technology' simply to mean extremely high-tech inventions that are predicated on and depend on an enormous global network of intense exploitation of all natural resources, including an exploited working class, mostly in the Third World.[45]

Thus, the issue is not reduced to a conflict between scientific/technological development on the one hand and environmental consciousness on the other hand, between selfish utilitarianism and responsible global awareness. This, according to Le Guin, is a false dichotomy, predicated on the mistaken assumption that any kind of "usefulness" runs contrary to the natural workings of the world. In fact, Heaven and Earth are bounteous and sustaining—"inexhaustibly giving"[46]—but the challenge is to engage their resources in a way that is conscientious and honorable: to utilize without exploiting, to take without taking away, to develop without depleting. And in the words of Le Guin's Laozi, "To have without possessing, / do without claiming, / lead without controlling: / this is mysterious power."[47]

Conclusion: The Question of Crosscultural Appropriation

To summarize, in Ursula K. Le Guin's lifelong engagement with the Laozhuang tradition, we have a fascinating encounter between the Western and the Daoist imaginations, and the result is a creative, responsible, and compelling vision that is worthy of both our scholarly attention and our moral ear. Still, the fact that Le Guin has executed her task so delicately does not entirely disarm the broader, much thornier matter of cross-cultural appropriation. As Russell Kirkland noted at the conference on Daoism and ecology, the borrowing of re-sources from other traditions, especially those of Native Americans and other groups not immediately or easily empowered to control the discourse, may in practice amount less to a faithful inter-religious dia-logue than to a kind of "cultural strip-mining." Modern sinologists are loath to replicate the intellectual "orientalism" of some of their prede-cessors, and they express some justifiable protectiveness when they encounter what might be termed "religious orientalism," whether in the form of unwitting insensitivity or predatory colonialism. It was suggested earlier that scholars do not really have the authority to dic-tate the directions of Western dialogue with Chinese religious and philosophical materials, but it is also not self-evident that they need remain silent when observing some kind of violation. The reality is that within the modern academic study of religion, the conversation on cross-cultural appropriation is still very much in its infancy, and many of us are struggling to develop viable criteria for distinguishing creative reinterpretation from problematic misappropriation and to establish reasonable norms for academic engagement with materials whose "ownership" may be a delicate or controversial matter. In short, it is one of the pressing challenges for the emerging generation of schol-ars to make intellectual sense of this issue and all of its concomitant implications. Much to her credit, Le Guin provides an exemplary role model in this endeavor, in her sincere and respectful encounters with Daoism, where she strives to take without taking away, to be both "passionately sympathetic" and "exceedingly conscious of possible transgression." Daoism has historically been a highly complex affair, as is the relationship between Daoist history and environmental con-cerns and the question of how the vast repository of Daoist wisdom may offer some helpful responses to modern ecological crises. Never-theless, I am confident that the Western transformation of Daoism—at least one corner of it—is in good hands.

Notes

1. This is quoted from the webpage of the Abode of the Eternal Tao. The journal is self-consciously ahistorical, as the editor resists including materials from the Daoist canon, arguing that "the last thing we need right now is another religion, even if it is based on Taoism." Solala Towler, "From the Editor," *The Empty Vessel*, spring 1994, 2.

2. Russell Kirkland, "Teaching Taoism in the 1990s," *Teaching Theology and Religion* 1, no. 2 (June 1998): 113.

3. Ibid., 114.

4. Norman Girardot, "My Way: Teaching The *Tao Te Ching* and Taoism at the End of the Millennium," forthcoming in *Teaching the Tao Te Ching*, ed. Warren Frisina (New York: Oxford University Press).

5. I am indebted to Masatoshi Nagatomi for emphasizing this point.

6. See Jonathan R. Herman, *I and Tao: Martin Buber's Encounter with Chuang Tzu*" (Albany: State University of New York Press, 1996).

7. The term "complement dualism" is actually lifted from Wilfred Cantwell Smith's perceptive introduction to Chinese religion in *The Faith of Other Men* (New York: Harper, 1962), 67–80.

8. See, for example, *Dancing at the Edge of the World* (New York: Grove, 1989).

9. Douglas Barbour, "The Lathe of Heaven: Taoist Dream," *Algol*, November 1973, 22.

10. Elizabeth Cummins Cogell, "Taoist Configurations," in *Ursula K. Le Guin: Voyager to Inner Lands and to Outer Space*, ed. Joe De Bolt (Port Washington: Kennikat Press, 1979), 154.

11. Ibid., 179.

12. Dena C. Bain, "The *Tao Te Ching* as Background to the Novels of Ursula K. Le Guin," *Extrapolation* 21, no. 3 (1980): 221.

13. Robert Galbreath, "Taoist Magic in the Earthsea Trilogy," *Extrapolation* 21, no. 3 (1980): 262–64.

14. John H. Crow and Richard D Erlich, "Words of Binding: Patterns of Integration in the Earthsea Trilogy," *Ursula K. Le Guin*, ed. Joseph D. Olander and Martin Harry Greenberg (New York: Taplinger, 1979), 238.

15. Cogell, "Taoist Configurations," 155.

16. Bain, "The *Tao Te Ching*," 209.

17. It is an interesting coincidence that Le Guin was invited to participate in the conference on Daoism and ecology *before* her translation of the *Daode jing* was published. It was only when she informed me of her "forthcoming version" of the text—"based on intuitive and poetic understanding . . . and a lot of chutspah"—that I considered my own project on her work.

18. Ursula K. Le Guin, *Lao Tzu, Tao Te Ching: A Book about the Way and the Power of the Way* (Boston: Shambhala, 1997), x.

19. Ibid., 108–10.

20. Paul Ricoeur, *Hermeneutics and the Human Sciences*, trans. John B. Thompson (Cambridge: Cambridge University, 1981), 178.

21. Le Guin, *Lao Tzu*, x.

22. Ibid., 35.

23. Ibid., 115.

24. Ibid., x.

25. Ibid., 47.

26. Ibid., 116, 122–23.

27. Ibid., 78.

28. Ibid., 77, 122.

29. Ibid., 38–39.

30. Ibid., 9.

31. Ibid., 20.

32. Ibid., 83.

33. Ibid., 7.

34. Ibid., 48.

35. Ibid., 17.

36. Ibid., 6.

37. "Lao Tzu thinks the material dualist, who tries to ignore the body and live in the head, and the religious dualist, who despises the body and lives for a reward in heaven, are both dangerous and in danger. So, enjoy your life, he says; live in your body, you are your body; where else is there to go? Heaven and earth are one. As you walk the streets of your town you walk the Way of heaven." Ibid., 100–01.

38. Ibid., 5, 91.

39. Ibid., 91.

40. Beverly Wildung Harrison, "The Power of Anger in the Work of Love," *Weaving the Visions: New Patterns in Feminist Spirituality*, ed. Judith Plaskow and Carol Christ (San Francisco: Harper, 1989), 221.

41. Le Guin, "Is Gender Necessary? Redux," from *Dancing at the Edge of the World*, 16.

42. Le Guin interview, cited from Jonathan White, *Talking on the Water: Conversations about Nature and Creativity* (San Francisco: Sierra Club Books, 1994).

43. Le Guin, *Lao Tzu*, 64.

44. This discussion, excerpted from Le Guin's answers to questions posed by her readers, appears on the website for Houghton Mifflin Company's McDougal Littell division. At present, the URL is http://www.mcdougallittell.com/lit/guest/le_guin/ukans.htm, though such internet addresses are notoriously transitory.

45. This excerpt from a Le Guin interview, the original source of which is not specified, appears in the preface to her short story, "Nine Lives," in *The Ascent of Wonder: The Evolution of Hard SF*, ed. David G. Hartwell and Kathryn Cramer (New York: Tor, 1994), 43.

46. Le Guin, *Lao Tzu*, 8.

47. Ibid., 66.

Sectional Discussion:
Daoism—A Vital Tradition for the
Contemporary Ecological Consciousness

JAMES MILLER

Introduction

A vital thread running throughout the Daoist tradition is that we must pay careful attention to the text of nature and not project onto it the false habits of our cultural imagination. For this reason Liu Ming asserts decisively that Daoism can have nothing to do with the idea of an "environmental crisis" inasmuch as this covertly describes a cosmic drama of sin, judgment, and redemption that stems from a false Christian conception of the relationship between the self and the world. If any form of environmentalism is acceptable to Daoism, it is the "nuclear environmentalism" where "change begins small." The essays in this section have all hinged in one way or another on how we come to perceive the Daoist tradition and how we, as scholars and practitioners, re-present Daoism in light of the contemporary situation. Liu Ming notes that when he was ordained as a Daoist priest he was presented with texts to which scholars have never had access since they do not form part of the Ming dynasty edition of the Daoist literary corpus. The proper functioning of the Daoist religion, therefore, depends upon a certain degree of mystery in that the lineages of transmission depend for their survival on guarding their texts and disclosing them only to those who have received the proper training. The problem that scholars of Daoism face in decoding these mysteries is that they are in a real sense disturbing the fabric of the tradition by doing so.

Jonathan Herman's essay is an intelligent treatment of this difficult question and points, in a decidedly Daoist fashion, toward the complementary relationship of production and destruction. Is the formation of a contemporary Western Daoism predicated on the "cultural strip-mining" of the historical Chinese tradition? Herman argues against both the sterile conservation of cultural traditions and their thoughtless exploitation in favor of an imaginative, "mutual appropriation." Perhaps this should be our Daoist model for a constructive hermeneutics and a pragmatic environmentalism. For sure, Daoism is right that these issues are intimately related.

I should like to present three strategies for "mutual appropriation" that try to relate the modern construction of Daoism to ecological concerns: 1) to use the ancient tradition to illuminate our self-understanding; 2) to learn to interpret the textuality of nature; and 3) to foster creative transgression.

Illuminate Oneself

The broad thrust of the declaration of the Chinese Daoist Association on Global Ecology is that the way to respect the universal Dao is to cultivate one's inner power or virtue (*de*). Conversely, the widespread failure of human beings to value their inner nature is a prime factor in the degeneration of our natural environment. Accordingly, the declaration revisits the ancient texts of the tradition in order to learn again what it means to be a human being who is modeled *after* the earth and the heavens. The greatest danger for nature is when the self-image of human beings is inflated. Consequently, the first step in any environmental program must be to illuminate oneself. The texts of the Daoist tradition therefore function as deep reflective pools in which we can wash away the sedimented layers of psychocultural delusions. When we have come to envision ourselves aright we cannot fail to be properly oriented with *respect* to our environment.

The danger for human understanding and for the natural environment comes when we project our own psychodramas into the text of nature. Nietzsche writes in *The Antichrist:* "Whatever a theologian feels to be true *must* be false: this is almost a criterion of truth. His most basic instinct of self-preservation forbids him to respect reality at any point or even to let it get a word in." If we understand the environmental crisis to be a crisis of human self-preservation, that is, an existential drama of apocalyptic consequence, then we run the risk of being "theologians" unable to respect reality, unable to decipher the text of nature.

Read the Text of Nature

In the Shangqing tradition, however, learning to "decipher the text of nature" does not mean abandoning the conventions of civilization in favor of the romantic romp of the *übermensch* through the wilderness. Rather, it means learning to see through religious texts to discern the cosmic textuality, or fabric, into which they have been woven. This takes place by following the practices detailed in the texts as a way to perceive, and therefore actualize, the cosmic complexity of biospritual life. By learning to visualize the cosmic landscape of the body, adepts learned to see that their *bios*, their biological nature, was concretely embedded in and inextricably folded into the cosmic complex. The medium through which this embedding takes place is the religious text.

Thus we are presented with a twofold hermeneutical problem. On the one hand people are able to engage the complex value of the natural world only through the mediation of Daoist religious texts and traditions. On the other hand a full, nonreductive appreciation of the religious texts is possible only by considering their cosmic construction. In the Shangqing tradition we are faced resolutely with the theological proposition that only by means of the mediation of the religious text shall we ever properly conceive of our relationship to our environment, and only by means of the proper conception of our relationship to our environment shall we be able to understand the religious texts and traditions. Nature, text, and person are wholly interlocked. One domain cannot be understood without the aid of the other two.

Foster Creative Transgression

A Daoist understanding of evolution is thus triadic in nature: the evolutionary strands of the human species, religious culture, and our universe cannot be separated one from another. This "dao," or way of understanding, is neither anthropocentric, cosmocentric, nor biocentric, but simply recognizes the reality of the interdependent transformations of these three threads. The danger of this book is that it has already turned the many strands of Daoist lineages into a reified entity that is capable of answering our questions. It seems, therefore, that in framing the question of Daoism and ecology, we have ourselves perpetrated some noticeable violence to the tradition. In fact, we have steered away from this danger by presenting Daoism and ecology as living conversation partners and not as lifeless bodies of scientific knowledge. In doing so we are acknowledging, but also fostering, an evolutionary process.

Ursula Le Guin recognizes her own role in the spread of Daoism in the West and adopts a model of stewardship that is useful both for environmentalists and Daoist scholars: be defensive and deferential, but accept that transgression and compromise is inevitable. There is, therefore, no turning back of the clock of intellectual history, nor is there any turning back of the clock of natural history. The strands of natural, human, and religious history are now most tightly interwoven in the impending collapse of our planetary life-support system. The Daoist, like the martial artist, must be alert, intelligent, flexible, and defensive. This way of being in the world is not a Daoist "ethic" that can be "applied" to "situations" of environmental concern, but rather a creative posture that merits adoption by scholars, scientists, politicians, and all who desire harmony with their cosmic environment. It is not intended as a substitute for the pragmatic Confucianism of the concerned intellectual, but as the necessary internal, spiritual accompaniment.

Epilogue: Dao Song

URSULA K. LE GUIN

J onathan Herman's discussion of the popular Western co-optation of Daoism (from the previous section), which I found both just and provocative, brought my thoughts to an area that I've had to think about with some of the same intensity a Chinese or a sinologist might bring to this one. That is the appropriation by non-Indians of Native American religious ideas and spiritual practices. Indians have a large and legitimate problem with this. It's one I've had to take sides on, because my father, as an anthropologist, was responsible for making Native cultural elements accessible to whites and because I myself have drawn on the writings both of anthropologists and of Indians for certain models of thought and behavior that inform my books, particularly *Always Coming Home*. And because when some white kid starts telling me that she's a "shawman" and knows what animal is her soul-guide and how good it makes her feel, something in me writhes in protest. To the one true shaman I knew, a Yurok, being a shaman wasn't a game but the heavy burden of an undesired lifetime commitment to learning and transmitting complex knowledge, involving a profound, continual, spiritual risk.

Feeling thus, that the Indians are legitimately defensive of their religions, I feel that as an outsider to Daoism, however passionately sympathetic, I must be exceedingly conscious of possible transgression. I tried to maintain a proper wariness while working on my version of Laozi.[1] Yet it is, to some degree, inevitably transgressive; and I quite expected to meet an equal degree of defensiveness here among practitioners and scholars of Daoism. Having taken upon oneself the transmission or translation of certain texts or ideas or ways of thought, one is responsible for them, protective of them. You want them to be understood rightly and respected for what they are, not trivialized and not misused. You feel toward them as the ecologist does toward a river or a desert: this is not to be abused, and if we use it we should do so very mindfully.

Defensiveness against cheapening and trivializing Daoism thus seems to me an inevitable, essential part of your work as scholars; and yet, like the ecologist, the conservationist, you don't have the luxury of being absolutely defensive. Compromise is also inevitable. People *will* use the river and the desert. Daoist texts *are* popular. The barbarians are inside the gates—here I am. I strongly support Jonathan Herman in saying that if we unscholarly types whose historical knowledge is gappy, who confuse hermeneutics with heuristics, who don't even know Chinese—if we amateurs are co-opting your texts, then perhaps your best move is to start co-opting ours. Use our efforts and our blunders, our naiveties and misunderstandings of Daoism, as signs, signals, guides to your own work: what most needs explaining, what keeps living and therefore changing in the tradition, what is translatable (in every sense of the word) and what is not. Not to declare war on foolishness and ignorance, but to use foolishness and ignorance as guides, seems quite in the spirit of Laozi.

And may I add a word here about the artist as the interpreter of religion and of thought. We have been mining our texts for ideas. But if the Laozi and Zhuangzi texts were aesthetically insignificant, important only intellectually, would they matter as they do? If they were not great works of art, would they be accessible to Western popular culture? Would they have their extraordinary influence? Daoism has expressed itself most directly (and appropriately) in art. Daoist painting teaches us to see the world; Daoist poetry gives us an understanding of the world.

And, as religion has always used art to speak itself, to show, to convince, so I believe ecology can and must use art. Art is *all* turtles, all the way down.

I will end with a small poem I wrote many, many years ago.

Tao Song

O slow fish
show me the way
O green weed
grow me the way

The way you go
the way you grow
is the way
indeed

O bright Sun
light me the way
the right way
the one
no one can say

If one can choose it
it is wrong
Sing me the way
O song:

No one can lose it
for long

Notes

1. [Editors' note based on material from J. P. Seaton and Ursula Le Guin]: Le Guin's version of Laozi is *Lao Tzu, Tao Te Ching: A Book About the Way and the Power of the Way* (Boston and London: Shambhala, 1997). She acknowledges J. P. Seaton as her "collaborator" for this work. As to the nature of this collaboration, Seaton says the following:

> Put a slice of ham between two pieces of bread and you have a ham sandwich. We conventionally name a sandwich for the goody in between the two slices of bread.
>
> When my children were little I used to put something good, like peanut butter and jam or cream cheese and cucumber (the cucumber on top, of course) on the two sides of a cracker and name the concoction, with a little twist of the convention, a cracker sandwich.
>
> Between Jonathan Herman's excellent analysis and Ursula K. Le Guin, I am the cracker in this particular cracker sandwich.
>
> As my response to Jonathan's presentation I'll try simply to clarify my role in Ms. Le Guin's translation of the *Tao Te Ching*. I am as delighted to have been able to take part in your conference as I was surprised, and honored, to be invited. Though I pride myself on being a sinologist and though many people who know me say, quite wrongly, that I "am a Taoist," I am not a scholar of comparative religions, nor am I, really, a *scholar* of Chinese literature. For thirty years I have taken "all (Chinese) poetry as my province" to very freely paraphrase Bacon . . . so I am to be damned as a generalist, in general, and as a literary translator, a translator of poetry, in specific. Having more or less willingly accepted the dictum that the translator is a traitor, I decided a couple of decades ago that if this be treachery, I'd make the most of it. My work and I are decidedly not the topic we are here to discuss, but it seems appropriate that I give you this much background, and perhaps just a little more, concerning myself, to allow you to understand my apparently rather enigmatic roll in Ms. Le Guin's *Tao Te Ching* translation.
>
> I read Lao Tzu first in the version of R. B. Blakney (paper, $.65) in 1963. I have since read it attentively in fourteen different English versions following Wang Pi, two that appear to follow the *Ho Shang Kung* commentary and two versions of the *Ma wang tui*. Until I undertook to assist Ms. Le Guin, I'd read the original piece by piece only as classroom demand required it. I had *never* read commentaries to Lao Tzu until Ms. Le Guin needed to see them. When we began working together, her knowledge of English language scholarship was both broader and deeper than mine, and I struggled to catch up. What I found in Hendrick and Lafargue, I found particularly illuminating.
>
> On the other hand, I had taught the work in English for twenty-five years to undergraduates at a beginning and a more advanced level (we don't have a graduate degree program in Chinese at Chapel Hill) and had pondered the various versions for personal edification. I found both encouragement and solace there. I have always identified the concept of *wei-wu-wei* most clearly with

artistic creativity. As a teacher of Chinese literature in translation for thirty years, I have often hoped to find a translation of the *Tao Te Ching* whose language would *excite* bright contemporary American students. That is the motive that brought me to encourage Ms. Le Guin in her project.

Simply put, in the process of Ms. Le Guin's translation I fulfilled the function of a research assistant. I checked Morohashi and the standard commentaries when asked, and occasionally when I felt the necessity myself. I discussed the sources and the scholarship when asked. Occasionally, with trepidation, I offered a suggestion of interpretation. I was given all the respect that every such assistant deserves, more than most get, and no more than necessary. Like a good research assistant, I learned more from my boss, Ms. Le Guin, in the process than she did from me. The translation (or *version* as she will continue to call it no matter how hard I try to convince her otherwise) is hers alone, an embodiment, an enunciation, of a lifetime of serious critical study and deep pondering, put forward with the grace and the power that only a real artist can command (or let flow), and there is not a word of it I would change, even if I were the sort to dare to try to change a word written by someone who has been known to make kings pregnant.

"In response to Dr. Seaton," Le Guin commented, "'research assistant' is hardly how I would describe the part he played in my version of the *Tao Te Ching*. I think of him more as a Sherpa. Here I am, the foolish foreign climber, and there's Lao-Tzu Mountain, and there's the Sherpa, the guide—the one who lives on the mountain, knows the mountain; the one who carries the oxygen; the one without whom I never would have got half way to the top."

Bibliography on Daoism and Ecology

JAMES MILLER, JORGE HIGHLAND, and LIU XIAOGAN
with the assistance of Belle B. L. Tan and Zhong Hongzhi

Allan, Sarah. *The Way of Water and the Sprouts of Virtue*. Albany: State University of New York Press, 1997.
  Based on her research in ancient Chinese language and culture, Allen demonstrates that the fundamental concepts of Chinese philosophy are organic metaphors derived from images of nature.

Ames, Roger T. "Taoism and the Nature of Nature." *Environmental Ethics* 8 (1986): 317–50.
  The crisis of environmental ethics requires more than an alternative metaphysics or "science of first principles." Daoism demonstrates an *ars contextualis* from which the nature of relatedness may be redefined as an "aesthetic cosmology."

Anderson, E. N. *Ecologies of the Heart*. New York: Oxford University Press, 1996.
  Anderson, a cultural ecologist, discusses the concrete application of fengshui and Chinese nutritional therapy in a wide-ranging survey of key themes in the relationship of ecology and religion.

Anderson, E. N., and Marja Anderson. *Mountains and Water: The Cultural Ecology of South Coastal China*. Taipei: Orient Cultural Service, 1973.

Callicott, J. Baird. *Earth's Insights: A Survey of Ecological Ethics from the Mediterranean Basin to the Australian Outback*. Berkeley and Los Angeles: University of California Press, 1994.
  Callicott displays a deep admiration for Daoism that rests chiefly on a philosophical exposition of the *Daode jing* that he earlier developed through his work with Roger Ames.

Callicot, J. Baird, and Roger T. Ames. *Nature in Asian Traditions of Thought.*
Albany: State University of New York Press, 1989.
An anthology of essays from Indian, Chinese, and Japanese perspectives. The Chinese section focuses primarily on Daoist and Confucian
philosophy.

Chen Rongbo. "Laozi de huanbao meixue" (Laozi's Aesthetics of Environmental Protection). *Zhexue zazhi* (Taipei) 7 (January 1994): 98–103.
Chen argues that the *Laozi* advocates the sanctification of the heart and
mind, respecting the relationship between humans and nature, and discarding the deep-rooted selfishness of human beings. These concepts
can be used as a basis for a Daoist aesthetic of environmental protection.

Chen Xiaoya. "Kuayue liangqian nian de woshou: Laozi zhexue yu lüse
zhengzhixue" (Laozi's Philosophy and Green Politics: A Meeting across
Two Millennia). *Xuexi* (Beijing) 10 (1995): 72–77.
The author suggests similarities and interchangeable ideas between
Laozi's doctrines and the green politics in the West, including *wuwei*
(non-action) and zero-growth theory, pacifism, and anti-extremisim.

Chen, Ellen M. "The Meaning of *Te* in the Tao Te Ching: An Examination of
the Concept of Nature in Chinese Taoism." *Philosophy East and West*
23 (1973): 457–70.

Cheng Chung-ying. "On the Environmental Ethics of the *Tao* and the *Ch'i*."
Environmental Ethics 8 (1986): 351–70.
Cheng proposes five basic axioms for an environmental ethic: 1) total
interpenetration; 2) self-transformation; 3) creative spontaneity; 4) a
will not to will; and 5) non-attaching attachment. These axioms are
rooted in a metaphysics of Dao and qi.

Clarke, J. J. "Environmentalism: New Ways?" In J. J. Clarke, *The Tao of the
West: Western Transformations of Taoist Thought*, 81–89. London and
New York: Routledge, 2000.

Cooper, David E. "Is Daoism 'Green'?" In *Morals and Society in Asian Philosophy*, ed. Brian Carr, 82–91. Richmond, Surrey: Curzon, 1996.

Elvin, Mark, and Liu Ts'ui-jung, eds. *Sediments of Time: Environment and
Society in Chinese History*. Cambridge: Cambridge University Press,
1998.
This collection of detailed historical investigations includes essays by

Dunstan, Finnane, and Santangelo that describe aspects of the representation of environment in Chinese cultural history.

Ge Rongjin, ed. *Daojia wenhua yu xiandai wenming* (Daoist Culture and Modern Civilization). Beijing: Zhongguo Renmin Daxue Chubanshe, 1991.
This is the first Chinese book to discuss the general issue of the relevance of Daoist culture to modern societies. In chapter 10, the authors investigate the relationship of Laozi's philosophy to an environmental consciousness, suggesting that it favors an ecological balance and sustainable development.

Girardot, N. J. *Myth and Meaning in Early Taoism*. Berkeley and Los Angeles: University of California Press, 1983.
Girardot investigates the themes of chaos and cosmos in early Daoist philosophy and religion, discovering the traces of their mythic structure.

Gottlieb, Roger S., ed. *This Sacred Earth: Religion, Nature, Environment*. New York: Routledge, 1996.
A comprehensive anthology of readings in the field of religion and ecology, but the representation of Daoism is limited to a paraphrase of the *Daode jing*.

Gu Linyu. "Laozi chongshang ziran de jiazhi quxiang" (Laozi's Value Orientation: Following Naturalness). *Xueshu yuekan* 7 (1989): 13–17.
The author argues that Laozi's advocacy of *ziran*, or naturalness, is demonstrated in three main areas, namely, epistemology, personality, and aesthetics.

Hall, David L. "On Seeking a Change of Environment." In *Nature in Asian Traditions of Thought*, ed. J. Baird Callicott and Roger T. Ames, 99–112. Albany: State University of New York Press, 1989.
The crisis in environmental ethics mirrors a broader crisis in philosophy. Modern views of philosophy as a rational enterprise call into question basic views of the nature of order and relatedness. Philosophical Daoism offers an alternative conception of the relationship of the particular person to his environment.

He Ping. "Shilun Laozi zhexue de ziran fanshi" (An Argument on the Natural Pattern in the Philosophy of Laozi). *Nankai xuebao* 2 (1993): 14–21.
The writer discusses three interrelated aspects of the concept of Dao, namely, nonbeing, oneness, and the principle of naturalness. The au-

thor argues that there are natural patterns of Heaven and Earth, and natural patterns of humankind. The relationship between the two is demonstrated in the structure of Laozi's philosophy.

Houten, Richard Van. "Nature and *tzu-jan* in Early Chinese Philosophical Literature." *Journal of Chinese Philosophy* 15 (1988): 33–49.
By means of a historical analysis of the semantics of the binome *ziran*, the author argues that it is only in the Han dynasty that Daoism develops an explicit connection to the world of nature.

Ip Po-keung. "Taoism and the Foundation of Environmental Ethics." *Environmental Ethics* 5 (1986): 335–43.
Daoist philosophy provides a metaphysical foundation for environmental ethics by establishing a link between ethics and science. The world is not a lifeless machine servicing selfish human desires but exists in a mutual ethical relationship with humankind.

Kinsley, David. *Ecology and Religion. Ecological Spirituality in Cross-Cultural Perspective*. Englewood Cliffs, NJ: Prentice Hall, 1995.
Kinsley relies on Tu Weiming for his treatment of the Chinese worldview, and confines his treatment of Daoism ("letting be") to the *Laozi* and the *Zhuangzi*.

Kirkland, Russell. "The Roots of Altruism in the Taoist Tradition." *Journal of the American Academy of Religion* 54 (1986): 59–74.
Kirkland examines the concept of altruism in Daoism with particular reference to the Daoism under the Tang. The section entitled "solicitude for non-human life in medieval Taoism" (pp. 72–74) presents historical evidence for a Daoist concern for nonhuman life.

———. "Taoism." In *Encylopedia of Bioethics*. 2d ed. New York: Macmillan, 1995.
Daoism assumes a universalistic ethic that extends not only to all humanity but also to the wider domain of all living things. This worldview encourages individuals and groups to engage in activities intended to promote the harmonious integration of the individual, society, and nature.

———. "Self-Fulfillment through Selflessness: The Moral Teachings of the *Daode jing*." In *Varieties of Ethical Reflection: New Directions for Ethics in a Global Context*, ed. Michael Barnhart. Forthcoming.

Lau, D. C., and Roger T. Ames, trans. *Yuandao: Tracing Dao to Its Source.* New York: Ballantine, 1998.
This translation of the first chapter of the *Huainanzi* is prefaced by an interpretive essay that draws together many themes of Daoist thinking about order, cosmos, situation, agency, and unity that are necessary for constructing a Daoist environmental philosophy.

Li, Huey-li. "A Cross Cultural Critique of Ecofeminism." In *Ecofeminism: Women, Animals, Nature*, ed. Greta Gaard. Philadelphia: Temple University Press, 1993.
Argues against the notion that transcendent dualism is the conceptual root of the oppression of women and the exploitation of nature, since both phenomena are as evident in Chinese history as in the West.

Liu Sifen and Liu Qiyu. "Daojia lunli yu dangdai Zhongguo de daode jiangou" (Daoist Ethics and the Moral Reconstruction in Modern China). *Xinhua Wenzhai* 11 (2000): 30–32.
The authors argue that some Daoist theories, including Daoist eco-ethics, should be integrated into the modern moral system of China.

Major, John S. *Heaven and Earth in Early Han Thought.* Albany: State University of New York Press, 1993.
A translation and analysis of Chapters 3, 4, and 5 of the *Huainanzi.* This is the key Han dynasty text that systematically established the Chinese cosmological worldview and the system of correlations between different dimensions of the cosmos.

Marshall, Peter H. *Nature's Web: An Exploration of Ecological Thinking.* London: Simon and Schuster, 1992.

Miller, James. "Daoism and Ecology." *Earth Ethics* 10, no. 1 (1998): 25–26.
This is the background paper prepared for the religions of the world and ecology series' culminating conference "Religion, Ethics, and the Environment" at the American Academy of Arts and Sciences. An amended version is incorporated into the introduction to this volume.

Neville, Robert C. "Units of Change—Units of Value." *Philosophy East and West* 37 (1987): 131–34.
Any adequate environmental ethics requires 1) a theory of how value is altered by change, which presupposes 2) an axiology of the value of actual things and the value of the relations obtained between things.

From this can be developed 3) a theory of how things are integrated in systems. The author examines the conceptual resources of Chinese philosophy for answering these questions.

Novak, Philip. "Tao How? Asian Religions and the Problem of Environmental Degradation." *ReVision* 16 (1993): 77–82. Reprinted from *ReVision*, 1985.
Novak contrasts the ideals of Daoist philosophy with the lack of environmental awareness in ancient China. In the face of environmental degradation, however, these ideals are being reappropriated by modern environmental ethics.

Palmer, Martin. "Saving China's Holy Mountains." In *People and the Planet*, 1996. <*http://www.oneworld.org/patp/vol5/feature.html*>

Paper, Jordan, and Li Chuang Paper. "Chinese Religions, Population, and the Environment." In *Population, Consumption, and the Environment: Religious and Secular Responses*, ed. Harold Coward, 173–91. Albany: State University of New York Press, 1995.
Insights from contemporary anthropology, sociology and comparative religion are interwoven in order to sketch the outlines of modern Chinese attitudes towards population and the environment.

Parkes, Graham. "Human/Nature in Nietzsche and Taoism." In *Nature in Asian Traditions of Thought*, ed. J. Baird Callicott and Roger T. Ames, 79–98. Albany: State University of New York Press, 1989.
Both the *Daode jing* and Nietzsche bemoan a human sickness that is the result of an anthropocentric misreading of the "text of nature" and advocate the reintegration of human beings into the "continuum of natural phenomena." Nietzsche's view of how this is possible, however, suggests a "wilder and alchemically tinged . . . opus contra naturam" than the early Daoist return to the "uncarved block."

Peerenboom, R. P. "Beyond Naturalism: A Reconstruction of Daoist Environmental Ethics." *Environmental Ethics* 13, no. 1 (1991): 3–22.

Rolston, Holmes, III. "Can the East Help the West to Value Nature?" *Philosophy East and West* 37 (1987): 172–90.
The bipolar complementarity of yang/yin is discussed as an empirical category, a descriptive metaphysics, and a moral prescription. These categories are related to ideas of biological integrity, organism, and ecosystem. The answer to the title question is, in the end, no.

She Zhengrong. "Laozi shengtai sixiang ji qi dui dangdai de qishi" (Laozi's Ecological Thought and Its Significance in the Contemporary Era). *Qinghai shehui kexue* 2 (1994): 43–50.
Laozi's ecological wisdom is expressed through his unique understanding of the relation of humankind and Heaven. The writer further stresses that this wisdom is derived from the intuition of a cosmic ontology, a simpler way to grasp the ecosystem in its entirety and complexity than through rational thinking.

Shi Baishi. "Zhuangzi sixiang yu xiandaihua wenming" (Zhuangzi's Thought and the Civilization of Modernization). *Anhui daxue xuebao*, June 1996, 72–77.
The writer discusses the modern significance of three major themes of Zhuangzi's philosophy in light of issues in modern society. One theme is Zhuangzi's principle of following naturalness and its significance for modern environmental protection and the preservation of nature.

Sivin, Nathan. *Medicine, Philosophy and Religion in Ancient China.* Aldershot, Hants.: Variorum, 1995.
A collection of essays on the comparative philosophy of science in China and the West.

Smil, Vaclav. *China's Environmental Crisis: An Inquiry into the Limits of National Development.* Armonk, N.Y.: M. E. Sharpe, 1993.
Smil, an environmental scientist, documents the ecological changes wrought by China's modernization and its population explosion. Although he makes occasional reference to classical philosophical and literary texts, these connections are not the subject of his book and thus are not pursued deeply.

Smith, Huston. "Tao Now: An Ecological Testament." In *Earth Might Be Fair: Reflections on Ethics, Religion, and Ecology,* ed. I. G. Barbour, 62–82. Englewood Cliffs, N.J.: Prentice-Hall, 1972.
Smith argues that the Western scientific method is fundamentally concerned with drawing distinctions. From this follow five epistemological goals of clarity, generalization, conceptualization, implication, and, ultimately, control. The Dao, on the other hand, is a realm of relativity and interdependence that engenders a spirit of simplicity and surrender.

Soulé, Michael E., and Gary Lease, eds. *Reinventing Nature? Responses to Postmodern Deconstruction.* Washington, D.C.: Island Press, 1995.
Contains a public survey of attitudes toward nature in Japan and America by Stephen Kellert that discounts the impact of cultural stereotypes of "Eastern" philosophy and the "Judaeo-Christian" tradition.

Stein, R. A. *The World in Miniature: Container Gardens and Dwellings in Far Eastern Religious Thought*. Trans. Phyllis Brooks. Stanford: Stanford University Press, 1990.
A classic work of sinology on the theme of microcosm and macrocosm in Chinese cultural practice.

Sylvan, Richard, and David Bennett. *Of Utopias, Tao and Deep Ecology*. Canberra: Australian National University, 1990.
The first half of this work is a philosophical investigation into utopia theory framed in terms of sistology (the theory of items, whether existent or nonexistent). The second half is an argument in the terms of item theory and object theory that the *Daode jing* offers the best example of deep ecological theory.

Tang Mingbang, and He Jianming. "Daojia wenhua de xiandai yiyi" (The Modern Significance of Daoist Culture). *Wuhan daxue xuebao* 1 (1991): 112–16.
Based on the analysis of the central idea of following naturalness, the paper discusses four topics: the principle of reversion in thinking, *wuwei* (non-action) in politics, venerating Dao and *de* in morality, and long life in the art of longevity.

Tuan Yi-fu. *Topophilia: A Study of Environmental Perception, Attitudes, and Values*. Englewood Cliffs, N.J.: Prentice Hall, 1974.

Tucker, Mary Evelyn. "Ecological Themes in Taoism and Confucianism." In *Worldviews and Ecology: Religion, Philosophy, and the Environment*, ed. Mary Evelyn Tucker and John Grim, 150–60. Maryknoll, N.Y.: Orbis Books, 1994.

Wang Qintian. *Shengtai wenhua* (Ecological Culture). Taipei: Yangzhi Cultural Career Company, 1995.
The emergence of ecology and its related culture of the present age can be understood as a revolution in the history of science and human intellect. The writer proceeds from the ecological situation in mainland China and proposes that the world should draw on the experience and wisdom of China or the East as a whole when constructing an new ecological culture.

Wei Yuangui. "Laozi zhexue de ziranguan yu huanbao xinling" (The View of Nature and the Mind of Environmental Protection in Laozi's Philosophy). *Zhexue Zazhi* 13 (July 1995): 36–55.
The writer discusses environmental concerns from the the Daoist understanding of *ziran*, or naturalness. Though Laozi did not address the

topics of environmental protection and preservation, his doctrine of the natural equilibrium and integral unity of the universe is, in fact, the way to uphold an ecological balance.

Yin Zhihua. "Daojiao jielü zhong de huanjing baohu sixiang" (The Thought of Environmental Protection in Daoist Commandments and Rules). *Zhongguo Daojiao* (China's Daoism) 2 (1996): 33–34.
The author briefly introduces Daoist rules for the protection of animals, plants, land, and water resources.

Zhang Jiacai. "Laozhuang daojia ziran zhuyi lungang" (An Outline of the Daoist Naturalism of Laozi and Zhuangzi). *Lanzhou daxue xuebao* 3 (1991): 87–94.
Zhang examines the word *ziran* and comes up with four interrelated perspectives: 1) The origin, development, and transformation of the myriad things are natural; 2) they follow natural laws; 3) these laws exist inherently in the myriad things; and 4) this disposition of the myriad things allows them to be self-fulfilling.

Zhang Jiyu, ed. *Dao fa ziran yu huanjing baohu* (The Way to Follow Naturalness and Environmental Protection). Beijing: Huaxia Chubanshe, 1998.
This is the first Chinese book on Daoism and environmental protection. The editor and co-author is a sixty-fifth generation descendant of Zhang Daoling, the founder of the Celestial Masters tradition, and vice president of the Chinese Daoist Association. The last chapter of the book introduces Daoist contributions to environmental protection, both theoretical and practical, from the ancient to modern.

Zhang Renwu, Ji Wenying, and Zhang Tong. "Zhongguo gudai pusu shengtai jingji guanian jiqi zai nongye shang de yingyong" (The Simple Economic View in Ancient China and Its Utilization in Agriculture). *Shengtai jingji* (Ecological Economy) 5 (1994): 27–33.
The authors demonstrate the eco-economic view in ancient China, based on historical documents, including Daoist classics, such as the *Huainanzi.*

Zhang Yunfei. *Tianren heyi: ruxue yu shengtai huanjing* (The Union of Heaven and Humanity: Confucianism and Ecological Environment). Chengdu: Sichuan renmin chubanshe, 1995.
The writer deals with three aspects of the Confucian consciousness, the ecological, philosophical and ethical. He argues that the present environmental crisis shows the significance of the concept of the union of Heaven and humanity.

Zhuang Qingxin. "Daojia ziranguan zhong de huanjing zhexue" (Environmental Philosophy in the Daoist View of Nature). *Zhexue zazhi* 13 (July 1995): 56–71.

Daoist philosophy requires us to act according to natural laws, to respect nature, and finally, to reach the state of the union of Heaven and human. This is equivalent to principles of environmental ethics in the West. To actualize these principles, we have to transcend ourselves and consciously cultivate ourselves.

Zimmerman, Michael E. *Contesting Earth's Future: Radical Ecology and Postmodernity*. Berkeley and Los Angeles: University of California Press, 1994.

A philosophical examination of deep ecology, social ecology, ecofeminism, chaos theory, and cyborgism that occasionally refers to the influence of Daoism as articulated in the modern West.

Notes on Contributors

Roger T. Ames is Professor of Chinese Philosophy at the University of Hawai'i. He is a translator of classics, such as *Sun-tzu: The Art of Warfare* (Ballantine Books, 1993), *Sun Pin: The Art of Warfare* (with D. C. Lau; Ballantine Books, 1996), *The Analects of Confucius: A Philosophical Translation* (with H. Rosemont, Jr.; Ballantine Books, 1998), and *Focusing the Familiar: A Translation and Philosophical Interpretation of the* Zhongyong (with David Hall; University of Hawai'i Press, 2001), and is the co-author of several interpretative studies of classical Chinese philosophy, including *Thinking through Confucius* (State University of New York Press, 1987), *Anticipating China* (State University of New York Press, 1995), and *Thinking from the Han* (State University of New York Press, 1998) (all with David Hall).

E. N. Anderson received his B.A. from Harvard College (1962) and his Ph.D. from the University of California, Berkeley (1967). He has been teaching at the University of California, Riverside, from 1966 to the present. His major works include *The Food of China* (Yale University Press, 1988) and *Ecologies of the Heart* (Oxford University Press, 1996). His professional focus is on cultural ecology and ethnobiology. His principal areas of fieldwork have been Hong Kong (1965–66 and 1974–75), Malaysia and Singapore (1970–71), British Columbia (1984–85), and Yucatan Peninsula of Mexico (1991, 1996).

Joanne D. Birdwhistell is Professor of Philosophy and Asian Civilization at the Richard Stockton College of New Jersey. She has published *Transition to Neo-Confucianism: Shao Yung on Knowledge and Symbols of Reality* (Stanford University Press, 1989), *Li Yong (1627–1705) and Epistemological Dimensions of Confucian Philosophy* (Stanford University Press, 1996), and a number of articles. She received her M.A. and Ph.D. from Stanford University and her B.A. from the University of Pennsylvania. Her research interests now focus on comparative philosophy, particularly in respect to environmental and gender issues.

Robert Ford Campany earned his doctorate from the University of Chicago in 1988. He is Associate Professor of Religious Studies and East Asian Languages and Cultures at Indiana University, where he has taught since 1988. His book on the origins of the Chinese marvel tales known as *zhiguai,* titled *Strange Writing: Anomaly Accounts in Early Medieval China,* was published in 1996 by the State University of New York Press. His book *To Live as Long as Heaven and Earth: A Translation and Study of Ge Hong's "Traditions of Divine Transcendents"* is forthcoming from the University of California Press.

Vincent F. Chu is the chief instructor at the Gin Soon Tai Chi Club in Boston's Chinatown. He started studying Classical Yang Family Style Tai Chi Chuan with his father, Master Gin Soon Chu, when he was seven years old. He has studied Tai Chi Chuan with Grandmaster Yang Sau-Chung, heir of the legendary Yang Cheng-Fu, and Master Ip Tai Tak of Hong Kong. He has a well-rounded training in Classical Yang Family fist forms (slow and fast forms), Push Hands, weapons forms (staff, broadsword, and spear), and *qigong* forms. He is a frequent contributor to martial arts magazines.

Edward Davis received his B.A. from Harvard College in 1976 and his M.A. and Ph.D. from the University of California at Berkeley (1981 and 1993). He is Associate Professor of Chinese History at the University of Hawai'i. His publications include the book *Society and the Supernatural in Song China* (University of Hawai'i Press, 2001). He is currently coediting the *Encyclopedia of Contemporary Chinese Culture* (Routledge, 2002).

Stephen L. Field received his Ph.D. from the University of Texas at Austin in 1985, and is Professor of Chinese at Trinity University in San Antonio, Texas. He has delivered papers on the classical origins of fengshui at international, national, and regional conferences since 1993 and has published two book chapters (in Chinese) and one journal article to date. His web page at Fengshui.com is one of the most popular fengshui sites on the Internet. It collects abridged versions of his research and provides an automatic fengshui reading for visitors based on the principles of the Compass School.

N. J. Girardot is University Distinguished Professor of the Comparative History of Religions at Lehigh University. His research has dealt with Daoism (*Myth and Meaning in Early Taoism* and other works), Chinese mythology, and the history of the study of Chinese religions. Other areas of interest include American visionary "folk" or "outsider" art and popular religious movements in the United States. His book entitled *The Victorian Translation of China: James Legge's Oriental Pilgrimage* is forthcoming from the University of California Press.

Russell B. Goodman studied at the University of Pennsylvania, Oxford, and Johns Hopkins. He is the author of *American Philosophy and the Romantic Tradition* (Cambridge University Press, 1990), *Pragmatism: A Contemporary Reader* (Routledge, 1995), *Wittgenstein and William James* (Cambridge University Press, 2002), and *Stanley Cavell: The Philosopher Responds to His Critics* (Vanderbilt University Press, 2002). He has published on Daoism and Hinduism and on the American philosopher Ralph Waldo Emerson. He was a Fulbright lecturer at the University of Barcelona in 1993. He is Professor and Chair of the Philosophy Department at the University of New Mexico.

Thomas H. Hahn received his M.A. from the University of Frankfurt in Main, Germany, in 1984. His thesis topic was Lu Dongbin and his commentaries on the *Lao Tzu*. From 1984 to 1987 he conducted research to investigate present-day Daoism in China (Shanghai, Chengdu, Beijing). From 1988 to 1990 he was at the University of Marburg where he was a researcher in Chinese religion and traditional Chinese education. He received librarian's training (1990–1998) at the University of Heidelberg. He completed his Ph.D. thesis in 1997 entitled "Chinese Mountains and Their Gazetteers." In 1998 he became an academic librarian at Memorial Library of the University of Wisconsin at Madison. In 2001, he became curator of the Wason Collection at Cornell University. He is working on a book project on Zhang Xiangwen, China's leading twentieth-century geographer.

David L. Hall is Professor of Philosophy at the University of Texas at El Paso. He received his Ph.D. in religious studies from Yale University in 1967. His principal research interests are American philosophy and comparative Chinese and Western philosophy. His books include: *Thinking From the Han: Self, Truth, and Transcendence in China and the West* (State University of New York Press, 1998), with Roger Ames; *Anticipating China: Thinking through the Narratives of Chinese and Western Culture* (State University of New York Press, 1995), with Roger Ames; *Richard Rorty: Prophet and Poet of the New Pragmatism* (State University of New York Press, 1994); and *Thinking through Confucius* (State University of New York Press, 1987), with Roger Ames.

Jonathan R. Herman is Associate Professor of Religious Studies in the Department of Philosophy at Georgia State University. He received his doctorate in Chinese religion from Harvard University in 1992. He is the author of *I and Tao: Martin Buber's Encounter with Chuang Tzu* (State University of New York Press, 1996) and several essays on Daoist philosophy, Neo-Confuciansim, comparative mysticism, and hermeneutics. He is currently an officer of the Society for the Study of Chinese Religions.

Russell Kirkland is Associate Professor of Religion at the University of Georgia. He earned his B.A. in religious studies and M.A. in Asian history from Brown University in 1976 and an M.A. in religious studies (1982) and Ph.D. in Chinese language and culture (1986) from Indiana University. He has taught Daoism and related subjects at the University of Rochester, the University of Missouri, Oberlin College, Stanford University, and Macalester College. He has published more than a dozen studies of Daoism and the history and religions of China, Tibet, Korea, and Japan. A frequent contributor to *Religious Studies Review*, he has been book review editor for the *Journal of Chinese Religions* since 1990 and serves on the executive board of the Society for the Study of Chinese Religions.

Terry F. Kleeman teaches in the Religious Studies and East Asian Languages and Literatures Departments at the University of Colorado at Boulder, and has taught at the College of William and Mary, the University of Minnesota, and the University of Pennsylvania. A graduate of the University of California at Berkeley (Ph.D.) and the University of British Columbia (M.A.), he has pursued research overseas at National Taiwan University, Taisho University (Tokyo), the École Pratique des Hautes Études (Sorbonne, Paris), and the University of Tokyo. His primary field is East Asian religion, especially Chinese popular religion and religious Daoism. He is the author of *A God's Own Tale: The Book of Transformations of Wenchang* (State University of New York Press, 1994) and *Great Perfection: Religion and Ethnicity in a Chinese Millennial Kingdom* (University of Hawai'i Press, 1998).

Livia Kohn is Professor of Religion and East Asian Studies at Boston University. A graduate of Bonn University, Germany, she has spent many years pursuing research at Kyoto University, Japan, and has also taught at the University of Michigan, the Stanford Center for Japanese Studies, Kyoto, and Eötvös Lorand University, Budapest. Her specialty is medieval Daoism. She has written *Taoist Mystical Philosophy: The Scripture of Western Ascension* (State University of New York Press, 1991), *Early Chinese Mysticism* (Princeton University Press, 1992), *The Taoist Experience* (State University of New York Press, 1993), *Laughing at the Tao* (Princeton University Press, 1995), *God of the Dao: Lord Lao in History and Myth* (University of Michigan, Center for Chinese Studies, 1998), and most recently the textbook *Daoism and Chinese Culture* (Three Pines Press, 2001). She has also edited *Taoist Meditation and Longevity Techniques* (University of Michigan, Center for Chinese Studies, 1989), *Lao Tzu and the Tao Te Ching* (with Michael LaFargue, State University of New York Press, 1998), and *Daoism Handbook* (Brill, 2000).

Michael LaFargue received his Th.D. in New Testament Studies from Harvard Divinity School 1978. Since then, he has specialized in applying methods developed in biblical studies to the interpretation of Asian religious texts. He has been a part-time professor in the Religious Studies Program at the University of Massachusetts-Boston since 1978 and has been the director of the Asian Studies Program there since 1995. He has also been a visiting professor at Wheaton College, Wellesley College, and Boston University. He has published four books: *Language and Gnosis: Form and Meaning in the Acts of Thomas* (Fortress, 1985); *The Tao of the Tao Te Ching* (State University of New York Press, 1992); *Tao and Method: A Reasoned Approach to the Tao Te Ching* (State University of New York Press, 1994); and *Lao Tzu and the Tao Te Ching*, co-edited with Livia Kohn (State University of New York Press, 1998).

Chi-tim Lai is Associate Professor of Religion at the Chinese University of Hong Kong, where he offers courses in the history of Daoism, Daoist scriptures, and ancient Chinese religions. He received his Ph.D. from the University of Chicago Divinity School, specializing in Daoism, history of religions, and psychology of religion. He is the editor of *Daoism and Popular Religion* (Center for the Study of Religion and Chinese Society, Chung Chi College, 1999). He has recently published articles on Six Dynasties Daoism in *Asia Major* and *Numen*. He is currently working on the early Celestial Master scriptures and the history of Hong Kong Daoism.

Ursula K. Le Guin was born in California, graduated from Radcliffe College, and earned her master's degree at Columbia University. She is one of the most distinguished living American authors and the winner of numerous literary awards, including the National Book Award, the Nebula, and the Hugo. She has written many novels and short stories, some of which have Daoist and ecological themes (e.g., *The Left Hand of Darkness*, *The Dispossessed*, *The Lathe of Heaven*, and *Always Coming Home*). She has recently published (with J. P. Seaton) *Tao Te Ching: A Book about the Way and the Power of the Way*, a new English version (Shambhala Press, 1997).

Li Yuanguo is a research fellow at the Sichuan Province Academy of Social Sciences' Institute of Philosophy and Culture and is the author of many books and articles, including *A Popular History of Daoist Religion in Sichuan* (*Sichuan Daojiao shihua*) and *Qigong and Longevity Practices in Daosim* (*Daojiao qigong Yangsheng xue*).

Liu Ming (Charles Belyea) is the founding director of Orthodox Daoism in America (O.D.A.), a nonprofit religious organization, begun in 1982, dedicated to the transmission of Chengyidao (orthodox Daoism) to the West. He holds a degree in Asian aesthetics and has received thirty years of training in

Tantric (Tibetan) Buddhism and orthodox Daoism, with particular interest in ritual meditation. In Taiwan (1977–78) he was adopted into a Chinese (Shanxi) Daoist family as a lineage-holding successor to their tradition of orthodox Daoism. Since completing a one-year solitary retreat (1981), he has been involved in the design of California state and U.S. national standards for graduate education in traditional Chinese medicine. Recently he has dedicated himself to teaching Daoism and designing a curriculum for the training of orthodox Daoist priests in America. O.D.A. is now in the process of publishing the *Willow Record* (a series of books on Chengyidao) and building an authentic Daotan, or Daoist shrine/temple. O.D.A. also publishes a quarterly newsletter called *Frost Bell*, which has subscribers worldwide.

Liu Xiaogan received his Ph.D. from Beijing University in 1985. Since then, he has taught and conducted research at Beijing, Harvard, Princeton, Singapore, and the Chinese University of Hong Kong. He is the author or contributor to books and journals, such as *Classifying the Zhuangzi Chapters, Our Religions, Lao-tzu and the Tao-te-ching, Journal of Chinese Philosophy, Taoist Resources, Taoism and Ecology, What Men Owe to Women: Men's Voices from World Religions, Essays on Religious and Philosophical Aspects of the Laozi,* and *Purity of Heart-Contemplation: A Monastic Dialogue between Christian and Asian Traditions,* in addition to Chinese books and papers on Daoism, ancient Chinese culture, modern Chinese intellectual issues, and comparative studies.

Weidong Lu is Professor of Chinese Medicine at the New England School of Acupuncture in Watertown, Massachusetts. He received his medical degree from Zhejiang College of Traditional Chinese Medicine in Hangzhou in 1983. He has been teaching Chinese medicine, including Chinese herbal medicine and acupuncture, in China and in the United States for more than ten years. He is currently the chair of the Department of Chinese Herbal Medicine at the New England School of Acupuncture. He is also the consultant of research projects at the Center for Alternative Medicine Research at Beth Israel Deaconess Medical Center in Boston. He serves as the vice chairman of the Committee on Acupuncture under the Massachusetts Board of Registration in Medicine. He has a private practice in Chinese herbal medicine and acupuncture in Watertown, Massachusetts.

Jeffrey F. Meyer received his Ph.D. in the history of religions from the University of Chicago in 1973. He is currently a professor in the Religious Studies Department, University of North Carolina at Charlotte. He has been awarded grants by the Fulbright Foundation, the American Council of Learned Societies, the Pacific Cultural Foundation, the UNCC Foundation, and the Cosmos Society. His publications on Beijing, Chinese architecture, and sacred places include: *Peking as a Sacred City* (Orient Cultural Service, 1976); *The Dragons of Tiananmen: Beijing as a Sacred City* (University of South Carolina Press, 1991); "Traditional Peking: The Architecture of Conditional Power," in *The City as a Sacred Center*, ed. Bardwell Smith and Holly Reynolds (E. J. Brill, 1987); "Feng-shui of the Chinese City," *History of Religions* 18, no. 2 (November 1978); "Chinese Buddhist Monastic Temples as Cosmograms," in *Sacred Architecture in the Traditions of China, Judaism and Islam*, ed. Emily Lyle (Edinburgh, 1992); and "The Eagle and the Dragon: Comparing Washington and Beijing," *Washington History* 8, no. 2 (fall/winter 1996–97). He has also published on moral education in China.

James Miller recently received his Ph.D. from Boston University and is now a postdoctoral research fellow in Daoism at Queen's University, Kingston, Ontario. His dissertation, "The Economy of Cosmic Power," is a cosmological theory of religious transaction and a comparative study of Shangqing Daoism and the Christian religion of Augustine of Hippo. He is a member of the advisory board of the Forum on Religion and Ecology (FORE), the consultant on Daoism for the FORE website at http://environment.harvard.edu/religion, and editor of http://www.daoiststudies.org.

René Navarro is a licensed acupuncturist and herbalist in Massachusetts and a senior instructor of the International Healing Tao Center. He has been an instructor of meditation and *qigong* at the Hawai'i College of Traditional Oriental Medicine in Maui and he has taught at the Tao Garden in Chiangmai, Thailand. He is also a faculty member of the Acupuncture Therapeutics and Research Center in the Philippines, where he teaches *qigong*, acupuncture, Tai Chi Chuan, and abdominal massage. He has edited and has been a contributing writer for Mantak Chia's *Greatest Enlightenment of Kan and Li* and *Chi Nei Tsang Internal Organs Chi Massage*. He has been appointed editor for the *Healing Tao* internal alchemy series presenting the Transmissions of White Cloud Hermit. His book of poetry, *Du-Fu's Cottage and Other Poems*, was published in 1997.

Jordan Paper is a professor in the East Asian and Religious Studies Programs, as well as an adjunct professor in the Faculty of Environmental Studies and the School of Women's Studies at York University (Toronto). He is also an associate research fellow at the Centre for the Study of Religion and Society at the University of Victoria. He has written books on Chinese religion, northern Native American religions, and female spirituality. Over the last few years, he has participated in several conferences on religion and the environment, focusing on China.

John Patterson taught philosophy at Massey University in New Zealand until his retirement in 2000. His publications include: a logic textbook, *Practical Logic*, the first book on Maori philosophy by a professional academic philosopher; *Exploring Maori Values* (Dunmore Press, 1992); *Back to Nature: A Taoist Philosophy for the Environment*; the novel *Pinkie's Problem*; and a book of poetry, *Cant Find Rest Room*. His latest book, *People of the Land: A Pacific Philosophy*, explores Maori environmental beliefs (Dunmore Press, 2000).

Lisa Raphals is Associate Professor of Chinese and Comparative Literature at the University of California, Riverside. She is the author of *Knowing Words: Wisdom and Cunning in the Classical Traditions of China and Greece* (Cornell University Press, 1992), *Sharing the Light and Representations of Women and Virtue in Early China* (State University of New York Press, 1998), and a range of studies in comparative philosophy and early Daoism.

Kristofer Schipper worked as a student in Paris with Max Kaltenmark and Rolf Stein. He was trained as a historian of religion and as a cultural anthropologist. After eight years of research at the Institute of Ethnology Academia Sinica, he became a professor of Chinese religions at the École Pratique des Hautes Études at the Sorbonne, Paris. He directs the Taoist Documentation Center of the EPHE, the Tao-tsang Project of the European Science Foundation, and the Peking Temple Project of the CNRS and the Dutch Academy of Sciences. He is editor of the journal *Sanjiao wenxian*. From 1992 to 1999 he was a professor of Chinese history at the University of Leiden.

Daniel Seitz received his B.A. and M.A.T. from the University of Chicago (1977, 1980) and J.D. from Boston University (1985). From 1992 to the present he has been president of the New England School of Acupuncture. He currently serves as chairman of the Accreditation Commission for Acupuncture and Oriental Medicine. From 1987 to 1990 he was chief of the Acupuncture Unit of the Massachusetts Board of Registration in Medicine. From 1977 to 1979 he served as a Peace Corps Volunteer in Malaysia.

Linda Varone is a fengshui practitioner with her own firm, Dragon and Phoenix Feng Shui, based in Cambridge, Massachusetts. She has consulted on residential and business applications of fengshui on the East Coast. She has given classes and workshops on her unique blend of East and West for the Peabody-Essex Museum in Salem, the FORTUNE 500 Cultural Program, and at the Boston Design Center, the New England School of Acupuncture, and the University of Vermont in Burlington. She has studied with Professor Lin

Yun, the foremost teacher in the West of Black Sect Tantric Buddhist Feng-shui, Steven Post, William Spear, and Roger Green. She has studied interior design at the Boston Architectural Center and holds an M.A. in psychology. She has appeared on WCBV-TV's *Chronicle* and in the *Sunday Boston Globe*.

Richard G. Wang received his B.A. and M.A. from Fudan University, Shanghai, China. From 1987 to 1990 he worked as an assistant research fellow in the Institute of Literature at the Shanghai Academy of Social Sciences. From 1991 to 1993 he studied at the University of Colorado at Boulder and received an M.A. in 1993. He then received his Ph.D. in 1999 from the University of Chicago. Since 1999 he has been an assistant professor at the Chinese University of Hong Kong.

Zhang Jiyu is a sixty-fifth-generation direct descendant of the founder of Daoist religion, the first Celestial Master, Zhang Daoling. Currently, he is vice president of the Chinese Daoist Association and editor-in-chief of the magazine *Chinese Daoism*, published in Beijing. He is the author of *A Brief History of the Way of the Celestial Masters* (*Tianshidao shilüe*) and co-author and editor of *Dao Follows Naturalness* and *Environmental Protection* (*Dao fa ziran* and *Huanjing baohu*).

Glossary of Chinese Characters

This glossary includes the Chinese characters for names and terms used in transliteration within this volume. While every attempt has been made to provide a comprehensive listing of these words, there may be some omissions. The glossary does not include titles of articles cited in endnotes and lengthier passages quoted from original Chinese texts. The editors greatly appreciate the contribution of Yang Yew Chong and Lim Moey Kia, students at the National University of Singapore, who helped compile the glossary, and Louis Komjathy at Boston University, who helped edit it.

Anhui	安徽
bagua zhang	八卦掌
bai yao	百藥
Baoji	寶雞
Baopuzi neipian	抱撲子內篇
Bencao jing	本草經
Bencao jing jizhu	本草經集注
benxing	本性
bian	變
bianhua zhi shu	變化之術
Bo	亳
Bohe	帛和
boshan	博山
Bowu zhi	博物志
Boyang fu	伯陽父
budai	不怠
buren	不仁

bushan ren	不善人
buwei shi	不爲始
buxian	不先
buyan	不言
buzheng	不爭
buzheng zhi de	不爭之德
Buzhou shan	不周山
Caijing	蔡經
Cao Cao	曹操
Chan (Zen)	禪
Chan, Wing-tsit	see Chen Rongjie
Changjiang	長江
chaoshan jingxiang	朝山敬香
chen	陳
Chen Rongbo	陳榮波
Chen Rongjie (Wing-tsit Chan)	陳榮捷
Chen Tuan	陳搏
Chen Xiaoya	陳小雅
Cheng Xuanying	成玄英
Cheng Zhongying	成中英
Chengdu	成都
chengfu	承負
chengli	成立
chenwei	讖緯
chuanze	川澤
Chuci	楚辭
chunpu	純樸
Chunqiu fanlu	春秋繁露
Chunqiu zuozhuan cidian	春秋左傳辭典
ci	慈
cun	存
da	大

Da wuliang shoujing	大無量壽經
Dadao jailing jie	大道家令戒
Dahuang	大荒
Dao (Tao)	道
dao (tread)	蹈
dao fa ziran	道法自然
Dao Zhang	道長
Daode jing	道德經
Daoji jing	道跡經
daojia	道家
daojiao	道教
daoqi	道氣
daosong	道頌
daotan	道壇
daoyin	導引
Daozang	道藏
daziran	大自然
de	德
Dedao jing	德道經
deyi	得一
di	地
dijun	帝君
Ding Tao	定陶
dixue	地學
dong (arid land)	垌
dong fu	洞府
Dong Zhongshu	董仲舒
dongfu	洞府
dongtian	洞天
dongyang	東洋
dou	斗
Du Guangting	杜光庭
Du Jiangyan	都江堰
Du Weiming (Tu Weiming)	杜維明

Dujiang Yan	都江堰
duli bugai	獨立不改
Dunhuang	敦煌
Dunhuang juanzi	敦煌卷子
Duren jing	度人經
Erlang Miao	二郎廟
ershisi zhi	二十四治
fa	法
fahui	法會
fajia	法家
fan	反
Fan Gaode	范高德
Fan Shangzhi	氾勝之
fang	方
feiding mingguan	非定命觀
Feihuangzi	飛黃子
fen	分
Fengdu	酆都
fengshui	風水
fojiao	佛教
fu (prose-poem)	賦
fu (talisman)	符
fu shan	浮山
Fu Weixun (Charles W. H. Fu)	傅偉勳
Fu Xi	伏羲
Fu, Charles W. H.	see Fu Weixun
fudi	福地
Fujian	福建
gai	改
Gan Ji	干吉
ganqing	感情
Gansu	甘肅
ganying	感應
Ge Hong	葛洪

Ge Rongjin	葛榮晉
Ge Xuan	葛玄
Geng Sangchu	庚桑楚
gengxiang chengfu	更相承負
gong	公
Gong Chong	宮崇
gonggong	共工
Gu Linyu	顧林玉
guan tianxing	觀天性
guanxi	關系
Guanyinzi	關尹子
Guanzi	管子
gui (ghost)	鬼
gui (return)	歸
guishen	鬼神
guopu	郭璞
Guoyu	國語
Han	漢
Han Chengdi	漢成帝
Han Jingdi	漢景帝
Han Wudi	漢武帝
He Bingdi (Ho Ping-ti)	何炳棣
He Jianming	何建明
He Ping	何平
He Shengdi	賀聖迪
hehe	和合
Henan	河南
Hengshan	衡山
Heshang gong	河上公
Ho Ping-ti	see He Bingdi
Hou Hanshu	後漢書
houtian	後天
hua	化
Huai	淮

Huainanzi	淮南子
huang (sovereign)	皇
huang (wild)	荒
Huangdi	黃帝
huangdi	荒地
Huangdi neijing	黃帝內經
Huangdi yinfu jing	黃帝陰符經
Huangdi yinfu jing shu	黃帝陰符經疏
Huang-Lao	黃老
Huangshanzi	黃山子
Huangting jing	黃庭經
huangye	荒野
Huashan	華山
Huashiyan	花石岩
Huayan	華嚴
Huizi	惠子
Huma	胡麻
hun	魂
hundun	混沌
Ji Cheng	計成
Ji Wenying	計文英
jian	儉
Jian Changchen	蹇昌辰
Jiangsu	江蘇
Jiangxi	江西
jiao (outskirts)	郊
jiao (teaching)	教
jiao (type of ritual)	醮
jiaowai	郊外
jiaxu	甲戌
jiayi	賈誼
jiazi	甲子
Jie Yu	接輿
jijiu	祭酒

Jin	晉
jindan	金丹
jing (essence)	精
jing (scene)	景
jing qi	精氣
jingshen	精神
jingzi	鏡子
Jinye jing	金液經
jiong	駉
jiu	酒
jiu zhou	九州
jiuzhen fa	九眞法
Jiuzhen zhongjing	九眞中經
jiuzhuan dan	九轉丹
ju	聚
juan	卷
jun	君
Junzhi	菌芝
junzi	君子
jusheng	巨勝
Kaishan zu	開山祖
kan	坎
keyi zizhu	可以自主
Kua Fu	夸父
Kunlun shan	昆侖山
Lai Chi-tim	see Li Zhitian
Lao Dan	老聃
Laojun	老君
Lao-Zhuang	老莊
Laozi	老子
Leishen shan	雷神山
li (a unit of distance)	里
li (ceremony)	禮
li (distant)	離

Li (Lady)	麗（姬）
Li Bing	李冰
Li Gen	李根
Li Quan	李筌
Li Yuanguo	李遠國
Li Zhitian (Lai Chi-tim)	黎志添
Li Zhuang	麗莊
liaojie quan yu	暸解全域
Liexian zhuan	列仙傳
Liezi	列子
Liji	禮記
lin	林
Lingbao	靈寶
Lingbao wufu	靈寶五符
Lingnan	嶺南
Liu An	劉安
Liu Bowen	劉伯溫
Liu Cuirong	劉翠溶
Liu Haichan	劉海蟾
Liu Ming	劉明
Liu Sifen	劉斯奮
Liu Xiaogan	劉笑敢
liwai	禮外
Longhu shan	龍虎山
longwen	龍文
Lu	魯
Lu Song	魯頌
Lu Weidong	陸衛東
Lu xiansheng daomen kelüe	陸先生道門科略
Lu Xiujing	陸修靜
Lüshi chunqiu	呂氏春秋
Ma Xiaohong	馬曉宏
Magu (Maid Ma)	麻姑
Mao Qiang	毛嬙

Maoshan	茅山
Mawangdui	馬王堆
meihua shu	梅花樹
Meng	蒙
meng	夢
Mengzi	孟子
miluan	迷亂
ming (bright)	明
ming (destiny; life)	命
mingjie	冥界
mingshan daze buyi feng	明山大澤不以封
mo (mai)	脈
Mojing	墨經
mou	牟
mu (mother)	母
mu (unit of area)	畝
Mutianzi zhuan	穆天子傳
nan	蛆
Nanyue xiaolu	南嶽小錄
neidan	內丹
Neijing tu	內經圖
Neiye	內業
nianming zai guxu zhi xia	年命在孤虛之下
Ningpa	寧杷
Niushan	牛山
niwan	泥丸
nongjia	農家
Nü Wa	女媧
Pan Gu	盤古
pei	沛
penjing	盆景
piaofeng	飄風
ping	平
pinyin	拼音

po	魄
pu	朴
Putuo shan	普陀山
qi	氣
Qian	遷
qiaowei	巧僞
qie	妾
qigong	氣功
Qilin	麒麟
Qin	秦
Qingcheng shan	青城山
Qinghai	青海
Qingshan si	青山寺
Qinwang ziying	秦王子嬰
Qitu chushu jue	起土出書訣
Quanzhen	全眞
qun	群
renjie	人界
rujia	儒家
san	散
sanhe	三和
Sanhuang neiwen	三皇內文
sanjue	三絕
Sanqing	三清
Santian neijie jing	三天內節經
santong	三統
sanxia	三峽
sha	煞
Shaanxi	陝西
shan	善
shan zhi	山志
Shang	商
Shang Qiu	商丘
Shangqing	上清

Shangqing dongzhen zhihui guanshen dajie wen	上清洞真智慧觀身大戒文
Shangqing taishang dijun jiuzhen zhongjing	上清太上帝君九眞中經
Shangqing taishang jiuzhen zhongjing jiangsheng shendan jue	上清太上九眞中經降生神丹訣
Shanhai jing	山海經
shanlin	山林
shanlin chuanze	山林川澤
shanren	善人
shanshui	山水
shanze suo chu	山澤所出
Shanzhi	山志
Shaolin	少林
She Zhengrong	佘正榮
shen (body)	身
shen (numinous)	神
sheng	生
shengqi	生氣
shengren	聖人
shengshu	聖書
shengtai huanjing	生態環境
Shengzhou	生洲
shenqi	神氣
shenren	神人
shenshu	神書
Shenxian zhuan	神仙傳
Shenyi jing	神異經
shi (begin)	始
shi (generation)	世
shi (influence)	勢
shi (pass away)	逝
shi (time)	時

Shi Baishi	史白事
Shi Guizhi	石桂芝
Shi Mizhi	石蜜芝
shi wu	十巫
Shiji	史記
shijie	世界
Shijing	詩經
shijing	寶景
Shisan jing zhushu	十三經注疏
Shizhou ji	十州記
shou yi	守一
Shu	蜀
shu (arts)	術
shu (reciprocity)	恕
Shun	舜
Shuowen jiezi	說文解字
si hai zhi da	四海之大
Si min yue ling	四民月令
Sichuan	四川
Sima Tan	司馬談
sishi	四勢
sixiang	思想
Song	宋
Songshan	嵩山
Soushen shan tu	搜神山圖
Su Shi	蘇軾
Sui	隋
Taibei (T'aipei)	台北
Taihe	太和
taiji	太極
taiji quan	太極拳
Taiji zhenren fu lingbao zhaijie weiyi zhujing yaojue	太極眞人敷靈寶齋戒威儀諸經要訣
Taiping	太平

Taiping jing	太平經
Taiping jing hejiao	太平經合校
Taiping qingling shu	太平清領書
Taiping yulan	太平御覽
Taiqing jing	太清經
Taishan	泰山
Taishang dongxuan lingbao sanyuan pinjie gongde qingzhong jing	太上洞玄靈寶三元品戒功德輕重經
Taishang Laojun	太上老君
Taishang Laojun jinglü	太上老君經律
Taiwan	台灣
taixi	胎息
taiyang	太陽
Taiyi	太一
taiyin	太陰
Taiyin qingjie	太陰清戒
Tang dynasty	唐
Tang Junyi	唐君毅
Tang Mingbang	唐明邦
Tang Taizong	唐太宗
Tao	陶
Tao Hongjing	陶弘景
Tao Qian	陶潛
ti	體
tian	天
tian ren heyi	天人合一
tian zhi dao	天之道
Tian'anmen	天安門
tiandi	天地
tiandi hehe	天地和合
tiandi ren	天地人
tiandi zhi jian	天地之間
tianjie	天界

tianjun	天君
Tianshi	天師
tiantan diyu	天談地語
tianwen	天文
tianxia	天下
tong (communication)	通
Tu Weiming	see Du Weiming
tujing	圖經
Tushan	禿杉
wang	王
Wang Bi	王弼
Wang Bing	王冰
Wang Chong	王充
Wang Chunwu	王純五
Wang Ni	王倪
Wang Qintian	王勤田
Wang Yangming	王陽明
Wang Yuan	王遠
wang[a]	亡
wang[a] wei	亡爲
wanwu	萬物
wanwu zhi guang	萬物之廣
wanwu zhi ziran	萬物之自然
Wei Huacun	魏華存
wei wuwei	爲無爲
Wei Yuangui	魏元圭
weiguo	爲國
Weihe	渭河
Weilei	畏壘
weilei da rang	畏壘大穰
weizhi	爲之
Wen wang (King Wen)	文王

Wing-tsit Chan	see Chan Rongjie
wu	物
wu buwei	無不爲
wu dingti guan	無定體觀
Wu wang (King Wu)	武王
wu zhi er wu	無之而無
wu[a]	无
wu[b]	無
wu[c]	舞
Wudang	武當
wuhua	物化
wuming	無名
Wushang biyao	無上秘要
wushen	五神
wushi	巫師
wuwei	無爲
wuxing	五行
wuyu	無欲
wuyue xianguan	五岳仙官
Wuyue zhenxing tu	五嶽眞形圖
Wuzhen pian	悟眞篇
wuzhi	無知
xi	息
xian	仙
xiandao	仙道
Xiangkai	襄楷
xiangtong	相通
xianguan	仙官
Xianmen	羨門
xianren (ancestor)	先人
xianren (transcendent; immortal)	仙人
xiantian	先天
xiao	孝
xiao tiandi	小天地

xieqi	邪氣
xin (heart)	心
xin (trust)	信
xing (form)	形
xing (nature)	性
xingchu	行廚
xingqi	行氣
Xiong Deji	熊德基
xionghuang shali	凶荒殺禮
xiudao	修道
Xuanxue	玄學
xue	穴
xueqi	血氣
Xunzi	荀子
Xushen	許慎
Yam Kah Kean	see Yan Jiajian
Yan Jiajian (Yam Kah Kean)	嚴家健
Yang (surname)	楊
yang (the yang of yin-yang)	陽
Yang Chengquan	楊誠泉
Yang Xi	楊羲
yangsheng	養生
Yangsheng zhu	養生主
Yangzi jiang	揚子江
Yao	堯
Yaoxiu keyi jielu chao	要修科儀戒律鈔
ye (wilderness; classical character)	野
ye (wilderness; pre-Han character)	埜
yeren	野人
yeyou	野游
yeyu	野虞
yi (city district)	邑

yi (so that)	以
Yi (the *Yijing*)	易
yi wei jiming	疑爲祭名
Yibaibashi jie	一百八十戒
Yijing	易經
yin	陰
Yin Zhihua	尹志華
Yinfu jing	陰符經
yinzhai	陰宅
yiwei jiming	疑爲祭名
Yiwen ji	異聞記
yong	用
you	有
you du	幽都
you er hou wu	有而後 無
you er si wu	有而似無
You wang (King You)	幽王
you wei	有爲
Yu	禹
yuan	遠
Yuan Ke	袁柯
Yuanlinre	園林熱
yuanqi	元氣
Yue	岳
Yuedu	樂都
yueling	月令
Yulizi	郁離子
Yunji qiqian	雲笈七籤
Yunnan	雲南
yunqi	運氣
Yupian	玉篇
zai yewai shali	在野外殺禮
zaibian guaiyi	災變怪異
zangshu	葬書

zhang	丈
Zhang Boduan	張伯端
Zhang Chao	張潮
Zhang Daoling	張道陵
Zhang Guangting	張光庭
Zhang Jiacai	張加才
Zhang Jiyu	張繼禹
Zhang Kunmin	張坤民
Zhang Ling	張陵
Zhang Lu	張魯
Zhang Renwu	張壬午
Zhang Tong	張彤
Zhang Yunfei	張云飛
Zhang Zai	張載
Zhang Zhenjun	張振軍
Zhang Zifang	張子房
Zhao Mu	昭穆
zhen (battle formation)	陣
zhen (perfect)	眞
Zhen'gao	眞誥
Zheng Yin	鄭隱
Zhengyi	正一
Zhengyi fawen tianshi jiao jieke jing	正一法文天師教戒科經
zhenren	眞人
zhenshan tu	眞山圖
zhenxing	眞形
zhexue zazhi	哲學雜志
zhi ("place of order")	治
zhi (branch)	支
zhi (fungus, exudation)	芝
Zhi (Robber Zhi)	(盜)跖
zhile	至樂
zhong	中

Zhongguo	中國
zhonghe	中和
zhonghe qi	中和氣
Zhongnan shan	終南山
zhongren	種人
Zhongyang Huanglao jun	中央黃老君
zhongyong	中庸
Zhou	周
Zhou Lingwang	周靈王
Zhou Zhixiang	周智響
zhouhou fang	肘後方
Zhouli	周禮
Zhouli zhushu	周禮注疏
Zhouyu	周語
Zhu Rongji	朱鎔基
Zhu Xi	朱熹
Zhuang Qingxin	莊慶信
Zhuang Zhou	莊周
Zhuangzi	莊子
ziran	自然
zong	宗
zongjiao	宗教
Zuigen pin	罪根品
zuowang	坐忘
Zuozhuan	左傳

Index